ANALOG CIRCUITS AND SIGNAL PROCESSING

Series Editors:
Mohammed Ismail. The Ohio State University
Mohamad Sawan. École Polytechnique de Montréal

For further volumes:
http://www.springer.com/series/7381

Raj Senani • D.R. Bhaskar • A.K. Singh
V.K. Singh

Current Feedback Operational Amplifiers and Their Applications

Raj Senani
Division of Electronics
and Communication Engineering
Netaji Subhas Institute
of Technology
New Delhi, India

D.R. Bhaskar
Jamia Millia Islamia
Electronics and Communication
Engineering
F/O Engineering and Technology
New Delhi, India

A.K. Singh
Electronics and Communication
Engineering
HRCT Group of Institutions
F/O Engineering and Technology
Mota, Ghaziabad, India

V.K. Singh
Department of Electronics Engineering
Institute of Engineering and Technology
Lucknow, India

ISBN 978-1-4939-0043-5 ISBN 978-1-4614-5188-4 (eBook)
DOI 10.1007/978-1-4614-5188-4
Springer New York Heidelberg Dordrecht London

Softcover reprint of the hardcover 1st edition 2013

Printed on acid-free paper

Springer is part of Springer Science+Business Media (www.springer.com)

Preface

In spite of all electronic systems prominently being dominated by digital circuits and systems, the analog circuits have neither become obsolete nor avoidable. In fact, despite the dominance of digital circuits, analog circuits and techniques continue to be indispensable and unavoidable in many areas since all real life signals are analog in nature. Thus, several types of processing of natural signals or interface of such signals with digital processing circuits has to be necessarily carried out by analog circuits. Also, many basic functions such as amplification, rectification, continuous-time filtering, analog-to-digital conversion and digital-to-analog conversion etc. need analog circuits and techniques.

Traditionally, the integrated circuit (IC) op-amp has usually been considered to be the workhorse of all analog circuit designs. However, over the years, it was found that there are many situations such as realization of voltage controlled current sources, current controlled current sources, instrumentation amplifiers, non-inverting integrators and non-inverting differentiators etc., where the traditional voltage mode op-amp (VOA)-based circuits suffer from two drawbacks namely employment of more than the minimum required number of passive components and requirement of perfect matching of several of them (due to which any mismatch may not only deteriorate the performance of the intended circuits but may also lead to instability in some cases). Furthermore, VOA-based amplifiers exhibit a gain bandwidth conflict and their frequency range of operation is limited by the effect of finite gain bandwidth product (GBP) of the op-amps on one hand and due to the slew-induced distortion (resulting due to finite slew rate of the op-amps) on the other hand. Consequently, there has been continuous search for alternative analog circuit building blocks to overcome these difficulties while still matching the versatility of the VOAs in realizing almost all kinds of analog functions.

During the past four decades, many alternative new analog circuit building blocks have been proposed out of which only the Operational Transconductance Amplifiers, Current Conveyors and Current Feedback Operational Amplifiers have been made available as of-the-shelf ICs and have therefore attracted the attention of educators, researchers and circuit designers worldwide who have explored their various applications. Among these building blocks, the current feedback operational amplifier (CFOA), sometimes also referred as operational trans-impedance amplifier, has received notable attention in literature because of its two very significant properties namely, a very high slew rate (theoretically infinite; practically as high as several thousand volts per μs as against a very modest 0.5 V/μs for the general purpose and most popular μA741 type op-amp) and its capability of offering gain bandwidth decoupling (thereby implying the feasibility of maintaining essentially a constant bandwidth and variable gain, for low to medium values of the gains). Though CFOAs have some limitations as compared to the traditional VOAs, their advantageous features coupled by their versatility and flexibility, particularly of a specific type which has its compensation pin accessible externally, overshadows their demerits in a number of applications.

This monograph is basically concerned with CFOAs and their applications and includes an extensive discussion about various types of CFOAs, the basic circuits realizable using them, their merits and demerits and their applications in the realization of continuous time analog filters, simulation of inductors and other type of impedances, synthesis of sinusoidal oscillators and miscellaneous linear and non-linear applications (including a variety of relaxation oscillators and chaotic circuits). Also covered are numerous examples of the use of CFOAs in realizing a number of other newly proposed active circuit building blocks. The monograph closes by giving a brief account of the recent developments in the design of bipolar and CMOS CFOAs, a discussion about various modified forms of CFOAs proposed in the recent literature from time to time, outlining the current directions of research in this area and including a supplementary list of references for further reading.

It is hoped that this monograph, which contains a comprehensive collection of over 200 CFOA-based analog circuits with their relevant theory and design/ performance details, should turn out to be a useful source of reference for academicians (both educators and students), practicing engineers and anybody interested in analog circuit design using CFOAs. Readers may also find a number of interesting and challenging problems worthy of further investigations, from the various suggestions given in the respective chapters of this monograph.

Contents

1 Introduction 1
1.1 Prologue 1
1.2 An Overview of Analog Circuits and Their Applications 2
1.3 The Ubiquitous Op-Amp: The Drawbacks and Limitations of Some Op-Amp Circuits 3
1.3.1 Op-Amp Circuits Which Employ More Than the Minimum Number of Resistors and Require Passive Component-Matching 3
1.3.2 The Gain-Bandwidth Conflict 7
1.3.3 Slew-Rate Based Limitations 8
1.4 A Brief Review of the Evolution of Alternative Analog Circuit Building Blocks 9
1.4.1 The Operational Transconductance Amplifiers 9
1.4.2 The Current Conveyors 11
1.4.3 The Current Feedback Op-Amp (CFOA) 14
1.4.4 The Operational Trans-resistance Amplifier 15
1.4.5 The Four-Terminal-Floating-Nullor 17
1.4.6 The Current Differencing Buffered Amplifier 18
1.4.7 The Current Differencing Transconductance Amplifier (CDTA) 19
1.5 The Necessity and the Scope of the Present Monograph 20
References 21

2 CFOAs: Merits, Demerits, Basic Circuits and Available Varieties 25
2.1 Introduction 25
2.2 AD844: The CFOA with *Externally-Accessible* Compensation Pin 25
2.3 The Merits and the Advantageous Features of the CFOAs 28
2.3.1 The Reason and the Origin of the High Slew Rate 28

2.3.2 De-coupling of Gain and Bandwidth: Realisability of Variable-Gain, Constant-Bandwidth Amplifiers 30
2.4 The Demerits and Limitations of CFOAs 31
2.4.1 Demerits 31
2.4.2 Difficulties with Capacitive Feedback 32
2.4.3 Effect of Stray Capacitances and Layout Issues 32
2.5 Basic Circuits Using CFOAs 32
2.5.1 VCVS Configurations 32
2.5.2 Instrumentation Amplifier Using CFOAs 34
2.5.3 VCCS, CCVS and CCCS Configurations 35
2.5.4 Unity Gain Voltage and Current Followers 36
2.5.5 Integrators and Differentiators 36
2.6 Commercially Available Varieties of CFOAs 42
2.6.1 The Mixed-Translinear-Cells (MTC) as Building Blocks of CFOAs 42
2.6.2 Elantec Dual/Quad EL2260/EL2460 44
2.6.3 Intersil HFA 1130 44
2.6.4 AD8011 from Analog Devices 45
2.6.5 THS 3001 from Texas Instruments Inc. 46
2.7 Concluding Remarks 47
References 48

3 Simulation of Inductors and Other Types of Impedances Using CFOAs 49
3.1 Introduction 49
3.2 An Overview of Op-Amp-RC Circuits for Grounded and Floating Inductor Simulation and Their Limitations 49
3.3 Realization of Gyrator and Grounded Impedances Using CFOAs 54
3.4 Single-CFOA-Based Grounded Impedance Simulators 56
3.4.1 Lossy Grounded Inductors/FDNRs 57
3.4.2 Single-CFOA-Based Grounded Negative Capacitance and Negative Inductance Simulators 60
3.5 Floating Inductors and Floating Generalized impedance Simulators Using CFOAs 60
3.6 Floating Inductance Circuits Employing Only Two CFOAs 65
3.6.1 Lossless/Lossy Floating Inductance Simulator 65
3.6.2 A Lossy Floating Inductance Simulator 67
3.7 Applications of Simulated Impedances in Active Filter Designs 68
3.7.1 Applications in the Design of Second Order Filters 68
3.7.2 Application in the Design of Higher Order Filters 69
3.8 Realization of Voltage-Controlled Impedances 71
3.8.1 Grounded Voltage Controlled Impedance Simulators 72
3.8.2 Floating Voltage Controlled Impedance Simulators 73
3.9 Concluding Remarks 77
References 78

4 Design of Filters Using CFOAs 81
4.1 Introduction 81
4.2 The Five Generic Filter Types, Their Frequency Responses and Parameters 82
4.3 Voltage-Mode/Current-Mode Biquads Using CFOAs 83
4.3.1 Dual Function VM Biquads 83
4.3.2 Single Input Multiple Output (SIMO) Type VM Biquads 84
4.3.3 Multiple Input Single Output (MISO) Type VM Biquads 91
4.3.4 MISO-Type Universal Current-Mode (CM) Biquads 99
4.3.5 Dual-Mode Universal Biquads Using Single CFOA 100
4.3.6 Mixed-Mode Universal Biquads 103
4.4 Active-R Multifunction VM Biquads 107
4.5 Inverse Active Filters Using CFOAs 110
4.6 MOSFET-C Filters Employing CFOAs 112
4.6.1 MOSFET-C Fully Differential Integrators 113
4.6.2 MOSFET-C Fully Differential Biquads 115
4.6.3 MOSFET-C Single-Ended Biquad 116
4.7 Design of Higher Order Filters Using CFOAs 118
4.7.1 Signal Flow Graph Based Synthesis of nth Order Transfer Function Using CFOAs 119
4.7.2 Doubly Terminated Wave Active Filters Employing CFOA-Based on LC Ladder Prototypes 119
4.7.3 Higher Order Modular Filter Structures Using CFOAs 119
4.8 Concluding Remarks 126
References 127

5 Synthesis of Sinusoidal Oscillators Using CFOAs 131
5.1 Introduction 131
5.2 The Evolution of Single Element Controlled Oscillators: A Historical Perspective 131
5.3 Advantages of Realizing Wien Bridge Oscillator Using CFOA vis-à-vis VOA 133
5.4 Single-Resistance-Controlled Oscillators (SRCO) Using a Single CFOA 134
5.4.1 A Novel SRCO Employing Grounded Capacitors 138
5.5 Two-CFOA-Two-GC SRCOs: The Systematic State Variable Synthesis 140
5.6 Other Two-CFOA Sinusoidal Oscillator Topologies 143
5.7 Design of Active-R SRCOs 148
5.7.1 Active-R Sinusoidal Oscillators Using CFOA-Pole 148

5.7.2 Low-Component-Count CFOA-Pole Based Active-R SRCOs . . . 149
5.7.3 Other Two-CFOA Based Active-R SRCOs . . . 150
5.7.4 CFOA-Pole-Based RC Oscillator . . . 150
5.7.5 A Simple Multiphase Active-R Oscillator Using CFOA Poles . . . 151
5.8 SRCOs Providing ***Explicit*** Current Output . . . 152
5.9 Fully-Uncoupled SRCOs Using CFOAs . . . 157
5.10 Voltage-Controlled-Oscillators Using CFOAs and FET-Based VCRs . . . 161
5.11 State-Variable Synthesis of Linear VCOs Using CFOAs . . . 161
5.12 Synthesis of Single-CFOA-Based VCOs Incorporating the Voltage Summing Property of Analog Multipliers . . . 168
5.13 MOSFET-C Sinusoidal Oscillator . . . 173
5.14 Concluding Remarks . . . 175
References . . . 176

6 Miscellaneous Linear and Nonlinear Applications of CFOAs . . . 181
6.1 Introduction . . . 181
6.2 Electronically-Variable-Gain Amplifier . . . 181
6.3 Cable Driver Using CFOA . . . 182
6.4 Video Distribution Amplifier . . . 182
6.5 Schmitt Triggers and Non-sinusoidal Waveform Generators . . . 183
6.6 Precision Rectifiers . . . 189
6.7 Analog Squaring Circuit . . . 190
6.8 Analog Divider . . . 191
6.9 Pseudo-exponential Circuits . . . 192
6.10 Chaotic Oscillators Using CFOAs . . . 193
6.11 Concluding Remarks . . . 198
References . . . 198

7 Realization of Other Building Blocks Using CFOAs . . . 201
7.1 Introduction . . . 201
7.2 Applications of the CFOAs in Realizing Other Building Blocks . . . 201
7.2.1 CFOA Realizations of Various Kinds of Current Conveyors (CC) . . . 202
7.2.2 CFOA-Realization of the Four-Terminal-Floating-Nullors (FTFN) . . . 204
7.2.3 CFOA Realization of Operational Trans-resistance Amplifier (OTRA) . . . 205

7.2.4 CFOA Realization of Current Differencing Buffered Amplifier (CDBA) Based Circuits 207
7.2.5 CFOA Realization of Circuits Containing Unity Gain Cells 208
7.2.6 Current Differencing Transconductance Amplifier (CDTA) 210
7.2.7 Current Follower Transconductance Amplifiers (CFTA) 211
7.2.8 Current Controlled Current Conveyor Transconductance Amplifier (CCCC-TA) 211
7.2.9 Differential Input Buffered Transconductance Amplifier (DBTA) 212
7.2.10 Voltage Differencing Differential Input Buffered Amplifier (VD-DIBA) 213
7.3 Concluding Remarks 213
References 214

8 Advances in the Design of Bipolar/CMOS CFOAs and Future Directions of Research on CFOAs 223
8.1 Introduction 223
8.2 Progress in the Design of Bipolar CFOAs 223
8.2.1 Bipolar CFOA with Improved CMRR 223
8.2.2 Bipolar CFOA with Higher Gain Accuracy, Lower DC Offset Voltage and Higher CMRR 224
8.2.3 Bipolar CFOA Architectures with New Types of Input Stages 225
8.2.4 Novel CFOA Architecture Using a New Current Mirror Formulation 227
8.3 The Evolution of CMOS CFOAs 227
8.3.1 CMOS CFOA with Rail-to-Rail Swing Capability 229
8.3.2 CMOS CFOA for Low-Voltage Applications 229
8.3.3 Fully-Differential CMOS CFOAs 229
8.3.4 CMOS CFOAs with Increased Slew Rate and Better Drive Capability 230
8.3.5 Other CMOS CFOA Architectures 231
8.4 Various Modified Forms of CFOAs and Related Advances 232
8.4.1 The Modified CFOA 232
8.4.2 Current-Controlled CFOA 232
8.4.3 Current Feedback Conveyor 233
8.4.4 The Differential Voltage Current Feedback Amplifier 233
8.4.5 Differential Difference Complementary Current Feedback Amplifier 235

8.5 Future Directions of Research on CFOAs and Their Applications 237
8.6 Epilogue 237
References 238

References for Additional Reading 241

About the Authors 243

Index 247

Acknowledgements

The motivation for writing this book came from the involvement of our research group in writing two short chapters for the Springer monograph *Integrated Circuits for Analog Signal Processing* (edited by Prof E. Tlelo-Cuautle) one of which was related to Current Feedback Operational Amplifiers (CFOA). During the process of writing these chapters, it dawned upon the first author that the topic of CFOAs and their applications deserved a full monograph by itself. Accordingly, a detailed proposal of the present monograph was submitted to Charles Glaser, Senior Editor Engineering, Springer US, who, after getting the proposal reviewed, gave us a go-ahead to prepare the proposed monograph.

The authors are thankful to the facilities provided by the Analog Signal Processing (ASP) Research Lab., Division of ECE, Netaji Subhas Institute of Technology (NSIT), New Delhi, where the first author works and where this entire project was carried out.

The authors gratefully thank their respective family members for their unflinching encouragement, moral support and understanding shown by them over several decades, in general and during the entire course of the preparation of this monograph, in particular.

The authors also take this opportunity to thank Charles Glaser, Rebecca Hytowitz and Susan Westendorf and in particular Shashi Rawat, who provided all necessary support in the preparation of the manuscript of the monograph. The authors would also like to thank the other colleagues from their research group namely, S.S. Gupta, R.K. Sharma and Pragati Kumar for their support and understanding.

The authors, all of whom are members of the research group at the ASP Research Lab. at NSIT, have also been involved in teaching a number of ideas contained in this monograph to their students in various courses related to Analog Integrated Circuit Design at their respective Institutes. A popular query from our students has been as to: *in which book the material taught to them could be found*? Their query has also been an important deriving force to write this monograph and we thank our numerous students for this and do hope that this monograph, at least partly, answers to their persistent query.

Abbreviations

A/D	Analog-to-digital
ABB	Active building block
AD	Analog devices
ADC	Analog-to-digital convertor
AM	Analog multiplier
BJT	Bipolar junction transistor
BW	Bandwidth
CB	Complementary bipolar
CC	Current conveyor
CCCC-TA	Current controlled current conveyor transconductance amplifier
CC-CFOA	Current controlled current feedback operational amplifier
CCCS	Current-controlled-current-source
CCIII	Third generation current conveyor
CCVS	Current controlled voltage source
CDBA	Current differencing buffered amplifier
CDTA	Current differencing transconductance amplifier
CE	Characteristic equation
CFC	Current feedback conveyor
CFOA	Current feedback operational amplifier
CFTA	Current follower transconductance amplifier
CMOS	Complementary metal oxide semiconductor
CMRR	Common mode rejection ration
CO	Condition of oscillation
CVC	Current voltage conveyor
D/A	Digital-to-analog
DBTA	Differential-input buffered transconductance amplifier
DDA	Differential difference amplifiers
DDCC	Differential difference current conveyor
DDCCFA	Differential difference complimentary current feedback amplifier
DOCC	Dual output current conveyor
DVCC	Differential voltage current conveyor

DVCC+	Differential voltage second generation current conveyor (positive-type)
DVCFA	Differential voltage current feedback amplifier
DVCFOA	Differential voltage current feedback operational amplifier
ECO	Explicit-current-output
ELIN	Externally linear but internally nonlinear
FDCC	Fully-differential current conveyor
FDCCII	Fully differential second generation current conveyor
FDCFOA	Fully differential current feedback operational amplifier
FDNC	Frequency-dependent-negative-conductance
FDNR	Frequency-dependent-negative-resistance
FET	Field effect transistor
FI	Floating inductance or floating impedance
FPBW	Full power band width
FTFN	Four-terminal floating nullor
GBP	Gain bandwidth product
GC	Grounded capacitor
GIC	Generalized impedance converter
GNIC	Generalized negative impedance converter
GNII	Generalized negative impedance inverter
GPIC	Generalized positive impedance converter
GPII	Generalized positive impedance inverter
IC	Integrated circuit
ICC	Inverting current conveyor
MCFOA	Modified current feedback operational amplifier
MTC	Mixed translinear cell
NE	Node equation
NMOS	N-type metal oxide semiconductor
OFC	Operational floating conveyor
OTA	Operational transconductance amplifier
OTRA	Operational trans-resistance amplifier
PMOS	P-type metal oxide semiconductor
SEC	Single element controlled
SR	Slew rate
SRC	Single resistance controlled
SRCO	Single resistance controlled oscillator
TAC	Transconductance and capacitance
THD	Total harmonic distortion
TI	Texas instruments
VCC	Voltage-controlled capacitance
VCCS	Voltage-controlled-current-source
VCFI	Voltage controlled floating impedance
VCL	Voltage controlled inductance
VCO	Voltage controlled oscillator

VCR	Voltage-controlled-resistor
VCVS	Voltage controlled voltage source
VCZ	Voltage-controlled impedance
VD-DIBA	Voltage differencing differential input buffered amplifier
VDTA	Voltage differencing transconductance amplifier
VLF	Very low frequency
VOA	Voltage-mode op-amp
WBO	Wien bridge oscillator

Chapter 1
Introduction

1.1 Prologue

Since all natural signals are analog, the analog circuits and techniques to process them are unavoidable in spite of almost everything going digital. In particular, several analog functions/circuits such as amplification, rectification, continuous-time filtering, analog-to-digital (A/D) and digital-to-analog (D/A) conversion are impossible to be performed by digital circuits regardless of the advances made in the digital circuits and techniques. Thus, analog circuits are indispensable in many applications such as processing of natural signals, digital communication, Disk-drive electronics, processing of signals obtained from optical and acoustical transducers and wireless and optical receivers, to name a few. Besides these applications, there are other areas like simulating artificial neurons, artificial neural networks and a number of applications in image processing and speech recognition which are better carried out by analog VLSI or mixed signal VLSIs than digital circuits. Realistically speaking, all electronic design is essentially analog; in fact, even high-speed digital design is basically analog in nature. In conclusion, the all-round proliferation of digital circuits and techniques has not made analog circuits and techniques obsolete rather, it has thrown more challenges to analog circuit designers to evolve new methods and circuits to design analog signal processing circuits compatible with concurrent digital technology.

This monograph focuses on Current feedback operational amplifiers (CFOA) and their applications.

Although most of the chapters of this book deal with various applications of CFOAs which take as the basis, the commercially available off-the-shelf IC CFOAs and hence, it would appear that all such circuits are essentially evolved for discrete circuit applications, however, with some changes, the basic circuit topologies can also be carried over to fully integratable circuit designs. For example, using bipolar CFOAs and the passive resistors realized by BJT-based translinear current-controlled resistances, the resulting circuits become suitable for implementation in bipolar IC technology. Similarly, when a CMOS CFOA is considered along with

R. Senani et al., *Current Feedback Operational Amplifiers and Their Applications*, Analog Circuits and Signal Processing, DOI 10.1007/978-1-4614-5188-4_1,

the resistors realized by CMOS voltage-controlled-resistors (VCR), the given CFOA configuration would be possible to be integrated as an IC in CMOS technology. It is interesting to note that in either case, the resulting integratable version can easily possess an additional property of electronic tunability which may usually not be available in the discrete counterpart. In fully-integrable versions of CFOA-based circuits, the various parameters of the realized circuits can be electronically adjusted through external DC bias currents in the former case and through external DC voltages in the latter case.

1.2 An Overview of Analog Circuits and Their Applications

In the world of analog circuits, it is widely believed that almost any function can be performed using the classical voltage-mode op-amp (VOA). Thus, on one hand, one can realize using op-amps, all linear circuits such as the four controlled sources (VCVS, VCCS, CCVS and CCCS), integrators, differentiators, summing and differencing amplifiers, variable-gain differential/instrumentation amplifiers, filters, oscillators etc., on the other hand, op-amps can also be used to realize a variety of non-linear functional circuits such as comparators, Schmitt trigger, sample and hold circuits, precision rectifiers, multivibrators, log-antilog amplifiers and a variety of relaxation oscillators. Though a large variety of op-amps are available from numerous IC manufacturers, the internally-compensated types, such as μA741 from Fairchild and (to some extent) LF356 from National Semiconductors can be regarded to be the most popular ones for general purpose applications. In view of this, therefore, it is not surprising that till about 1990 or so, analog electronic circuit design was heavily dominated by VOAs.

Although the 'current feedback operational amplifier' (CFOA), sometimes also called 'operational trans-impedance amplifier', had been in existence since around 1985 or so, it actually started receiving attention of the analog circuit designers only when it was recognized that the circuits built using CFOAs can exhibit a number of advantages in analog circuit design such as, gain-bandwidth independence, relatively higher slew rate and consequently higher frequency range of operation and advantage of requiring a minimum number of external passive components without component-matching in most of the applications; see [1–10] and the reference cited therein.

While several dozens of books by various publishers have been published on traditional operational amplifiers and their applications, to the best knowledge of the authors, no such treatment has so far been given to its close relative—the CFOA. It is this reason which necessitated the writing of this monograph which is exclusively devoted to the CFOAs and their applications which are currently available only in research papers published in various international journals over the past two decades.

This chapter gives a brief overview of analog circuits and their applications, outlines some difficulties and limitations of certain types of op-amp circuits, surveys the state of the art of some prominent alternative building blocks and outlines the necessity and scope of the present monograph which deals with the CFOAs and their applications in modern analog circuit design and signal processing.

1.3 The Ubiquitous Op-Amp: The Drawbacks and Limitations of Some Op-Amp Circuits

Whereas the fact that the traditional VOA is a time-proven building block need not be emphasized in view of its wide spread recognition as the work horse of analog circuit design for several decades now, a comparably less acknowledged fact is that there are several applications in which the use of VOA does not lead to very appealing circuits. Some such VOA-based circuits are brought out by the examples which follow:

1.3.1 Op-Amp Circuits Which Employ More Than the Minimum Number of Resistors and Require Passive Component-Matching

There are a number of basic op-amp circuits which not only require more number of passive components than necessary but also call for the use of a number of matched resistors or require certain conditions/constraints to be fulfilled for realizing the intended functions. Some examples are as follows.

1.3.1.1 Voltage-Controlled-Current-Sources (VCCS) and Current-Controlled-Current-Sources (CCCS)

Consider two well-known VCCS configurations shown in Fig. 1.1a, b.

A straight forward analysis of the circuits of Fig. 1.1 shows that the relation between the output current and input voltage (assuming ideal op-amp) for the circuit of Fig. 1.1a is given by

$$I_0 = \frac{V_{in}}{R_1} + V_0\left(\frac{R_3}{R_2R_4} - \frac{1}{R_1}\right) \tag{1.1}$$

whereas for circuit of the Fig. 1.1b it is:

$$I_0 = -\frac{V_{in}R_2}{R_1R_3} + V_0\left(\frac{R_2}{R_1R_3} - \frac{1}{R_4}\right) \tag{1.2}$$

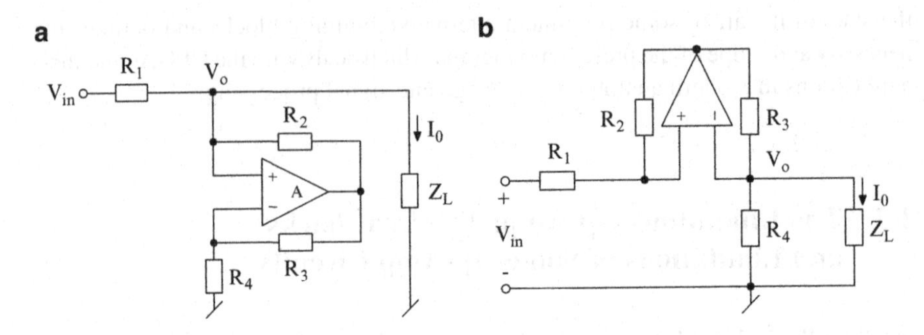

Fig. 1.1 VCCS. (**a**) Non-inverting VCCS, (**b**) inverting VCCS

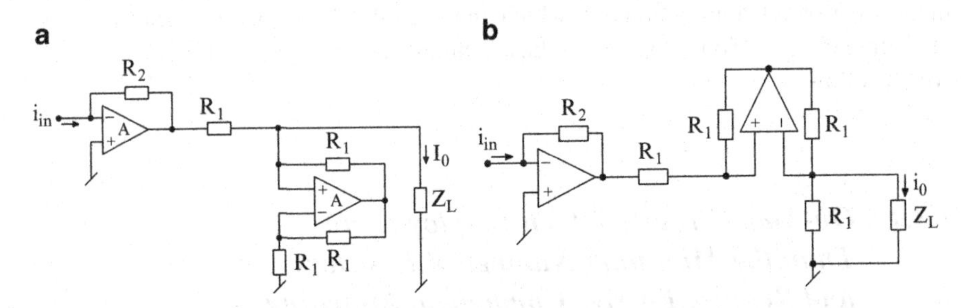

Fig. 1.2 CCCS. (**a**) Non-inverting CCCS, (**b**) inverting CCCS

From the above, it may be seen that to realize a VCCS, the op-amp circuits not only require more than the minimum number of resistances necessary[1] but also require that all the four resistors should have either a relationship $R_1 = R_2R_4/R_3$ or else all the four resistors be equal-valued and matched so that the output current becomes independent of the output voltage and depends only on the input voltage, as required. Thus, any mismatch in resistor values from the intended ones would degrade the performance of the circuit.

Figure 1.2a shows the realization of a non-inverting CCCS while the circuit of Fig. 1.2b realizes an inverting CCCS.

Assuming ideal op-amps, the expressions for the output current in terms of input current for the two circuits are given by

$$I_0 = -\left(\frac{R_2}{R_1}\right)i_{in} \text{ and } I_0 = \left(\frac{R_2}{R_1}\right)i_{in} \tag{1.3}$$

respectively. Thus, in these cases also as many as four resistors need to be equal-valued and matched and a total of five resistors are needed whereas (1.3) indicates

[1] The minimum number of resistors necessary to realize VCCS and CCVS is *one*.

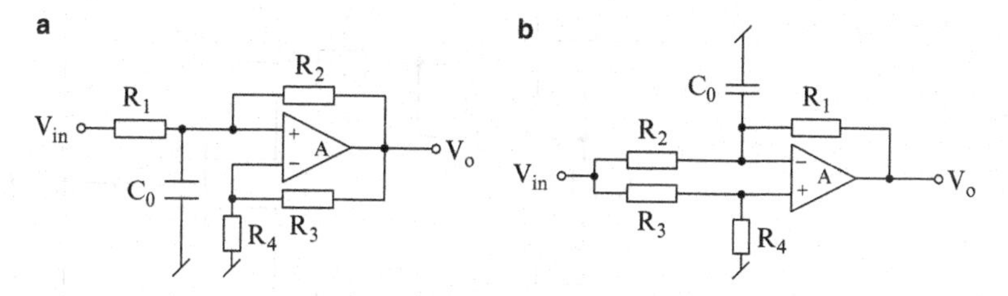

Fig. 1.3 (**a**) Non-inverting integrator, (**b**) non-inverting differentiator

that theoretically two resistors should be the minimum number of resistors necessary for realizing a CCCS.

1.3.1.2 Non-inverting Integrator/Differentiator Using a Single Op-Amp

Figure 1.3a shows a non-inverting integrator popularly known as Deboo's integrator [11] realized with a single op-amp whereas Fig. 1.3b shows non-inverting differentiator using exactly the same number of passive components. This circuit was independently proposed by Horrocks [12] and Ganguli [13] separately. In retrospection, the circuits of Fig. 1.3a, b are also derivable from each other by inverse transformation of Rathore [14].

A straight forward analysis of the first circuit reveals that its transfer function is given by

$$\frac{V_0}{V_{in}} = \frac{\left(\frac{R_3 + R_4}{R_1}\right)}{sC_0R_4 + \left(\frac{R_4R_2 - R_3R_1}{R_1R_2}\right)} \tag{1.4}$$

$$for\ R_1R_3 = R_2R_4;\ \frac{V_0}{V_{in}} = \frac{(R_3 + R_4)}{sC_0R_1R_4} \tag{1.5}$$

On the other hand, the transfer function of the non-inverting differentiator is given by

$$\frac{V_0}{V_{in}} = \frac{sC_0R_4 + \left(\frac{R_4R_2 - R_3R_1}{R_1R_2}\right)}{\left(\frac{R_3 + R_4}{R_1}\right)} \tag{1.6}$$

$$for\ R_1R_3 = R_2R_4;\ \frac{V_0}{V_{in}} = \frac{sC_0R_4R_1}{(R_3 + R_4)} \tag{1.7}$$

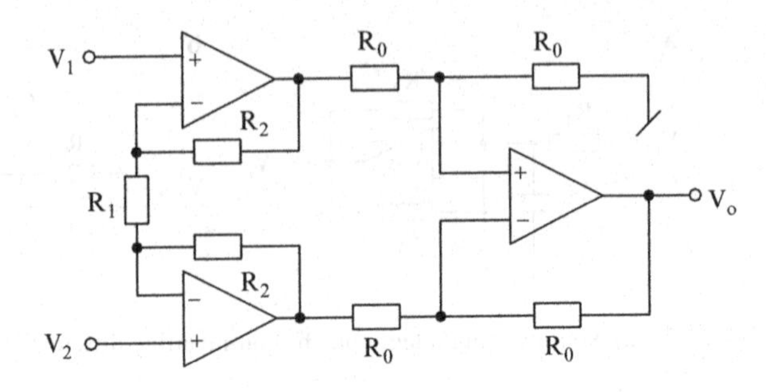

Fig. 1.4 Instrumentation amplifier

In both the cases, for ease of design, one normally takes all identical resistors (though not necessary) i.e. $R_1 = R_2 = R_3 = R_4 = R$. Thus, the transfer function of the integrator is given by

$$\frac{V_0}{V_{in}} = \frac{2}{sC_0R} \tag{1.8}$$

while that of the differentiator is given by

$$\frac{V_0}{V_{in}} = \frac{sC_0R}{2} \tag{1.9}$$

Thus, in both the cases, the circuits require more than the minimum required number (only *one*) of resistors. Furthermore, any mismatch of the resistor values may lead to the difference term $(R_2R_4 - R_1R_3)$ as in (1.4) and (1.6) becoming negative. This mismatch in case of the integrator may lead to instability since the pole of the transfer function would move into the right half of the s-plane. On the other hand, in case of the differentiator, the mismatch would degrade the performance, since it will not remain an ideal differentiator any more.

1.3.1.3 Instrumentation Amplifier

The conventional instrumentation amplifier is another circuit which uses more number of resistors than the minimum number required. This circuit is shown in Fig. 1.4.

The input output equation of this circuit is given by

$$V_0 = \left(1 + \frac{2R_2}{R_1}\right)(V_1 - V_2) \tag{1.10}$$

Note that while for realizing a variable gain, two resistors should be the minimum number of resistors necessary, this circuit employs as many as seven

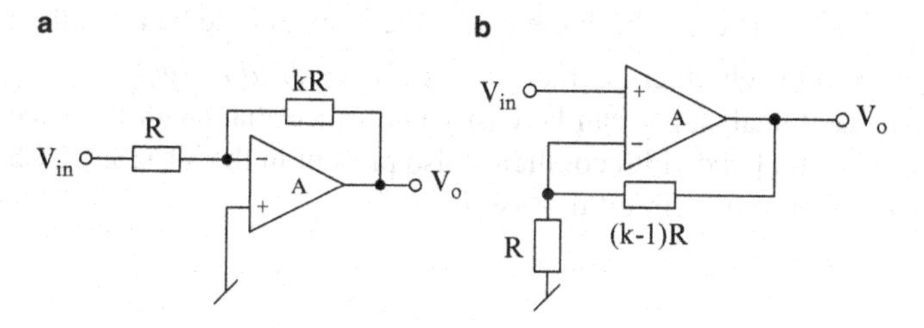

Fig. 1.5 K-gain amplifiers. (**a**) Inverting amplifier, (**b**) non-inverting amplifier

resistors out of which four must be either perfectly matched (or the constraint between them required to get $V_O \propto (V_1 - V_2)$ must be exactly satisfied).

1.3.2 The Gain-Bandwidth Conflict

A major demerit of the various controlled source implementations (except CCVS) using the traditional VOAs is the so-called 'gain-bandwidth-conflict'. This can be explained as follows.

Consider the well-known realization of the K-gain non-inverting and inverting amplifiers using op-amps, shown in Fig. 1.5.

Consider the one pole model of the op-amp as

$$A = \frac{A_0\omega_p}{s+\omega_p} \cong \frac{A_0\omega_p}{s} \cong \frac{\omega_t}{s} \; for \; \omega >> \omega_p \tag{1.11}$$

where $\omega_t = A_0\omega_p$ is the gain bandwidth product of the op-amp.

For the non-inverting amplifier, the non-ideal transfer function is given by

$$\frac{V_O}{V_{in}} = \frac{\omega_t}{s+\frac{\omega_t}{K}} \tag{1.12}$$

whereas the non-ideal gain function of the inverting amplifier of gain $-$K is found to be

$$\frac{V_O}{V_{in}} = -K\frac{\frac{\omega_t}{(K+1)}}{s+\frac{\omega_t}{(K+1)}} \tag{1.13}$$

It is, thus, seen that in the former case, maximum gain is K (at DC) and its 3-dB bandwidth is $\frac{\omega_t}{K}$ whereas in the latter case, the maximum gain is $-$K but the 3-dB bandwidth is $\frac{\omega_t}{(K+1)}$. Thus, in both the cases, the gain and the bandwidth cannot be set independent of each other i.e. *there is a gain-bandwidth conflict*.

By a non-ideal analysis, it can be easily confirmed that (with the exception of the CCVS) this gain bandwidth conflict is also present in the VCCS, CCCS and the instrumentation amplifier circuit discussed earlier.

1.3.3 Slew-Rate Based Limitations

Another factor, which limits the application of VOA-based circuits in higher frequency ranges, is the finite slew rate of the op-amp which is defined as the maximum rate of change of the output voltage with respect to time i.e. Slew Rate $(SR) = \frac{dV_0}{dt}\Big|_{max}$. Internally-compensated type op-amps have the first stage as a differential transconductance-type amplifier followed by a high gain intermediate stage, with the frequency compensating capacitor C_c connected across the intermediate stage such that it is charged by the current delivered by the input transconductance stage. When a large differential input is applied to an op-amp configuration (such as to an op-amp configured as a voltage follower), the input stage gets saturated and delivers a constant maximum current equal to the dc bias current I_{bias} of this stage by which the compensation capacitor is charged. Thus, the voltage across the compensating capacitor (which is equal to the output voltage of the op-amp) can change with a maximum rate of change equal to the finite and fixed dc bias current of the input transconductance stage divided by the value of the compensating capacitor and hence, the $SR = I_{bias}/C_c$ and is, therefore, limited. Thus, at large input voltages or high frequencies or a combination of the two, the output voltage fails to respond with the same speed as the input (due to finite maximum SR) and this results in slew-induced distortion. Conversely, to avoid slew-induced distortion, the input voltages and their frequencies are constrained to be kept small.

Thus, the finite slew rate affects both the dynamic range of the op-amp circuits as well as the maximum frequency of the input signal which can be applied without causing noticeable distortion in the output waveform. It may, however be kept in mind that the operational frequency range of an op-amp circuit or the maximum frequency of the input signal which can be applied to an op-amp circuit is also limited by the finite gain-bandwidth product of the op-amp (which results in finite close loop 3-dB bandwidth as explained earlier).

The maximum frequency f_{max} up to which an op-amp can operate without being slew-rate limited is a function of both the frequency and peak amplitude V_{op} of the output. This f_{max} is given by

$$f_{max} = \frac{SR}{2\pi V_{op}} \tag{1.14}$$

As the output voltage peak amplitude increases, the maximum frequency at which slew-rate-limiting occurs decreases. The frequency at which the op-amp becomes slew-rate-limited is called full power band width (FPBW) and is same as in (1.14) above. It is interesting to note that FPBW of a given op-amp amplifier circuit can be considerably less than the small-signal bandwidth of the same circuit.

1.4 A Brief Review of the Evolution of Alternative Analog Circuit Building Blocks

The various methods of linear analog circuit design, encompassing the classical as well as the modern approaches, can be broadly classified in two major categories:

(a) *Building block approach:* In this approach, first an ideal building block is postulated and synthesis/realization methods are formulated around such building blocks which are then realized using BJTs or MOSFETs.
(b) *Transistor-level approach*: In this approach, BJTs and FETs are used directly as non-linear elements and synthesis/realization methods are developed to achieve the required functions such that resulting circuits are externally linear but internally nonlinear (ELIN).

A vast majority of developments in the analog circuit design belong to the first category, whereas the so-called translinear, log domain and square root domain circuits constitute the second category.

In the following, we outline a number of popular and prominent analog circuit building blocks, which have been extensively investigated as alternatives to the classical VOAs over the last four decades and have been shown to offer a number of significant features and advantages over VOAs and VOA-based circuits in various analog signal processing/signal generation applications.

1.4.1 The Operational Transconductance Amplifiers

Because of inability of the traditional op-amp-RC filters in making it possible to realize precision fully-integratable filters in monolithic form and because of the need for having fully-integratable continuous-time filters in both bipolar and CMOS technologies, the operational transconductance amplifier (OTA)-C or g_m-C circuits had been widely investigated by a number of research groups throughout the world during eighties and nineties and were found to be useful in numerous applications. Because of the electronic controllability of their transconductance, OTAs have been extensively used for designing a variety of linear and non-linear electronically-controllable signal processing and signal generation circuits. The symbolic notation of the OTA is depicted in Fig. 1.6 and is characterized by the equations $i_1 = 0 = i_2$, $i_0 = g_m(v_1 - v_2)$; $g_m = f(I_B)\ or\ f(V_B)$.

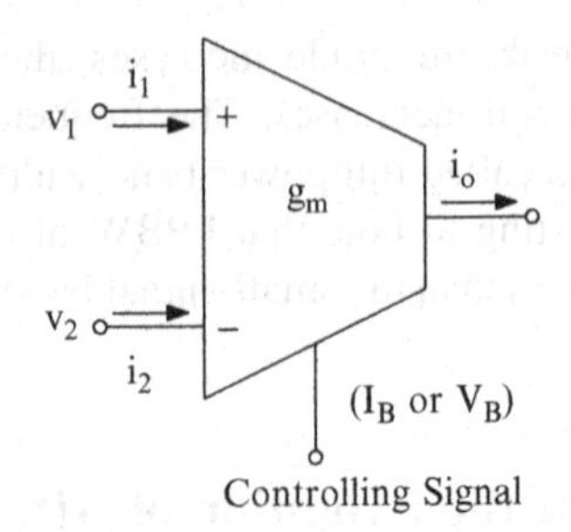

Fig. 1.6 Symbolic notation of the OTA

The OTA-C circuits (also known as transconductance and capacitance (TAC) circuits) employ only transconductors and capacitors to build various functional circuits and thus, generally do not require any external resistors. Furthermore, since the internal circuits of OTAs also can be designed without using any resistors, the resulting circuits are completely resistor-less. Since the transconductance of an OTA is electronically controllable through an external DC bias voltage/current, the OTA-C circuits are suitable for realizing electronically-controllable functions and are amenable to integration in both bipolar and CMOS technologies. Thus, the period 1985–1995 witnessed a phenomenal research activity on the various aspects of OTAs and OTA-C circuits.

A variety of OTAs is commercially available from a number of manufacturers out of which 3080 type and 13600/13700 type are most popular ICs which have been extensively used by several researchers for discrete implementation of OTA-C circuits. The circuit schematic of these two popular commercial OTAs are given below in Figs. 1.7 and 1.8, whereas an exemplary circuit schematic of a CMOS OTA is shown in Fig. 1.9.

Among various applications of the OTAs, that of realizing fully integratable and/or electronically-tunable filters has received major attention in literature. Together with the filters, the use of the OTAs, in conjunction with only capacitors as passive elements, to synthesise sinusoidal oscillators has also been extensively investigated in literature. Developments in OTA-C oscillators were motivated by the resulting features of electronic-controllability of the oscillation frequency *linearly* through the external DC bias currents of the OTAs. Such OTA-C oscillators were extensively investigated by a number of researchers, for instance, see [15–20]. Comprehensive catalogues of all possible OTA-C sinusoidal oscillators realisable with only three/four OTAs and two capacitors were also made available in [21–25]. The feasibility of implementing these circuits in CMOS was also demonstrated in a number of works; see [26–29] and the references cited therein.

It may be mentioned that the continued publication of improved CMOS implementations of OTAs and OTA-based application circuits in various technical journals even now (for instance see [30]) shows that all possible ideas related to the design of OTAs (bipolar, CMOS and Bi-CMOS [30–32]) as well as application circuits using OTAs, have still not been completely exhausted.

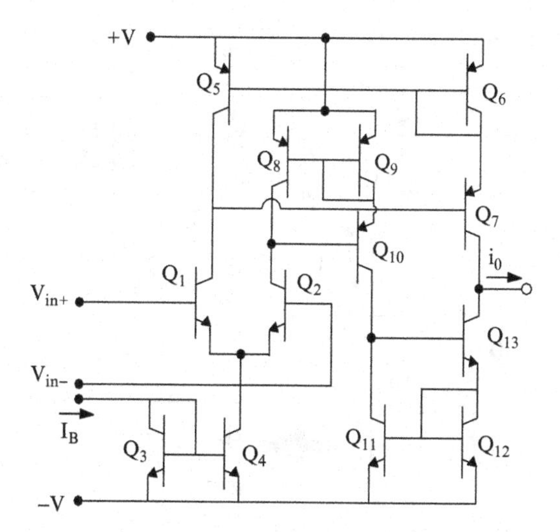

Fig. 1.7 Simplified Schematic diagram of 3080 type IC OTA

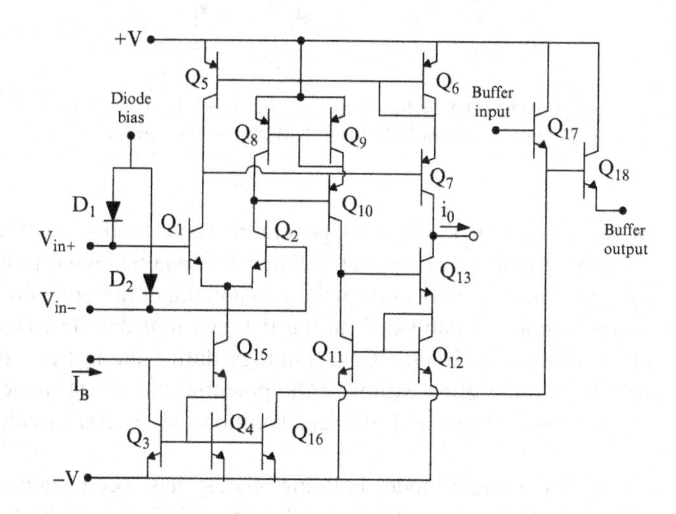

Fig. 1.8 Simplified Schematic diagram of 13600/13700 type IC OTA

1.4.2 The Current Conveyors

Historically, the progress in the current-mode circuits/techniques can be visualized to be considerably stimulated due to two major developments. The first one of these was the proposition of new building blocks known as *Current Conveyors* by Smith and Sedra during 1968–1970 [33, 34] and the second one was the introduction of the so-called *translinear circuits* by Gilbert in 1975 [35].

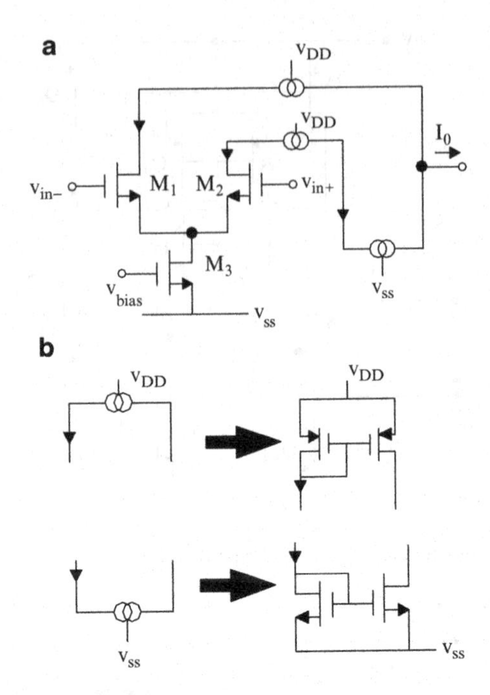

Fig. 1.9 An exemplary CMOS OTA architecture [28]. (**a**) The schematic of the CMOS OTA, (**b**) Symbolic notation and the circuits of PMOS and NMOS current mirrors

The current-mode techniques, in spite of generating some controversy [36], have indeed given way to a number of interesting/important analog signal processing/ signal generating circuits as is evident from the vast amount of literature on current-mode circuits and techniques published during the past four decades. Due to the advances made in integrated circuit (IC) technology during the last two decades, circuit designers have quite often exploited the potential of current-mode analog techniques for evolving elegant and efficient solutions to several circuit design problems.

The most popular current-mode building block has been, undoubtedly, the Current Conveyor (CC) introduced by Smith and Sedra as *first generation Current Conveyor* or CCI in 1968 [33] and later refined to *second generation Current Conveyor* in 1970 by Sedra and Smith [34]. Because of the extensive work done by researchers for more than four decades, the CCs and other current-mode circuits notably the current feedback op-amps (CFOA) have begun to emerge as an important class of building blocks with properties and capabilities that enable them to rival their voltage-mode counterparts (e.g. the traditional voltage mode op-amp) in a wide range of applications.

The first generation and second generation Current Conveyors (popularly known as CCI and CCII respectively), are 3-port active elements for which the symbolic notations are shown in Fig. 1.10.

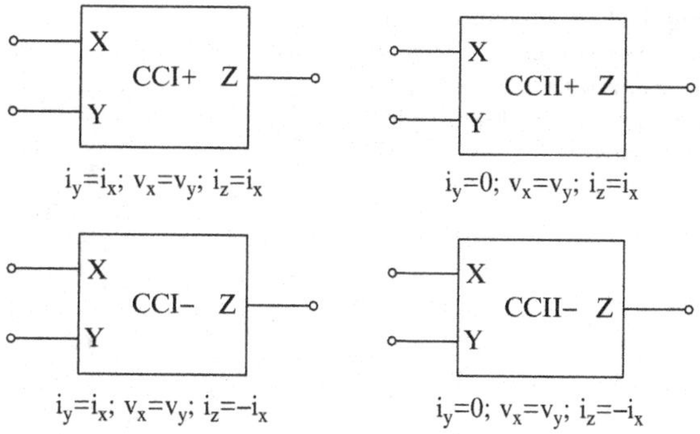

Fig. 1.10 CCI+, CCI−, CCII+ and CCII− Current Conveyors

A CCI (±) is characterized by the hybrid matrix

$$\begin{bmatrix} i_y \\ v_x \\ i_z \end{bmatrix} = \begin{bmatrix} 0 & 1 & 0 \\ 1 & 0 & 0 \\ 0 & \pm 1 & 0 \end{bmatrix} \begin{bmatrix} v_y \\ i_x \\ v_z \end{bmatrix} \tag{1.15}$$

while, a CCII (±) is characterized by

$$\begin{bmatrix} i_y \\ v_x \\ i_z \end{bmatrix} = \begin{bmatrix} 0 & 0 & 0 \\ 1 & 0 & 0 \\ 0 & \pm 1 & 0 \end{bmatrix} \begin{bmatrix} v_y \\ i_x \\ v_z \end{bmatrix} \tag{1.16}$$

In current-mode circuits, the operating variable becomes current rather than the voltage such that voltage swings could be kept small while permitting large swings in the signal currents. Motivated by the attractiveness of current-mode approach, extensive research has been carried out on Current Conveyors and translinear circuits implementable in both bipolar and CMOS technologies during the past four decades. This research has led to the developments by which applications of Current Conveyors (and now CFOAs) have been found in almost all the domains which were once dominated by the traditional voltage-mode op-amps.

These developments have resulted in a large number of integratable bipolar and CMOS implementations of a wide variety of current conveyors and other related building blocks such as current voltage conveyors (CVC), dual output CC (DOCC), differential voltage CC (DVCC), differential difference CC (DDCC), third generation CC (CCIII), inverting CC (ICC), fully-differential CC (FDCC), Operational floating conveyor (OFC), Operational Transresistance Amplifier (OTRA), Current differencing buffered amplifier (CDBA), Current differencing transconductance amplifiers (CDTA), Voltage differencing transconductance amplifier (VDTA), Current follower transconductance amplifier (CFTA), Four terminal floating nullor (FTFN) etc.

In the following, we give a brief account of the prominent developments on the evolution of only some of these building blocks which are closely related to CCs.

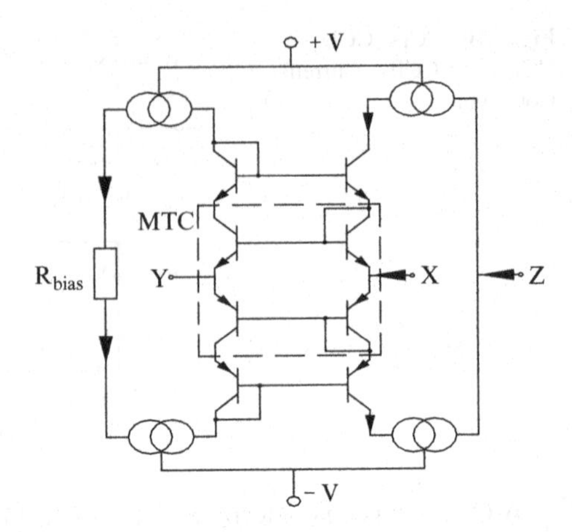

Fig. 1.11 A simplified form of Fabre-Normand Translinear CCII⊕ (adapted from [37, 38] © 1985 Taylor & Francis)

In 1985, Fabre [37] and Normand [38] independently proposed a Current Conveyor implementation based on a mixed translinear cell (MTC), a simplified form of which is as shown in Fig. 1.11.

The Mixed Translinear cell (MTC) and mirror arrangements force the current out of Z-terminal to be equal to the current out of X-terminal, while the voltage at X-terminal will be equal to the voltage at Y-terminal; with no current flowing into the Y-terminal, thereby exhibiting exactly the properties of a CCII ⊕.

Of all the building blocks evolved as alternatives to the classical op-amp, no other building block has received as much attention as the CC even when an IC CC was not available. In fact, there have been at least four different IC CCs produced at different times: PA630 from Phototronics Ltd, Canada in 1989; CCII01 from LTP Electronics in 1991, AD844 *disguised as a high slew rate op-amp* but containing a CCII+ inside and more recently, Max 4223 from MAXIM introduced in 2010. However, the most popular of these has been undoubtedly AD844 which can realize both CCII+ and CCII− (apart from its normal use as a current feedback op-amp); the other varieties have simply not taken off! It is worth mentioning that the number of publications on the proposals on hardware realization of CCs runs into several hundreds and so do the number of papers dealing with the applications of the CCs.

1.4.3 The Current Feedback Op-Amp (CFOA)

A Current Feedback Op-amp is essentially a translinear Current Conveyor (CCII+) followed by a translinear voltage buffer (see Fig. 1.12a for the symbolic notation and b for a typical bipolar implementation). One of the most popular CFOA namely, the AD844 from Analog Devices, is a 4-terminal building block characterized by the following equations:

$$i_y = 0, \ v_x = v_y, \ i_z = i_x \text{ and } v_w = v_z \tag{1.17}$$

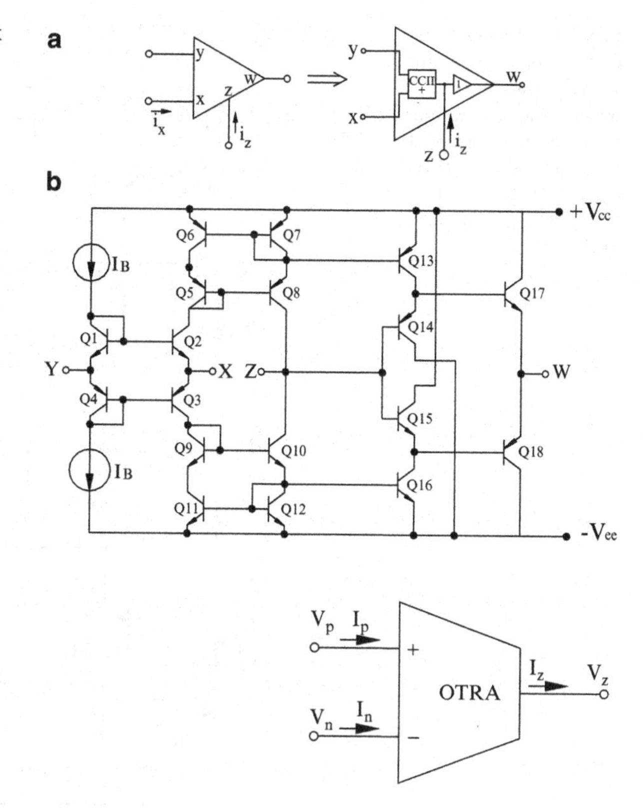

Fig. 1.12 Current feedback operational amplifier. (**a**) Notation and internal constituents, (**b**) bipolar implementation (adapted from [1] © 1990 Analog Devices, Inc.)

Fig. 1.13 Symbolic notation of the operational transresistance amplifier

CFOAs have attracted prominent attention in analog circuit design due to their two significant properties namely, the gain-bandwidth independence and very high slew rates together with their commercial availability as off-the self ICs from almost all leading IC manufactures.

Since this monograph is primarily about CFOAs, further aspects of the CFOAs would be elaborated in more details in the subsequent chapters.

1.4.4 The Operational Trans-resistance Amplifier

Operational trans-resistance amplifier (OTRA) has attracted considerable attention of analog designers in the context of recent developments in current-mode analog integrated circuits. The symbolic notation of the OTRA is given in Fig. 1.13.

An OTRA is characterized by the matrix equation

$$\begin{bmatrix} V_p \\ V_n \\ V_z \end{bmatrix} = \begin{bmatrix} 0 & 0 & 0 \\ 0 & 0 & 0 \\ R_m & -R_m & 0 \end{bmatrix} \begin{bmatrix} I_p \\ I_n \\ I_z \end{bmatrix} \tag{1.18}$$

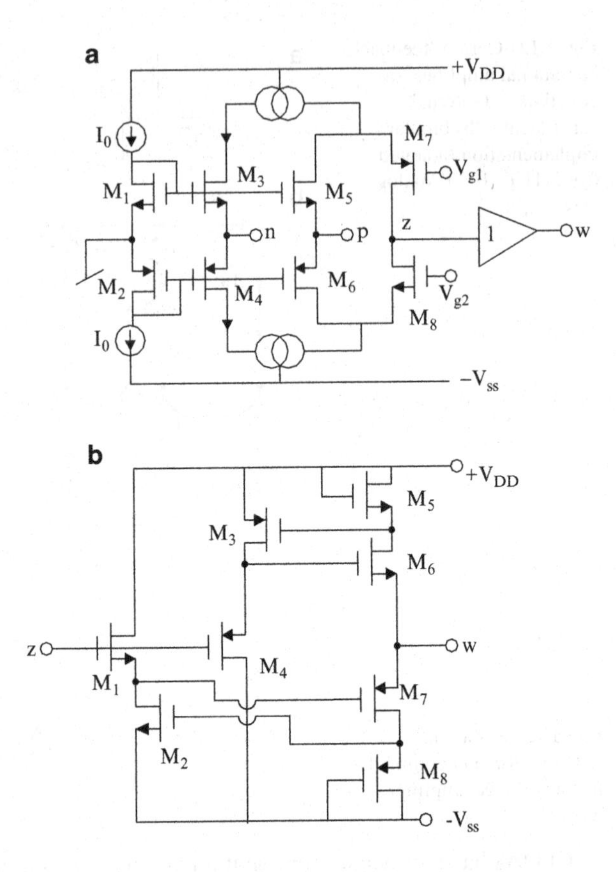

Fig. 1.14 An exemplary CMOS implementation of the operational transresistance amplifier by Toker et al. (adapted from [39] © 2000 IEEE). (**a**) The schematic of the CMOS OTRA. (**b**) Schematic of CMOS buffer

An exemplary CMOS OTRA advanced by Toker et al. [39] is shown in Fig. 1.14a whereas the voltage follower employed therein is shown in Fig. 1.14b.

OTRAs have so far been used in the realization of MOSFET-C filters, immittance simulators, square wave generators etc. for instance, see [40–43, 57] and the references cited therein.

From the literature survey, it has been found that the possible applications of OTRAs are clearly in infant stage and newer applications are still continuously being explored. Also, the work towards improving the CMOS circuit design of OTRAs is still continuing. Conceptually, an OTRA[2] has two low-impedance inputs and one low-impedance output. Since both input and output terminals of OTRA are characterized by low-impedance they offer the advantage of eliminating response limitations incurred by parasitic capacitances.

[2] Although a specific type of OTRA namely, the so called Norton amplifier had been commercially available since long from several manufacturers such as LM3900 from National Semiconductors, these commercial realizations do not provide virtual ground at the input terminals and they allow the input current to flow only in one direction. The former disadvantage limits the functionality of the Norton amplifier whereas the later calls for the use of external DC bias circuits leading to complex and clumsy designs even for simple functions.

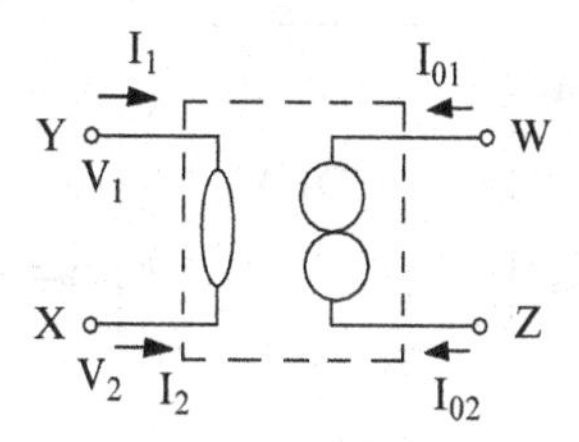

Fig. 1.15 Symbolic notation of the four terminal floating nullor

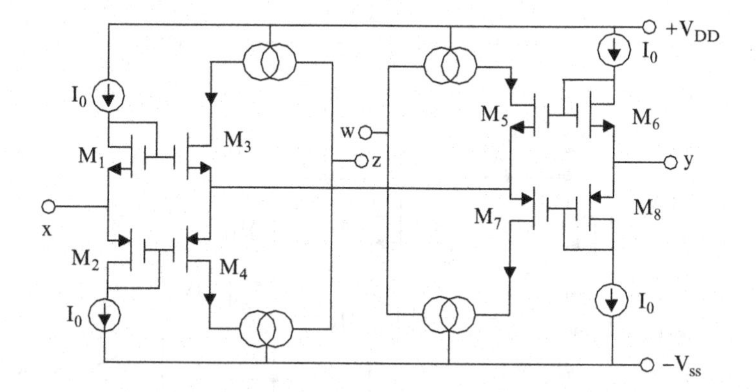

Fig. 1.16 CMOS implementation of the FTFN (adapted from [49] © 2000 IEE)

1.4.5 *The Four-Terminal-Floating-Nullor*

It was demonstrated in [44] and [45] (see also [46] and the references cited therein) that the Four terminal floating nullor (FTFN)[3] is a very general and flexible building block compared to the other active elements such as voltage-mode op-amps. This led to a growing interest in the design of amplifiers, gyrators, inductance simulators, oscillators and current-mode filters using FTFN as the active element [47]. FTFNs have been implemented using either a supply current sensing method with an op-amp and current mirrors [48] or using two CCIIs (as suggested in [44]) or two current feedback op-amp ICs AD844 from Analog Devices. An FTFN can be considered to be high gain transconductance amplifier with floating input and output terminals and can also be called an operational floating amplifier (OFA) [58]. The ideal nullor notation of the FTFN is shown in Fig. 1.15 whereas, a typical CMOS implementation of this building block, proposed by Cam et al. [49] which is an embodiment of the interconnection of two CMOS CCII+, is given in Fig. 1.16 and is characterized by $V_1 = V_2$, $I_{01} = -I_{02}$, $I_1 = 0 = I_2$, V_w and V_z being arbitrary.

Though nullors have been regarded in the circuit theory literature as universal elements and have found numerous applications as well as several integratable FTFN architectures have been evolved but a *perfect* FTFN implementation is still elusive.

[3] It may be mentioned that acronym 'FTFN' was first coined explicitly in [44] and [45].

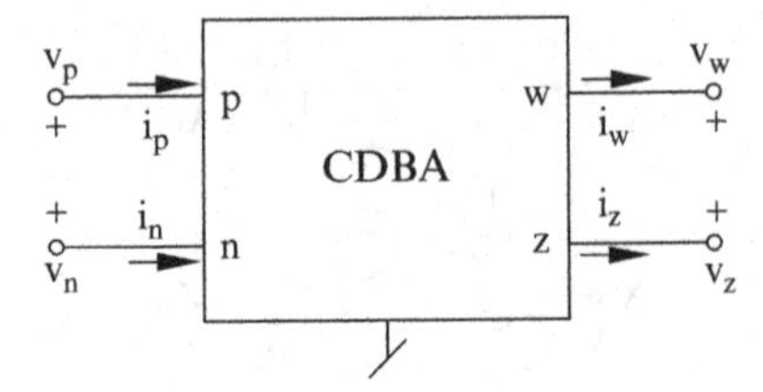

Fig. 1.17 Symbolic notation of the current differencing buffered amplifier

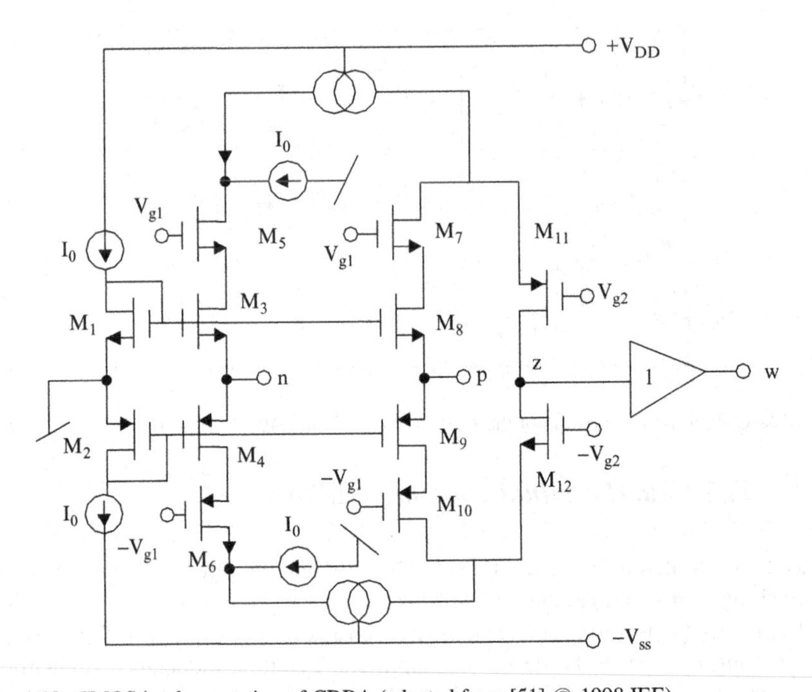

Fig. 1.18 CMOS implementation of CDBA (adapted from [51] © 1998 IEE)

1.4.6 The Current Differencing Buffered Amplifier

The Current Differencing Buffered Amplifier (CDBA) was introduced by Acar and Ozoguz [50]. The symbolic notation of the CDBA, is given in Fig. 1.17 and an exemplary realization of CDBA advanced by Ozoguz et al. [51], is reproduced here in Fig. 1.18., and is characterized by $I_z = I_p - I_n$, $V_p = 0 = V_n$ and $V_w = V_z$. The buffer shown in Fig. 1.18 is same as in Fig. 1.14b.

From the available literature on CDBAs (for instance see, [50–52] and the references cited therein), it is found that the advantages of CDBAs have not been fully exploited in available applications so far and work is still continuing in this area.

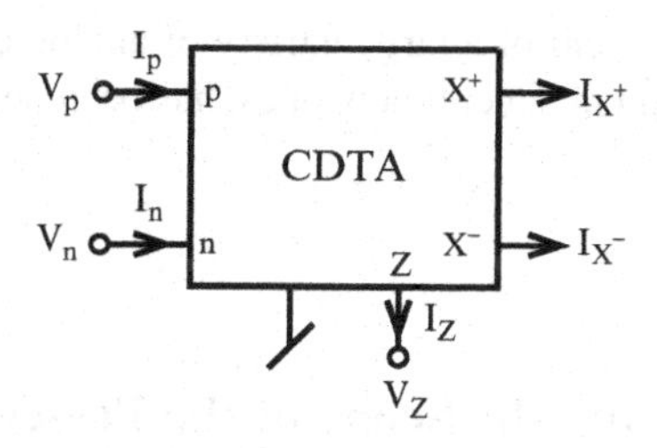

Fig. 1.19 Symbolic notation of the CDTA

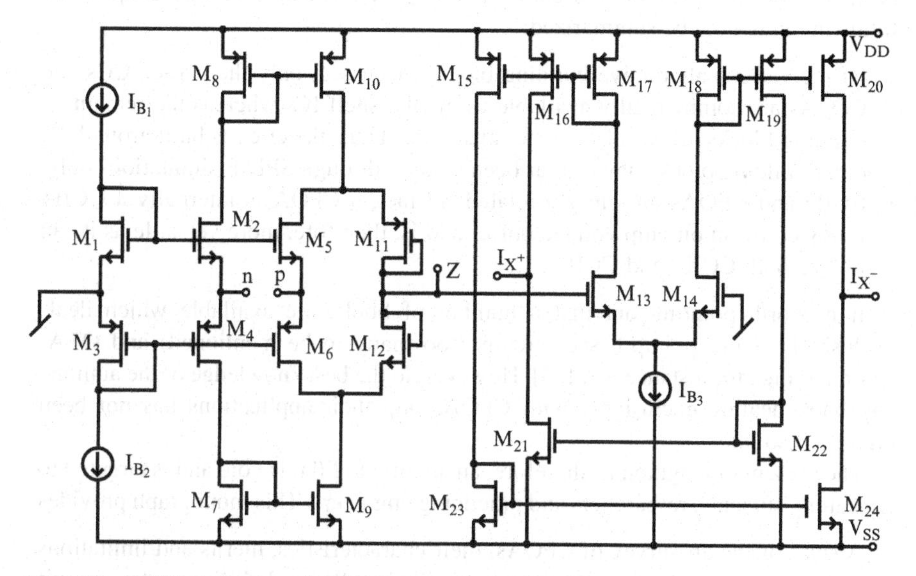

Fig. 1.20 CMOS realization of CDTA (adapted from [56] © 2006 Elsevier)

1.4.7 *The Current Differencing Transconductance Amplifier (CDTA)*

The CDTA was introduced in [53] by Biolek. The CDTA is a five terminal active element and has been shown to be a useful building block for the realization of a class of analog signal processing circuits [59, 60]. The symbolic notation of CDTA is shown in the Fig. 1.19. The CDTA can contain an arbitrary number of X-terminals, providing currents I_x of both directions. The port relations characterizing CDTA are given by

$$V_p = V_n = 0,\ I_z = I_p - I_n, \text{and } I_X = \pm g\ V_Z \tag{1.19}$$

where g is the transconductance.

A representative CMOS implementation of the CDTA is shown in Fig. 1.20.

A comprehensive treatment of a large variety of analog circuit building blocks, along with the introduction of several new ones, has been dealt in a recent paper by Biolek et al. [54].

1.5 The Necessity and the Scope of the Present Monograph

From the brief exposition presented in the earlier sections of this chapter the following may now be summarized:

- Of the various alternative building blocks discussed, only the OTAs, CCs and CFOAs are commercially available as off-the-shelf ICs whereas the remaining building blocks are, as of now, not available. Thus, the circuits built around the other building blocks have so far been studied through SPICE simulations only.
- CCIIs and CFOAs are closely related; in fact, a CFOA is internally a CCII+ followed by an on-chip voltage buffer and is, therefore, more versatile as it can realize both CCII+ and CCII−.

It is worth pointing out that a number of books are available which deals exclusively with op-amp-based circuits (too many to be mentioned) and OTA-based circuits, for instance, see [55]. However, to the best knowledge of the authors, any book dealing exclusively with CFOAs and their applications has not been published so far.

The present monograph is, therefore, an attempt to fill this void and is targeted to educators, students, researchers and practicing engineers. This monograph provides

- A state-of-the art survey of CFOAs, their characteristics, merits and limitations and various types of commercially-available off-the-shelf integrated circuit CFOAs
- A repertoire of prominent application circuits using CFOAs (covering both linear and no-linear applications) at a single place, with critical comments on the merits and demerits of various configurations (instead of being required to search a vast amount of literature published in various professional journals over the last more than 15 years).
- An appraisal of recent advances made in the design of bipolar and CMOS CFOAs and their variants
- A number of open problems and ideas for research for more advanced research-oriented readers
- A comprehensive list of references on Current feedback operational amplifiers and their applications (including those referred in the text as well as those suggested for further reading).

References

1. Analog Devices (1990) Linear products data book. Analog Devices Inc., Norwood, MA
2. Fabre A (1992) Gyrator implementation from commercially available transimpedance operational amplifiers. Electron Lett 28:263–264
3. Svoboda JA, McGory L, Webb S (1991) Applications of commercially available current conveyor. Int J Electron 70:159–164
4. Toumazou C, Payne A, Lidgey FJ (1993) Current-feedback versus voltage feedback amplifiers: history, insight and relationships. IEEE Int Symp Circuits Syst 2:1046–1049
5. Franco S (1993) Analytical foundations of current-feedback amplifiers. IEEE Int Symp Circuits Syst 2:1050–1053
6. Bowers DF (1993) The so-called current-feedback operational amplifier-technological breakthrough or engineering curiosity? IEEE Int Symp Circuits Syst 2:1054–1057
7. Toumazou C, Lidgey FJ (1994) Current-feedback op-amps: a blessing in disguise? IEEE Circ Devices Mag 10:34–37
8. Soliman AM (1996) Applications of the current feedback operational amplifiers. Analog Integr Circ Sign Process 11:265–302
9. Lidgey FJ, Hayatleh K (1997) Current-feedback operational amplifiers and applications. Electron Commun Eng J 9:176–182
10. Senani R (1998) Realization of a class of analog signal processing/signal generation circuits: novel configurations using current feedback op-amps. Frequenz 52:196–206
11. Deboo GJ (1967) A novel integrator results by grounding its capacitor. Electron Des 15:90
12. Horrocks D (1974) A non-inverting differentiator using a single operational amplifier. Int J Electron 37:433–434
13. Ganguly US (1976) Precise noninverting operator realization with high-resistive input impedance. Proc IEEE 64:1019–1021
14. Rathore TS (1977) Inverse active networks. Electron Lett 13:303–304
15. Abuelma'atti MT, Almaskati RH (1987) Active-C oscillator. Electron Wireless World 93:795–796
16. Linares-Barranco B, Rodriguez-Vazquez A, Huertas JL, Sanchez-Sinencio E, Hoyle JJ (1988) Generation and design of sinusoidal oscillators using OTAs. Proc IEEE Int Symp Circ Syst 3:2863–2866
17. Abuelma'atti MT, Almaskati RH (1989) Two new integrable active-C OTA-based linear voltage (current)-controlled oscillations. Int J Electron 66:135–138
18. Senani R (1989) New electronically tunable OTA-C sinusoidal oscillator. Electron Lett 25:286–287
19. Abuelma'atti MT (1989) New minimum component electronically tunable OTA-C sinusoidal oscillators. Electron Lett 25:1114–1115
20. Senani R, Amit Kumar B (1989) Linearly tunable Wien bridge oscillator realised with operational transconductance amplifiers. Electron Lett 25:19–21
21. Senani R, Tripathi MP, Bhaskar DR, Amit Kumar B (1990) Systematic generation of OTA-C sinusoidal oscillators. Electron Lett 26:1457–1459, also see (1991) *ibid*, 27:100–101
22. Senani R, Amit Kumar B, Tripathi MP, Bhaskar DR (1991) Some simple techniques of generating OTA-C sinusoidal oscillators. Frequenz 45:177–181
23. Bhaskar DR, Tripathi MP, Senani R (1993) A class of three-OTA-two-capacitor oscillators with non-interacting controls. Int J Electron 74:459–463
24. Bhaskar DR, Tripathi MP, Senani R (1993) Systematic derivation of all possible canonic OTA-C sinusoidal oscillators. J Franklin Inst 330:885–900
25. Bhaskar DR, Senani R (1994) New linearly tunable CMOS-compatible OTA-C oscillators with non-interacting controls. Microelectron J 25:115–123
26. Rodriguez-Vazquez A, Linares-Barranco B, Huertas JL, Sanchez-Sinencio E (1990) On the design of voltage-controlled sinusoidal oscillators using OTAs. IEEE Trans Circ Syst 37:198–211

27. Linnares-Barranco B, Rodriguez-Vazquez A, Sanchez-Sinencio E, Huertas JL (1989) 10 MHz CMOS OTA-C voltage-controlled quadrature oscillator. Electron Lett 25:765–767
28. Linnares-Barranco B, Rodriguez-Vazquez A, Sanchez-Sinencio E, Huertas JL (1991) CMOS OTA-C high-frequency sinusoidal oscillators. IEEE J Solid State Circ 26:160–165
29. Sanchez-Sinencio E, Silva-Martinez J (2000) CMOS transconductance amplifiers, architectures and active filters: a tutorial. IEE Proc Circ Devices Syst 147:3–12
30. Guo N, Rout R (1998) Realisation of low power wide-band analog systems using a CMOS transconductor. IEEE Trans Circ Syst II 45:1299–1303
31. Wilson G (1992) Linearized bipolar transconductor. Electron Lett 28:390–391
32. Lee J, Hayatleh K, Lidgey FJ (2002) Linear Bi-CMOS transconductance for Gm-C filter applications. J Circ Syst Comput 11:1–12
33. Smith KC, Sedra AS (1968) The current conveyor—a new circuit building block. Proc IEEE 56:1368–1369
34. Sedra AS, Smith KC (1970) A second generation current conveyor and its applications. IEEE Trans Circ Theory 17:132–134
35. Gilbert B (1975) Translinear circuits: a proposed classification. Electron Lett 11:14–16
36. Schmid H (2003) Why 'Current Mode' does not guarantee good performance. Analog Integr Circ Sign Process 35:79–90
37. Fabre A (1985) Translinear current conveyors implementation. Int J Electron 59:619–623
38. Normand G (1985) Translinear current conveyors. Int J Electron 59:771–777
39. Toker A, Ozoguz S, Cicekoglu O, Acar C (2000) Current-mode all-pass filters using current differencing buffered amplifier and a new high-Q bandpass filter configuration. IEEE Trans Circ Syst II 47:949–954
40. Chen JJ, Tsao HW, Liu SI (2001) Voltage-mode MOSFET-C filters using operational transresistance amplifiers (OTRAs) with reduced parasitic capacitance effect. IEE Proc Circ Devices Syst 148:242–249
41. Cam U, Kacar F, Cicekoglu O, Kuntman H, Kuntman A (2004) Novel two OTRA-based grounded immittance simulator topologies. Analog Intger Circ Sign Process 39:169–175
42. Gupta A, Senani R, Bhaskar DR, Singh AK (2012) OTRA-based grounded-FDNR and grounded-inductance simulators and their applications. Circuits Syst Sign Process 31:489–499
43. Hou CL, Chien HC, Lo YK (2005) Square wave generators employing OTRAs. IEE Proc Circ Devices Syst 152:718–722
44. Senani R (1987) A novel application of four-terminal floating nullors. Proc IEEE 75:1544–1546
45. Senani R (1987) Generation of new two-amplifier synthetic floating inductors. Electron Lett 23:1202–1203
46. Huijsing JH (1990) Operational floating amplifier. IEE Proc 137:131–136
47. Kumar P, Senani R (2002) Bibliography on nullors and their applications in circuit analysis, synthesis and design. Analog Integr Circ Sign Process 33:65–76
48. Higashimura M (1991) Realization of current-mode transfer function using four-terminal floating nullor. Electron Lett 27:170–171
49. Cam U, Toker A, Kuntman H (2000) CMOS FTFN realization based on translinear cells. Electron Lett 36:1255–1256
50. Acar C, Ozoguz S (1999) A new versatile building block: current differencing buffered amplifier suitable for analog signal-processing. Microelectron J 30:157–160
51. Ozoguz S, Toker A, Acar C (1998) Current-mode continuous-time fully integrated universal filter using CDBAs. Electron Lett 35:97–98
52. Pathak JK, Singh AK, Senani R (2011) Systematic realization of quadrature oscillators using current differencing buffered amplifiers. IET Circ Devices Syst 5:203–211
53. Biolek D (2003) CDTA-building block for current-mode analog signal processing. Proc ECCTD Poland III: 397–400
54. Bilolek D, Senani R, Biolkova V, Kolka Z (2008) Active elements for analog signal processing: classification, review, and new proposals. Radioengineering 17:15–32

55. Deliyannis T, Sun Y, Fidler JK (1999) Continuous-time active filter design. CRC, Boca Raton, FL
56. Keskin AU, Bilolek D, Honcioglue E, Biolkova V (2006) Current-mode KHN filter employing current differencing transconductance amplifiers. Int J Electron Commun (AEU) 60:443–446
57. Cakir C, Cam U, Cicekoglu O (2005) Novel all pass filter configuration employing single OTRA. IEEE Trans Circ Syst II 52:122–125
58. Huijsing JH (1993) Design and applications of operational floating amplifier (OFA): the most universal operational amplifier. Analog Integr Circ Sign Process 4:115–129
59. Prasad D, Bhaskar DR, Singh AK (2010) New grounded and floating simulated inductance circuits using current differencing transconductance amplifiers. Radioengineering 19:194–198
60. Prasad D, Bhaskar DR, Singh AK (2008) Realisation of single-resistance-controlled sinusoidal oscillator: a new application of the CDTA. WSEAS Trans Electron 5:257–259

Chapter 2
CFOAs: Merits, Demerits, Basic Circuits and Available Varieties

2.1 Introduction

Current feedback op-amps (CFOA) started attracting attention of the analog circuit designers and researchers when it was realized that one can design amplifiers exhibiting a characteristic which was the most significant departure from the characteristics exhibited by well-known VOA-based realizations in that CFOA-based circuits could realize variable-gain and yet constant bandwidth, as against the unavoidable gain-band-width-conflict in case of the VOA-based designs (as explained in Chap. 1). Furthermore, it was recognized that due to much higher slew rates of the order of several hundred to several thousand V/μs (which can be as large as 9,000 V/μs for modern CFOAs), as compared to a very modest 0.5 V/μs for the general purpose and most popular μA741-type VOA, CFOAs could lead to circuits capable of operating over much wider frequency ranges than those possible with VOAs.

In this chapter, we focus on the merits and demerits of CFOAs; discuss the various basic analog circuits realizable with CFOAs and highlight a variety of commercially available IC CFOAs from the various leading IC manufacturers.

2.2 AD844: The CFOA with *Externally-Accessible* Compensation Pin

Although in view of the popularity of the CFOAs they have been manufactured as integrated circuits by a number of IC manufacturers, there are two varieties which are in use. There are CFOAs which are pin-compatible to VOAs and do not have externally accessible compensation pin. On the other hand, AD 844-type CFOA from Analog Devices [1] has the option that its compensation pin (number 5) is externally-accessible while still maintaining pin-capability with VOAs.

R. Senani et al., *Current Feedback Operational Amplifiers and Their Applications*, Analog Circuits and Signal Processing, DOI 10.1007/978-1-4614-5188-4_2,

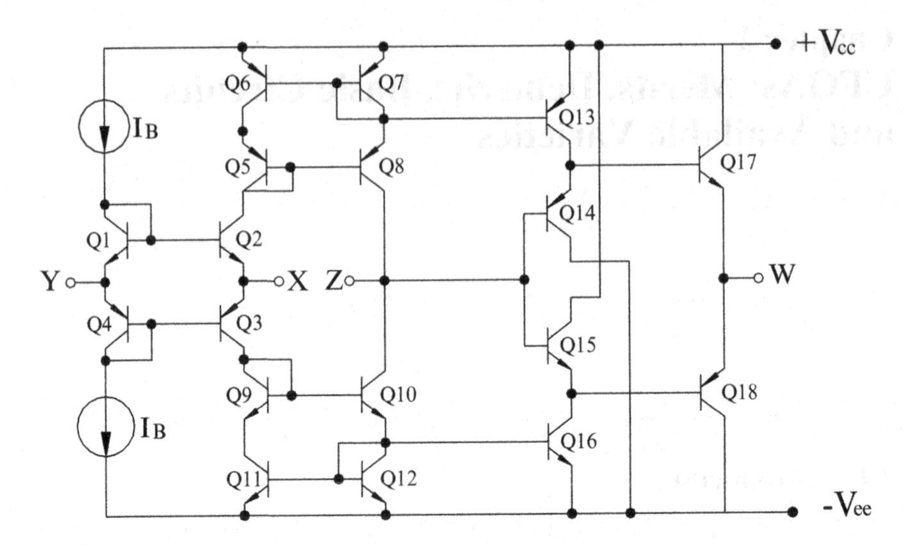

Fig. 2.1 A simplified schematic of the CFOA AD844 (adapted from [1] © 1990 Analog Devices, Inc.)

The AD844 from Analog Devices is a high speed monolithic (current feedback) op-amp which has been fabricated using junction- isolated complementary bipolar (CB) process. It has high bandwidth (around 60 MHz at gain of -1 and around 33 MHz at gain of -10) and provides very fast large signal response with excellent DC performance. It has very high slew rate, typically, 2,000 V/μs. Although it is optimized for use in current to voltage conversion applications and as inverting amplifier, it is also suitable for use in many non-inverting and other applications. Typical applications recommended by the manufacturers include Flash ADC input amplifiers, High speed current DAC interfaces, Video buffers and cable drivers and pulse amplifiers.

The AD844 can be used for replacement of traditional VOAs but due to its current feedback architecture results in much better AC performance, high linearity and excellent pulse response. The off-set voltage and input bias currents of the AD844 are laser- trimmed to minimize DC errors such that drift in the offset voltage is typically 1 μV/°C and bias current drift is around 9 nA/°C. AD844 is particularly suitable for video applications and as an input amplifier for flash type analog-to-digital convertors (ADC). A simplified schematic of the AD844 CFOA [1] is shown in Fig. 2.1.

It is interesting to point out that due to AD844 being sold, *disguised* as a large bandwidth, high slew-rate op-amp, initially it almost got unnoticed that its internal architecture, is, in fact, a translinear second generation plus type Current Conveyor[1]

[1] The *Current Conveyors* were introduced as new circuit building blocks by Sedra and Smith in [2, 3]; the first generation Current Conveyor (CCI) in [2] and the more versatile, the second generation Current Conveyor (CCII±) in [3].

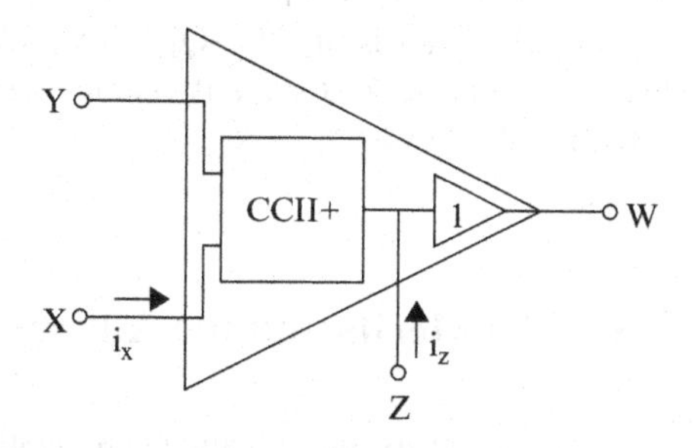

Fig. 2.2 A block diagram of the internal architecture of CFOA AD844

(CCII+) followed by a (translinear) voltage buffer. Its simplified symbolic diagram showing this identification is shown in Fig. 2.2.

Since the internal architecture of AD844 consists of a CCII + followed by a voltage buffer, this flexibility was later found to be useful in allowing the AD844 to be used as a CCII + and CCII− (using two CCII+), as pin-by-pin replacement of a VOA (with Z-pin left open) and lastly, as a 4-terminal building block in its own right.

In view of its front end being a CCII + and the back end being a voltage follower, the terminal equations of the CFOA can be written as

$$i_y = 0,\ v_x = v_y,\ i_z = i_x \quad \text{and} \quad v_w = v_z \tag{2.1}$$

In the internal architecture of the CFOA, transistors Q_1–Q_4 are configured as a mixed translinear cell (MTC) while the collector currents of transistors Q_2 and Q_3 are sensed by two modified p-n-p and n-p-n Wilson Current Mirrors consisting of transistors Q_5–Q_8 and Q_9–Q_{12} respectively to create a replica of current i_x at the terminal- Z thereby yielding $i_z = i_x$. The two constant current sources, each equal to I_B, force equal emitter currents in transistors Q_1 and Q_4 thereby forcing input current $i_y = 0$ when a voltage V_y is applied at the input terminal Y. It can be easily proved that with $i_x = 0$, $V_x = V_y$ and the Z-port current i_z will be zero. However, for the case of $i_x \neq 0$, an exact analysis [4] of the circuit using exponential relations between collector currents and base-emitter voltages for the transistors Q_1–Q_4 yields

$$I_z = I_x = -2I_B \operatorname{Sinh}\left(\frac{V_y - V_x}{V_T}\right) \tag{2.2}$$

from which an approximate relation between V_x, V_y and r_x (for $I_x << 2I_B$) can be expressed as follows

$$V_x \cong V_y + r_x i_x \quad \text{where } r_x = \frac{V_T}{2I_B} \tag{2.3}$$

If terminal-Z is terminated into an external impedance/load Z_L, a voltage V_z is created which passes through the voltage follower made from another MTC composed of transistors Q_{13}–Q_{18} for which transistors Q_{13} and Q_{16} provide the DC bias currents. The last stage is characterized by an equation similar to (2.3) which provides $V_w \cong V_z$.

2.3 The Merits and the Advantageous Features of the CFOAs

Two major merits and advantageous features of the CFOAs are (1) its very high (theoretically infinite) slew rate and (2) its capability of realizing amplifiers exhibiting gain-bandwidth decoupling. In the following, we elaborate these two characteristics of the CFOAs.

2.3.1 *The Reason and the Origin of the High Slew Rate*

In this sub-section we explain the origin and the reason for a very high slew rate of CFOAs as compared to conventional op-amps [5]. Figure 2.3a shows a simplified schematic of an internally compensated type IC op-amp exhibiting the differential transconductance stage consisting of transistors Q_1-Q_2-Q_3-Q_4, the intermediate gain stage (normally made from a cascade of CC-CE stages) having an inverting gain $-A_{v2}$ and the output stage which is a class AB type push-pull amplifier having both complementary transistors in emitter follower mode providing a voltage gain A_{v3} close to unity.

A straight forward analysis of the first stage reveals that the output current I_{out} is given by

$$I_{out} = I_B \tanh\left(\frac{V_{id}}{2V_T}\right) \tag{2.4}$$

A graphical representation of the above equation is shown in Fig. 2.3b. From this characteristic, it is seen that the output current i_o saturates to $+ I_B$ when V_{id} is large and positive while i_o saturates to $-I_B$ when V_{id} is large and negative. Thus, the maximum current available to charge the compensating capacitor C_c, is $\pm I_B$.

If such an op-amp is configured as a voltage follower by a feedback connection from V_{out} to the inverting input terminal of the op-amp and a large step signal is applied to the non-inverting input terminal at $t = 0$. This forces the transistor Q_1 into saturation and Q_2 into cut off due to which $I_{out} = I_B$ and thus, the capacitor C_c is charged *linearly* through constant current I_B.

In view of the high gain of the intermediate stage, for simplicity, its input node can be treated to be at *virtual ground* potential in which case one can write

$$I_{out} = C_c \frac{dV_{out}}{dt} \tag{2.5}$$

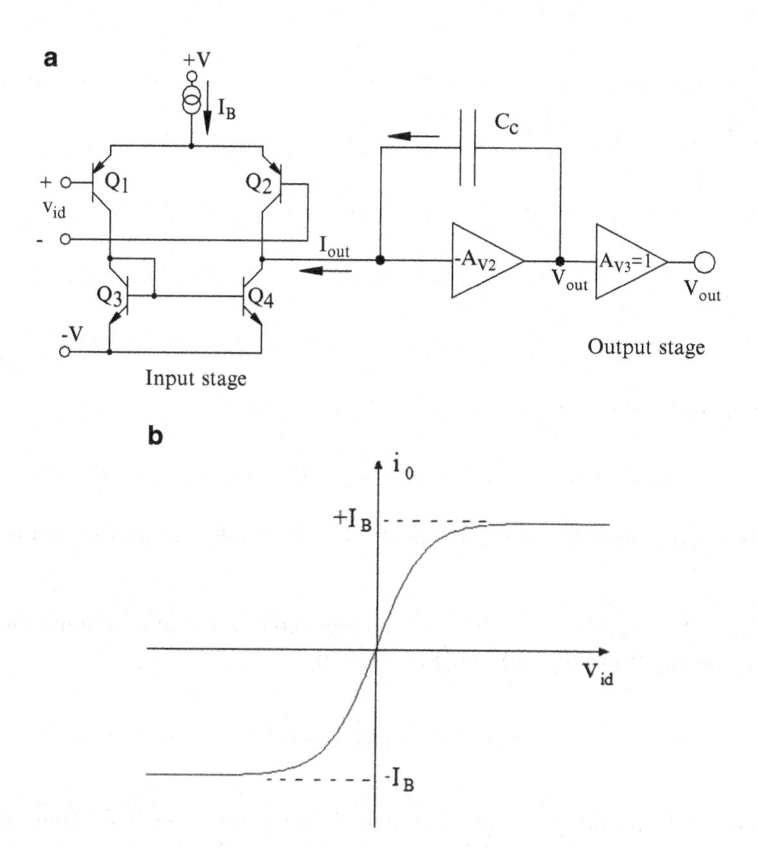

Fig. 2.3 (**a**) Simplified model of an internally compensated IC op-amp. (**b**) The tanh-characteristics of the input differential transconductance stage

Hence, the slew-rate (SR) is given by

$$\mathrm{SR} = \left.\frac{dV_{out}}{dt}\right|_{\max} = \frac{I_{out}}{C_c} = \frac{I_B}{C_c} \tag{2.6}$$

With $C_c = 30$ pF and $I_B = 19$ μA (as applicable to a μA741 type op-amp biased with ±15 V DC power supplies), the above figure turns out to be around 0 .63 V/μ s which is close to the data sheet value of 0.5 V/μs. For a sinusoidal output $V_0 = V_m \sin \omega t$, it can be shown that the maximum frequency ω_{max}, for which the limitation imposed by the finite slew rate will not come into play, is given by

$$\omega_{\max} = \frac{SR}{V_m} \tag{2.7}$$

Consider now a simplified schematic of the CFOA shown in Fig. 2.4a. An analysis of the input stage of the CFOA, which is made from MTC consisting of

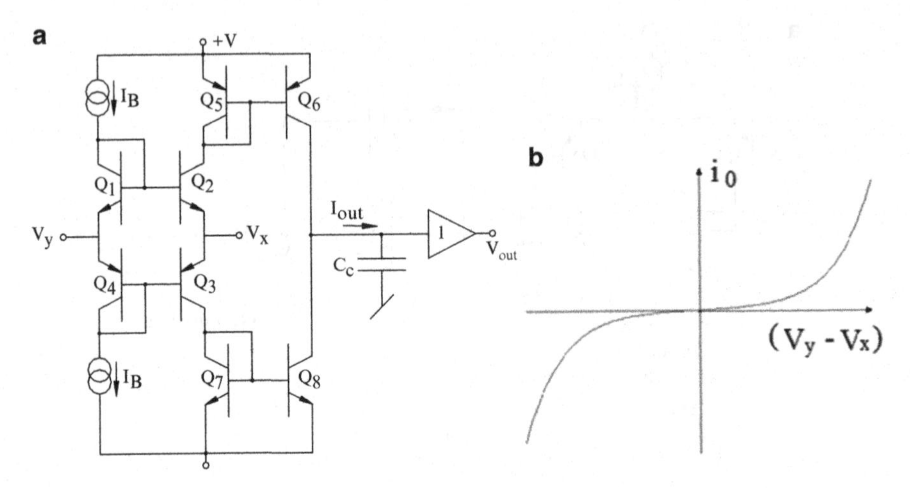

Fig. 2.4 (**a**) Simplified model of the CFOA. (**b**) The transfer characteristic between i_o and $(V_y - V_x)$

transistors $Q_1 - Q_4$ shows that the current output coming out of Z- terminal (which charges the compensating capacitor) is given by

$$I_{out} = 2I_B \,\mathrm{Sinh}\left(\frac{V_y - V_x}{V_T}\right) \tag{2.8}$$

A plot of the resulting transfer characteristic is shown in Fig. 2.4b. Thus, in this case, it is found that for a large differential input voltage (V_y–V_x), the output current which is the charging current of the compensating capacitor would be theoretically infinite. Thus, in contrast to VOAs, CFOAs have ideally infinite slew rate. In practice, slew rates from several hundred V/μs to as high as 9,000 V/μs are attainable. Consequently, a CFOA implementation of a circuit will not have the same kind of limitations on the maximum operational frequency range as prevalent in the corresponding VOA-based circuit. In other words, a CFOA-based circuit would operate satisfactorily over a frequency range much larger than possible for a VOA circuit realizing the same function.

2.3.2 De-coupling of Gain and Bandwidth: Realisability of Variable-Gain, Constant-Bandwidth Amplifiers

It has been explained in the previous chapter that all VOA-based controlled sources suffer from the drawback of gain- bandwidth-conflict. An important advantage of employing CFOAs is that this gain bandwidth conflict can be overcome due to the *current feedback* prevalent in the same configurations realized with CFOAs

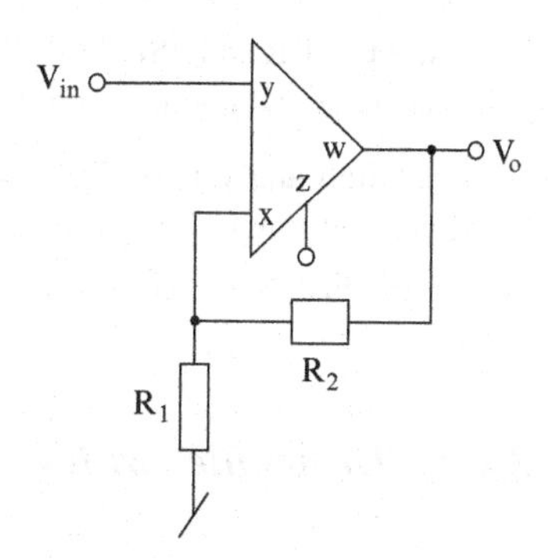

Fig. 2.5 The non-inverting amplifier using a CFOA

(Interestingly, we will see that even two alternative ways of realizing VCVS from CFOAs are also free from the gain-bandwidth-conflict).

Consider now the CFOA-based non-inverting amplifier of Fig. 2.5.

From an analysis of this circuit, taking CFOA characterization as $i_y = 0, v_x = v_y$, $i_z = i_x$ and $v_w = v_z = -i_z Z_p$ where Z_p is the parasitic impedance looking into the Z-terminal and consists of a resistance R_p (typically, around 3 MΩ) in parallel with a capacitance C_p (typically in the range 4.5 − 5.5 pF), the maximum gain of the circuit is found to be

$$\frac{V_0}{V_{in}} = \frac{\left(1 + \frac{R_2}{R_1}\right)}{\left(1 + \frac{R_2}{R_p}\right)} \tag{2.9}$$

whereas the −3 dB bandwidth is given by

$$\omega_{3-dB} = \frac{1}{C_p R_2}\left(1 + \frac{R_2}{R_p}\right) \cong \frac{1}{C_p R_2}; \quad \text{for } R_2 << R_p \tag{2.10}$$

It is, thus, seen that the bandwidth of the circuit can be fixed by setting the feedback resistor R_2 while the gain can be still varied through the variable resistor R_1 and therefore, the gain and bandwidth have become de-coupled and it has become possible to realize a constant-bandwidth, variable gain amplifier.

2.4 The Demerits and Limitations of CFOAs

2.4.1 Demerits

Despite its significant advantages over traditional VOAs, as explained in previous section, CFOAs generally have the following demerits:

- Relatively inferior DC precision.
- Relatively poor DC offset voltage due to the use of both PNP and NPN transistors.

- Lower CMRR and PSRR than VOAs due to the unsymmetrical complimentary-pair input stage and unequal and un-correlated input bias currents.

A detailed analysis of the input DC current, input offset voltage and maximum input voltage range for the input stage of a CFOA is given in [6] while a comprehensive analysis of output stage has been dealt in [7].

2.4.2 *Difficulties with Capacitive Feedback*

It should be kept in mind in devising CFOA-based circuits that a capacitive feedback between X and W is not recommended as it often leads to instability. Therefore, an inverting Miller integrator cannot be realized with a CFOA in the same way as the conventional op-amp-based Miller integrator.

2.4.3 *Effect of Stray Capacitances and Layout Issues*

Another important practical consideration to be taken care of is to take care of the stray capacitances on the inverting input node (X-input) and across the feedback resistor which invariably lead to peaking or ringing in the output response and sometimes even to oscillations. In view of this, appropriate care has to be taken in making an appropriate PCB layout and eliminate any stray capacitances. The performance of a CFOA-based circuit can be improved considerably with a good layout, good decoupling capacitors and low inductance wiring of the components.

2.5 Basic Circuits Using CFOAs

We now show how a number of basic analog circuits such as the four controlled sources, the voltage and current followers, the instrumentation amplifier and the integrators and differentiators can be realized in a number of advantageous ways using CFOAs sans the disadvantages associated with VOA-based realizations of the same functions.

2.5.1 *VCVS Configurations*

Consider now the various other VCVS realizations depicted in Fig. 2.6a–c.

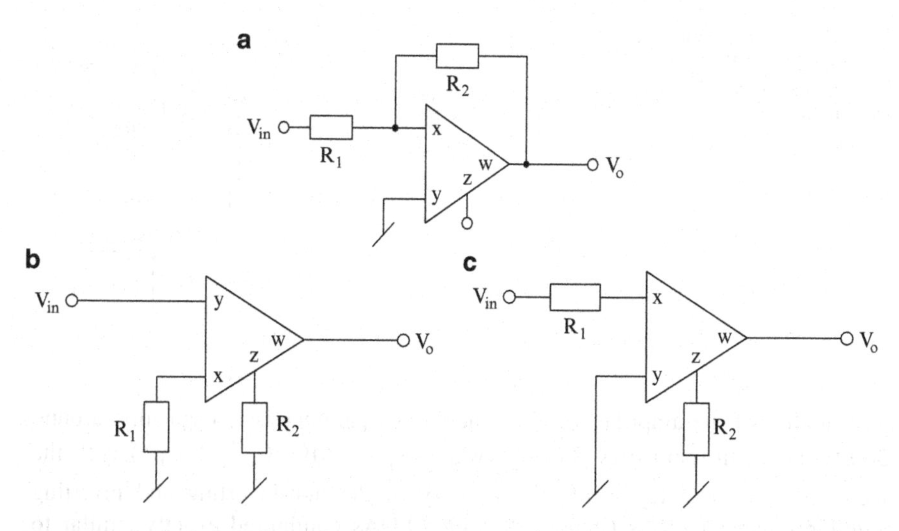

Fig. 2.6 Realization of various other VCVS circuits using a CFOA (**a**) inverting VCVS, (**b**) alternative non-inverting VCVS, (**c**) alternative inverting VCVS

A non-ideal analysis of all the three circuits reveals their non-ideal gains as:

$$\frac{V_0}{V_{in}} = -\frac{\frac{R_2}{R_1}}{\left(1 + \frac{R_2}{R_p}\right)} \text{ for the circuit of Fig. 2.6a} \tag{2.11}$$

$$\frac{V_0}{V_{in}} = \frac{\frac{R_2}{R_1}}{\left(1 + \frac{R_2}{R_p}\right)} \text{ for the circuit of Fig. 2.6b} \tag{2.12}$$

$$\frac{V_0}{V_{in}} = \frac{-\left(\frac{R_2}{R_1}\right)}{\left(1 + \frac{R_2}{R_p}\right)} \text{ for the circuit of Fig. 2.6c} \tag{2.13}$$

whereas the 3-dB bandwidth in all cases is given by the same value as in (2.10). Thus, in all the cases, the bandwidth can be set by the feedback resistor R_2 after which the gain can still be made variable through a single variable resistance R_1. Thus, the gain bandwidth conflict is not present in any of the four circuits. It is, therefore, possible to design constant-bandwidth variable-gain amplifiers using CFOAs which unfortunately cannot be done with the same topologies such as those of Figs. 2.5 and 2.6a realized with a traditional VOA.

However, it must be kept in mind that, in practice, constant bandwidth is achievable only for low to medium gains (typically, 1–10). Furthermore, the feedback resistor R_2 also cannot be chosen arbitrarily since this critically affects

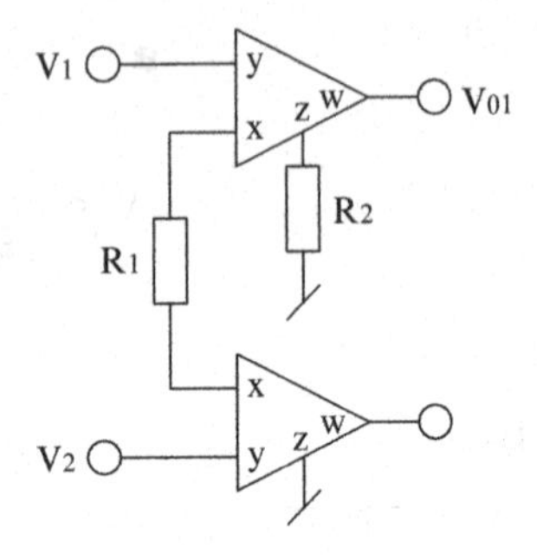

Fig. 2.7 An instrumentation amplifier using CFOAs (CFOA-version of Wilson's CCII-based circuit [8])

the stability of the amplifier. In fact, the CFOA parameters r_x (typically, around 50 Ω) and Z-pin parasitics $R_p \| \frac{1}{sC_p}$ (where $R_p = 3M\Omega$; $C_p = 4.5$ pF) with the feedback resistance R_2 decide the stability of the non-inverting and inverting amplifiers using CFOAs (if realized with CFOAs configured exactly similar to their VOA-counterparts). The manufacturer determines the optimum value of the feedback resistor R_2 during the characterization of the IC. Normally, lowering R_2 decreases stability whereas increasing R_2 decreases the bandwidth.

2.5.2 *Instrumentation Amplifier Using CFOAs*

We now show that, contrary to the traditional instrumentation amplifier which requires three VOAs and as many as seven resistors out of which four are required to be completely matched, the use of CFOAs makes it possible to realize a variable gain instrumentation amplifier with no more than two CFOAs along with a minimum number of only two resistors. Such a circuit is readily evolved from a known CCII-based circuit proposed by Wilson [8] and is shown here in Fig. 2.7.

Considering the finite input resistance looking into terminal-X of the CFOA as r_x and taking parasitic output impedance looking into terminal-Z as a resistance R_P in parallel with capacitance C_P, the maximum gain of this circuit is found to be:

$$\frac{V_0}{V_1 - V_2} = \frac{R_2}{(R_1 + 2r_x)\left(1 + \frac{R_2}{R_p}\right)} \tag{2.14}$$

whereas its 3-dB bandwidth is given by the some expression as in (2.10). Thus, it is seen that the bandwidth of the amplifier can be fixed at a constant value by fixing R_2 while the gain can be made variable by changing R_1. Thus, CFOA-based instrumentation amplifier also does not have the gain-bandwidth-conflict while employing a minimum possible number of passive components for realizing a variable gain.

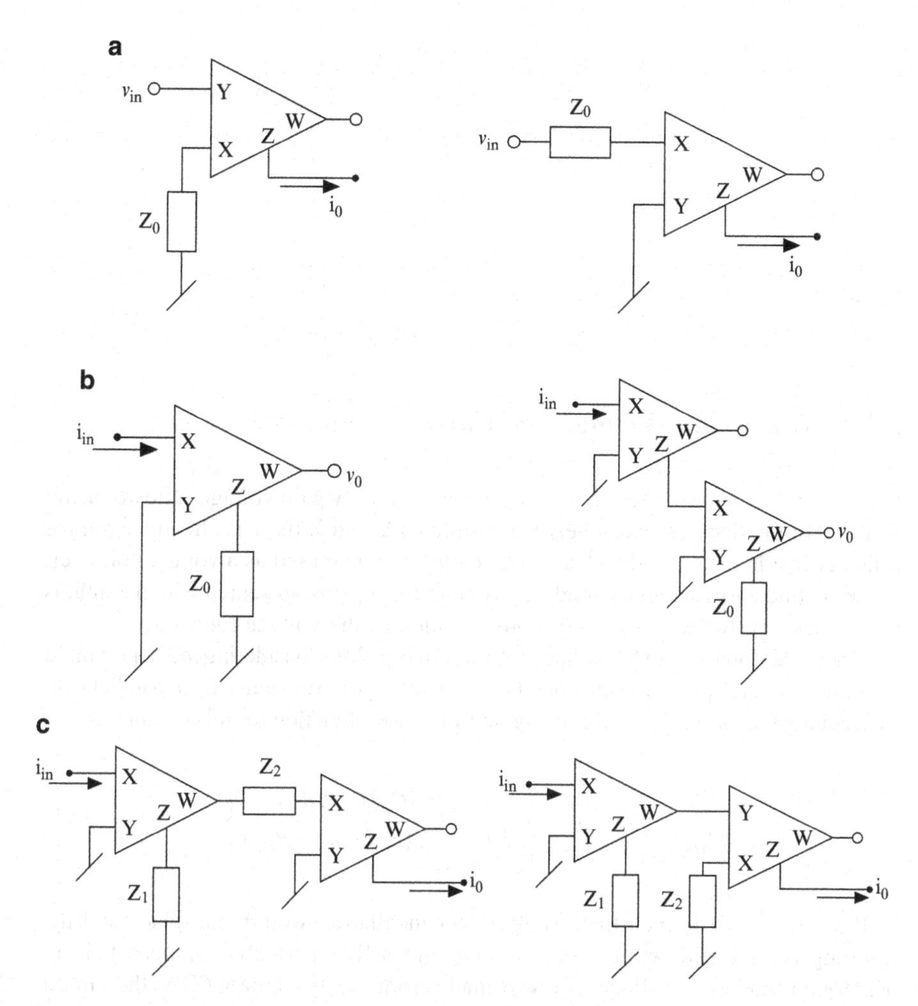

Fig. 2.8 Various controlled sources (**a**) Voltage controlled current sources. (**b**) Current controlled voltage sources. (**c**) Current controlled current sources

2.5.3 VCCS, CCVS and CCCS Configurations

In Fig. 2.8 we show the CFOA-based realization for non-inverting and inverting VCCS, CCVS and CCCS circuits. It may be noted that contrary to VOA-based circuits for VCCS and CCCS requiring as many as four identical resistors the corresponding realizations using CFOAs as in Fig. 2.8a–c employ a minimum possible number of passive components namely only one in case of Fig. 2.8a, b and two in case of Fig. 2.8c respectively thus, no component matching whatsoever is needed. Furthermore, it is straight forward to verify that all these circuits possess the most notable property of CFOA-based circuits i.e. no gain-bandwidth-conflict in the realization of any controlled sources.

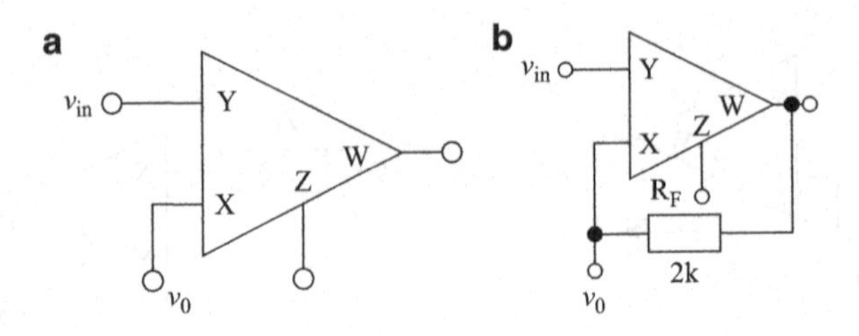

Fig. 2.9 Unity gain voltage followers using CFOA

2.5.4 *Unity Gain Voltage and Current Followers*

Figure 2.9 shows two different ways of realizing a unity gain voltage follower using CFOAs. In the first case since between terminals Y and X there is already a voltage follower inside the chip, the same voltage buffer can be used as a voltage follower. In the second case, a slightly modified version from [9] is presented which contains a feedback resistor R_F for the self-compensation of the voltage follower.

A non-ideal analysis of the voltage follower of Fig. 2.9b considering the X-port input resistance r_x and Z-port parasitic impedance consisting of a resistance R_p in parallel with a capacitance C_p, reveals the following non-ideal gain function for this circuit

$$\frac{V_0}{V_{in}} = \frac{\left(1+\frac{R_F}{R_p}\right)}{\left(1+\frac{r_x+R_F}{R_p}\right)}\left\{\frac{1+sC_p(R_p//R_F)}{1+sC_p(R_p//(r_x+R_F))}\right\} \tag{2.15}$$

If $R_F >> r_x$, it is seen that a pole-zero cancellation would take place and the resulting voltage gain will be close to unity and will be perfectly compensated for. It is found that for a voltage follower made from AD844-type CFOA, the circuit works quite well with $R_F = 2\ k\Omega$ [9].

The two possible realizations for unity gain current follower are shown in Fig. 2.10. As expected, none of the two circuits requires any resistors and both the circuits offer ideally zero input resistance and ideally infinite output resistance.

2.5.5 *Integrators and Differentiators*

In this subsection we first explain some integrators and differentiators [10] realizable similar to their VOAs counter parts. Due to the reason spelt out earlier an inverting integrator with a CFOA is not feasible. Since a capacitive feedback from

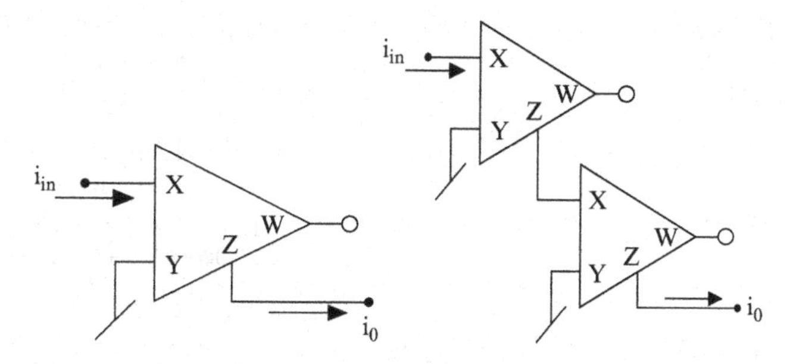

Fig. 2.10 Unity gain current followers using CFOA

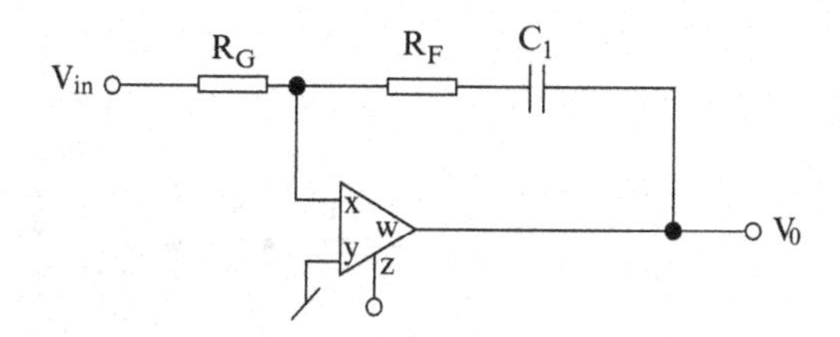

Fig. 2.11 An inverting integrator using a CFOA [10]

W to X leads to instability. However, a slightly modified version with an additional resistance incorporated in the feedback path is still possible as shown in Fig. 2.11.

Addition of resistor R_F is acceptable since at high frequency the resistor is dominant and hence feedback impedance would never drop below the resistor value. The transfer function of this circuit is given by

$$\frac{V_0}{V_{in}} = -\left(\frac{R_F}{R_G}\right)\left[\frac{s+\frac{1}{R_F C_1}}{s}\right] \tag{2.16}$$

$$\approx \frac{-1}{sC_1R_G}; \quad \text{for } \omega << \frac{1}{R_F C_1} \tag{2.17}$$

On the other hand, to realize a non-inverting integrator, one can make Deboo's integrator [11] almost in the same manner as is done with a VOA (see Fig. 2.12) however; this circuit suffers from the drawback of requiring four identical resistors and also has to fulfill a condition to ensure stable operation.

This circuit is characterized by the following transfer function.

$$\frac{V_0}{V_1} \cong \left(\frac{1+\frac{R_F}{R_G}}{sR_1C_1}\right) \tag{2.18}$$

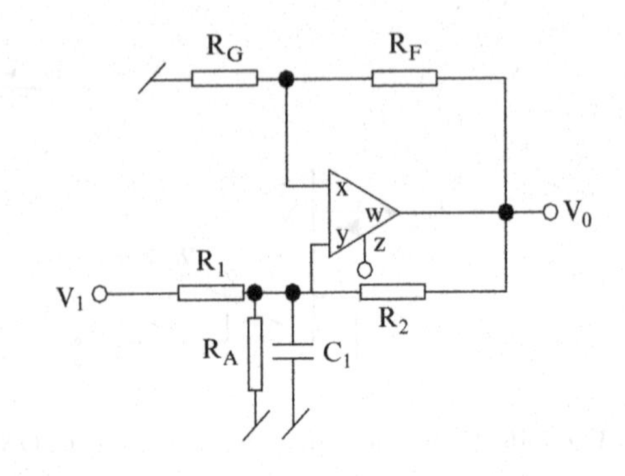

Fig. 2.12 CFOA-version of non-inverting Deboo's integrator [10, 11]

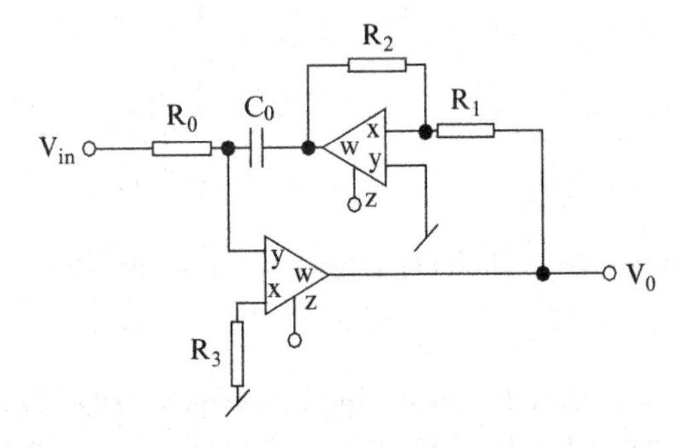

Fig. 2.13 Active-compensated non-inverting CFOA integrator

whereas the condition required for stable operation is

$$\frac{R_2}{R_1//R_A} \geq \frac{R_F}{R_G} \tag{2.19}$$

To circumvent the above problems, in Fig. 2.13 we show an alternative circuit for creating non-inverting integrator using two CFOAs [12]. This circuit has an in-built compensation for the non-ideal effects of the CFOA parasitic impedances.

The circuit of Fig. 2.13 realizes a non-inverting integrator since its transfer function is given by

$$\frac{V_0}{V_{in}} = \frac{1}{sT} \quad \text{where } T = \frac{C_0R_0R_2}{R_1} \tag{2.20}$$

Considering the Z-port parasitic impedance $Z_p = R_p//\frac{1}{SC_P}$ for both the CFOAs, a non-ideal analysis reveals

$$\frac{V_0}{V_{in}} \cong \frac{R_1}{sC_0R_0R_2}\varepsilon(s) \tag{2.21}$$

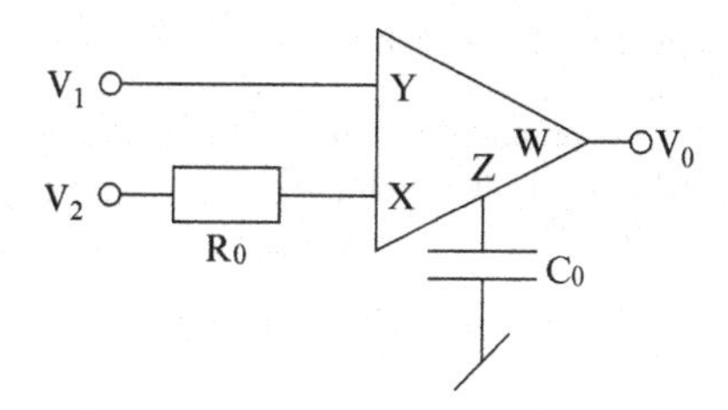

Fig. 2.14 A differential integrator using a CFOA

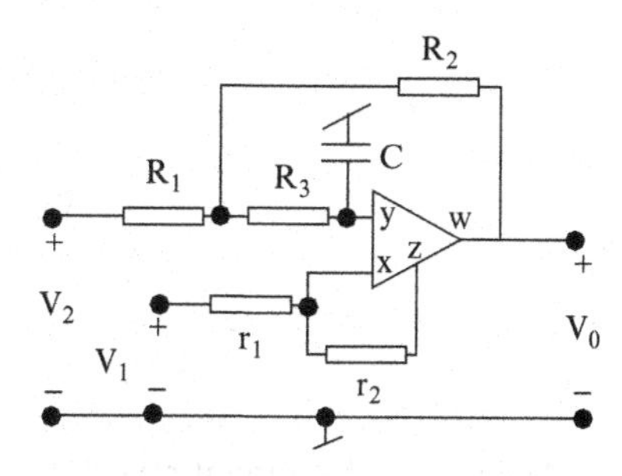

Fig. 2.15 Dual input integrator proposed by Lee and Liu (adapted from [13] © 1999 IET)

for $R_i << R_{pi}$, i = 0–3. The error function $\varepsilon(s)$ is given by

$$\varepsilon(s) = \frac{1 + sT_2}{1 + sT_1 + s^2 T_1 T_2} \quad \text{with } T_1 = C_P R_2; T_2 = C_P R_1 R_3 / R_2 \tag{2.22}$$

From the above, the phase error is given by

$$\phi \cong \omega(T_2 - T_1) - \omega^3 T_1^2 T_2 \tag{2.23}$$

Hence, for negligible phase error, one requires $T_1 = T_2$ which gives the required condition as $R_3 = R_2^2 / R_1$.

From the above, it is seen that with $R_3 = R_2^2 / R_1$, the phase error is minimized and active-compensation is achieved.

In the above cases, the circuits devised using CFOAs are exactly similar to their VOA counterparts. However, since a CFOA has an in built CCII+, there is an alternative way of realizing inverting/non-inverting integrators. A general circuit to realize an integrator in an alternative manner is shown in Fig. 2.14. An analysis of this circuit shows that the output voltage is given by

$$V_0 = \frac{1}{sC_0 R_0}(V_1 - V_2) \tag{2.24}$$

Thus, both inverting and non-inverting integrators can be realized from this circuit as special cases by grounding V_1 or V_2 respectively. A differentiator is obtainable from the same circuit by interchanging the resistor and the capacitor.

We now show a circuit which can perform the operation of dual input integrator using a single CFOA proposed by Lee and Liu [13] (Fig. 2.15).

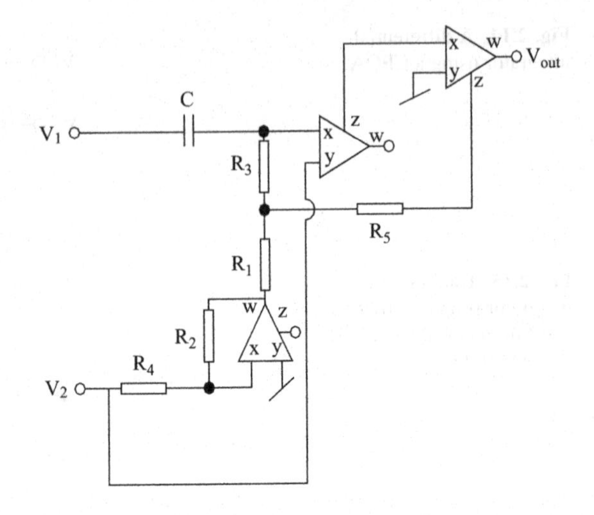

Fig. 2.16 Dual input differentiator using CFOAs proposed by Lee and Liu (adapted from [13] © 1999 IET)

Analysis of this circuit reveals

$$V_{out} = \frac{V_2 \frac{R_2}{R_1} - V_1 \frac{r_2}{2r_1}}{sC\left(R_3\left(1+\frac{R_2}{R_1}\right)+R_2\right) - \left(\frac{r_2}{2r_1} - \frac{R_2}{R_1}\right)} \tag{2.25}$$

If we choose $R_2/R_1 = r_2/2r_1$ the circuit realizes a dual input integrator with output voltage given by

$$V_{out} = \frac{1}{s\tau}(V_2 - V_1) \tag{2.26}$$

$$\tau = CR_1\left[1 + R_3\left(\frac{1}{R_1} + \frac{1}{R_2}\right)\right] \tag{2.27}$$

From equations (2.26) and (2.27) it is seen that the time constant of the integrator can be varied by changing the resistor R_3. The circuit operates well within the frequency range of 450 Hz to 1 MHz with a phase error of 5 °.

A dual-input differentiator [13] is shown in Fig. 2.16. The input of this circuit with $R_4 = (R_2 + R_3)$ and $R_5 = R_2$, is given by

$$V_{out} = V_2\left(\alpha - \frac{R_2}{R_3}(1-\alpha)\right) + sC(V_1 - V_2)\left[R_2 + R_1\left(1 + \frac{R_2}{R_3}\right)\right] \tag{2.28}$$

If $\alpha = R_2/(R_2 + R_3)$, (2.28) reduces to

$$V_{out} = sCR_2(V_1 - V_2)\left(1 + R_1\left(\frac{1}{R_2} + \frac{1}{R_3}\right)\right) \tag{2.29}$$

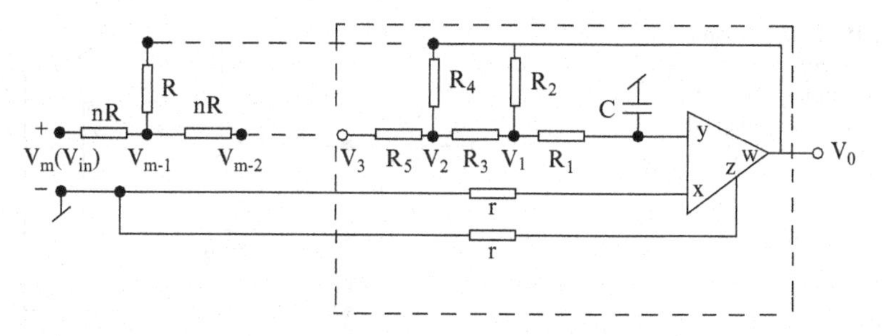

Fig. 2.17 Integrator with time constant multiplication proposed by Lee and Liu (adapted from [14] © 2001 IET)

Hence, the time constant can be varied by changing R_1. Over an operating frequency range of 1–100 kHz, this circuit works well with a phase error of the order of ±10 °.

In Fig. 2.17 we show another integrator circuit which was proposed by Lee and Liu in [14] and has the facility for time constant multiplication. Analysis of this circuit, as in [14], shows that its transfer function is given by

$$\frac{V_0}{V_{in}} = \frac{1}{snRC\frac{1}{2^m\sqrt{n^2+4n}}\left[\left(n+2+\sqrt{n^2+4n}\right)^m - \left(n+2-\sqrt{n^2+4n}\right)^m\right]+1} \tag{2.30}$$

By appropriate selection of **m** and **n**, a desired multiplication factor can be achieved. For instance, if we take m = 3 and n = 10 with V_3 as the input, it is possible to achieve a multiplication factor of 143.

The transfer function of the differentiator circuit of Fig. 2.18 is given by

$$V_0 = sC\frac{R_5\left(1+\frac{R_2}{R_1}\right)+R_2\left(1+\frac{R_5}{R_4}\right)\left(\frac{R_3}{R_1}+\frac{R_3}{R_2}+1\right)}{1-\frac{R_2R_5}{R_1R_4}}\,V_{in} \tag{2.31}$$

If $R_2 = R$, $R_4 = 2R$, $R_1 = R_3 = R_5 = 2nR$ the above equation can be expressed as

$$V_0 = 4sRC\left[n\left(1+\frac{1}{2n}\right)+(n+1)(n+1)\right] \tag{2.32}$$

As an example, if we select $R_2 = 1$ kΩ, $R_4 = 2$ kΩ, $R_1 = R_3 = R_5 = 10$ kΩ then the multiplication factor turns out to be 166.

In reference [14], it has been demonstrated that the circuit of Fig. 2.17 works well with in the frequency range 200Hz to 1 MHz with a phase error of less than 12° whereas for the differentiator of Fig. 2.18, the operating frequency range has been found to be 100 Hz to 10 kHz with a phase error less than 6 °.

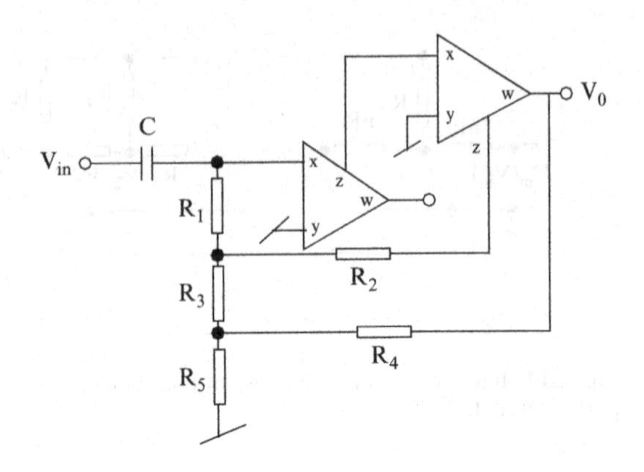

Fig. 2.18 Differentiator with time constant multiplication proposed by Lee and Liu (adapted from [14] © 2001 IET)

It must be mentioned that yet another differential integrator implemented from two AD 844-type CFOAs and capable of operating up to several MHz without encountering any stability problem was presented by Maundy et al. in [15].

2.6 Commercially Available Varieties of CFOAs

Although a wide varieties of CFOAs are available from various IC manufacturers, optimized with respect to a chosen parameter, it is interesting to note that the key building blocks used are two types of *mixed translinear cells.* In the following, we identify these two basic blocks and then briefly describe the internal architecture and characteristics/parameters of some exemplary IC CFOAs available from leading analog IC manufacturers.

2.6.1 The Mixed-Translinear-Cells (MTC) as Building Blocks of CFOAs

Most of the CFOA architectures have the internal structure of a CCII + followed by a voltage buffer. Since a CCII + itself has a voltage follower between its Y and X terminals, it, therefore, follows that a typical CFOA architecture would have two voltage followers (VF): one between Y and X terminals and the other between Z and W terminals. Furthermore, there has to be a mechanism of sensing the current flowing into the low-input impedance terminal X of the input VF, creating a copy of the same and making it available at the high output impedance Z-terminal where a compensating capacitor can be connected either internally or externally. Two standard configurations for realizing VFs are the two *mixed translinear Cells*

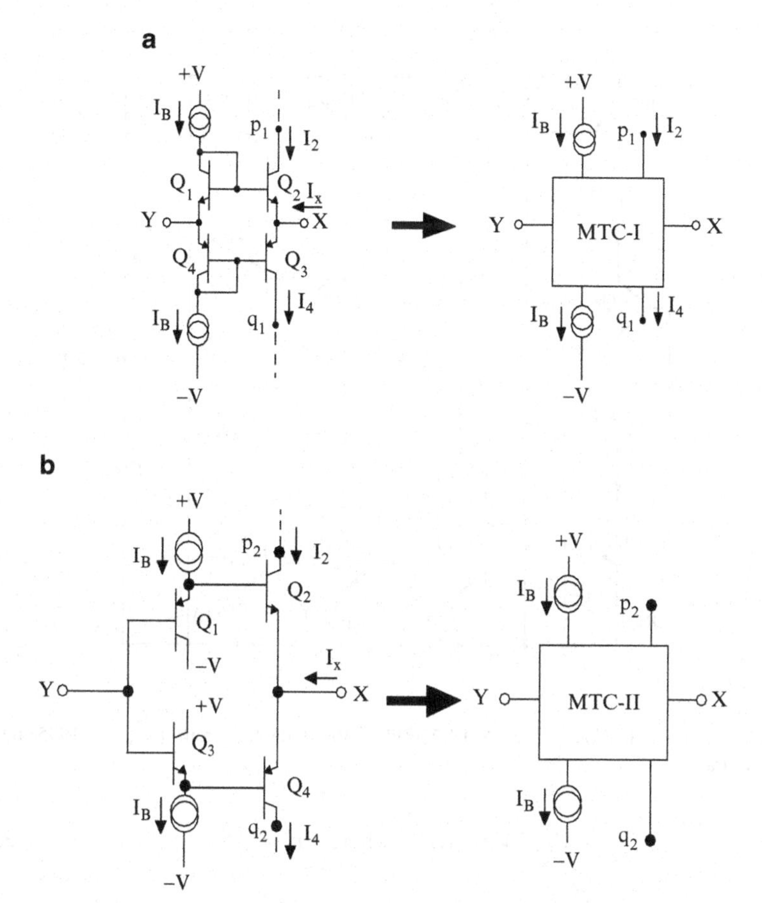

Fig. 2.19 The two types of mixed-translinear cells (MTC) (**a**) MTC-I, (**b**) MCT-II (adapted from [4] ©1997 Taylor & Francis)

(MTC) [4, 16] shown in Fig. 2.19a, b. An analysis of the type-I MTC reveals that the current I_x and differential input (V_y–V_x) are related by the following equation:

$$I_x = 2I_B \sinh\left\{\frac{V_y - V_x}{V_T}\right\} \tag{2.33}$$

Incidentally, type-II MTC of Fig. 2.19b, although has a different topology, it is also governed by the same equation [4]. This equation can be re-arranged as:

$$\frac{V_y - V_x}{V_T} = \sinh^{-1}\left(\frac{I_x}{2I_B}\right) \cong \frac{I_x}{2I_B}; \text{for } I_x << 2I_B \tag{2.34}$$

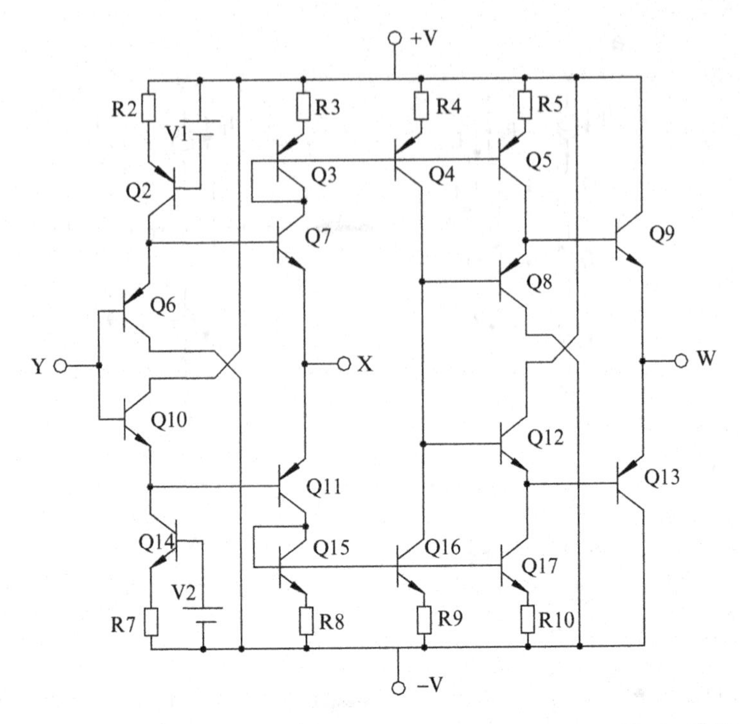

Fig. 2.20 Elantec dual/quad CFOA EL2260/EL2460 (adapted from [17] © 1995 Intersil American Inc.)

$$V_x \cong V_y - I_x r_x \quad \text{where} \quad r_x = \frac{V_T}{2I_B} \tag{2.35}$$

Note that when r_x is zero, one gets $V_x = V_y$ (as it should be, in the ideal case).

2.6.2 *Elantec Dual/Quad EL2260/EL2460*

Figure 2.20 shows a simplified schematic of Elantec dual/quad 130 MHz CFOA EL 2260/EL 2460 [17]. As can be seen, this architecture has both input and output buffers as type- II MTC and no compensating lead is available externally. This CFOA provides 130 MHz 3-dB band width (for a gain of +2) with a slew rate of 1,500 V/μs.

2.6.3 *Intersil HFA 1130*

Intersil HFA1130 (Fig. 2.21) CFOA is an ideal choice for applications requiring output limiting which allows the designer to set the maximum positive and negative output levels thereby protecting the later stages from damage or input saturation [18].

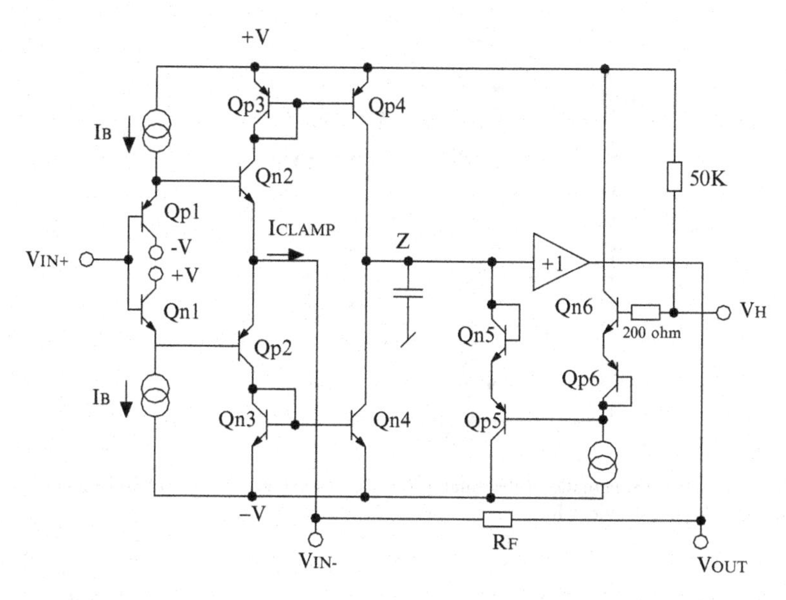

Fig. 2.21 Intersil HFA1130 output-limiting low-distortion CFOA (adapted from [18] © 2005 Intersil American Inc.)

The mechanism of high clamp (V_H circuit) can be explained as follows. The unity gain buffer made from type-II MTC forces V_{IN-} to track V_{IN+} and sets up a slewing current = $(V_{-IN} - V_{OUT})/R_F$. This current through the mirror action of the current mirrors Q_{p3}–Q_{p4} and Q_{N3}–Q_{N4} creates a replica of this current at the high impedance node Z. The base voltage of Q_{p5} is $2V_{BE}$ (Q_{N6} and Q_{P6}) less than V_H which permits the conduction of Q_{p5} whenever the voltage at the Z node reaches a voltage = Q_{p5}'s base $+2V_{BE}$ (Q_{p5} and Q_{N5}) in this manner the transistor Q_{p5} clamps node Z whenever Z reaches to a voltage level = V_H. The resistance R_1 acts as a pull-up resistance to ensure functionality with the clamp input floating. There is similar circuit (not shown in this diagram) which provides a symmetrical low clamp control by voltage V_L.

HFA1130 has a slew rate of the order of 2,300 V/μs and −3 dB bandwidth of 850 MHz and is capable of provide a high output current of the order of 60 mA and is recommended for applications in the design of residue amplifier, video switching and routing, pulse and video amplifiers, Flash A/D Driver, RF/IF signal processing and Medical imaging systems.

2.6.4 AD8011 from Analog Devices

Figure 2.22 shows a simplified schematic of the two-stage CFOA AD8011 from Analog Devices [19]. The input stage is a type-I MTC with a complementary second gain stage created from the pair of transistors Q_5 and Q_6. The circuit

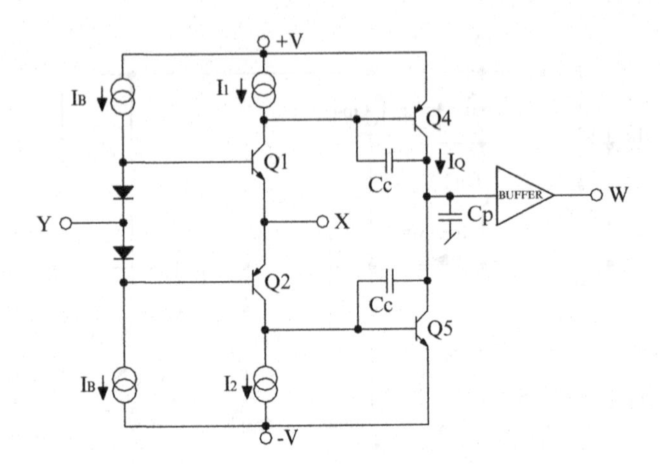

Fig. 2.22 Simplified schematic of the Analog Devices two-stage CFOA AD8011 (adapted from [19] © 1995 Analog Devices Inc.)

provides low distortion; high speed and high current drive while running on low quiescent currents. This CFOA has a −3 dB bandwidth of 57 MHz, slew rate of 3,500 V/μs, output current of 30 mA with quiescent power of 12 mW.

2.6.5 THS 3001 from Texas Instruments Inc.

Figure 2.23 shows the CFOA THS3001 from TI has 420 MHz 3-dB bandwidth for gain of +1, and has slew rate of 6,500 V/μs with current output drive as high as100mA. The simplified schematic of this CFOA is shown in Fig. 2.23.

This CFOA is built by using a 30-V dielectrically isolated, complementary bipolar process with NPN and PNP transistors possessing f_T of several GHz. This configuration implements an exceptionally high performance CFOA having wide bandwidth, high slew rate, fast settling time (40 ns) and low distortion (THD −80 dBc at 10 MHz).

Lastly, it may be pointed out that a wide varieties of CFOAs optimized for enhancement of one or more of the several specific performance features such as higher slew rate, increased output current drive capability, wider bandwidth etc. are available from leading IC manufacturers. For further information, the readers are referred to the datasheets of various IC manufacturers. Lastly, it may be pointed out that CFOAs with slew rate as high as 9,000 V/μs (such as THS3202 from Texas Instruments Inc.) are available as off-the-shelf items.

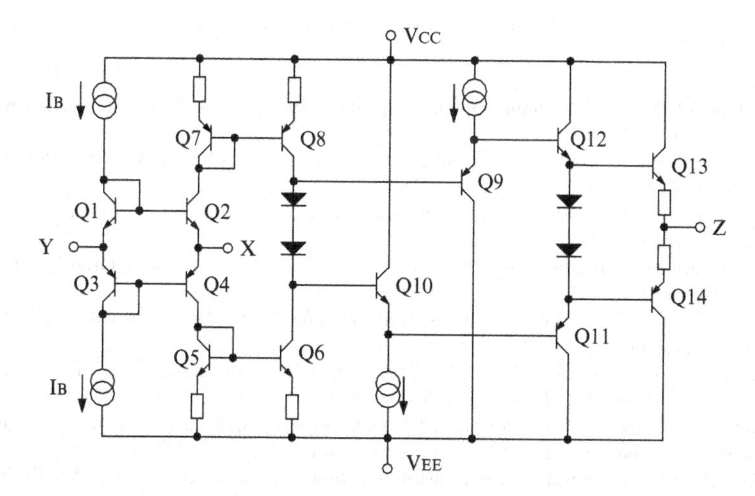

Fig. 2.23 A simplified equivalent of 420-MHz, high-speed CFOA THS 3001 type CFOA (adapted from [10] © 2009Texas Instruments Inc.)

2.7 Concluding Remarks

In this chapter, we have outlined the distinct merits of CFOAs over VOAs particularly the mechanism leading to a high (theoretically infinite) slew rate and the resolution of the gain-bandwidth conflict resulting in the notable property of the CFOA-based circuits (particularly VCVS structures) of providing constant-bandwidth with variable gains. We have also outlined the various de-merits of the CFOAs [5] namely, their inferior CMRR, unsymmetrical input bias dc currents, high input offset voltage and lower PSRRs etc.

Various basic analog circuit building blocks using CFOAs were outlined and a number of examples of commercially available CFOAs from leading IC manufacturers were highlighted.

In spite of their limitations, CFOAs are quite useful for numerous applications which can be carried out more efficiently with CFOAs than with VOAs, with one or more of the following advantages: employment of smaller number of external passive components, elimination of passive component-matching requirements in several cases and higher operational frequency range. In fact, the nature of many high frequency applications of CFOAs is such that the very high slew rate puts the CFOA in the spotlight [20].

In view of the above, it must be emphasized that the focus of the subsequent chapters of the present book would be primarily on those applications where the CFOAs are found to provide significant advantages and/or resulting in novel circuits—the type of which cannot be realized with conventional VOAs.

References

1. 60 MHz 2000 V/μs Monolithic op-amp AD844 (1990) Analog Devices, Inc. Norwood, MA 02062-9106, USA
2. Smith KC, Sedra A (1968) The current conveyor—a new circuit building block. Proc IEEE 56:1368–1369
3. Sedra A, Smith KC (1970) A second-generation current conveyor and its applications. IEEE Trans Circ Theor 17:132–134
4. Abuelma'atti MT, Al-Zaher HA (1997) Nonlinear performance of the mixed translinear loop. Int J Electron 83:467–471
5. Lidgey FJ, Hayatleh K (1997) Current-feedback operational amplifiers and applications. Electron Commun Eng J 9:176–182
6. Hayatleh K, Tammam AA, Hart BL (2010) Analysis of the input stage of the CFOA. Int J Electron Commun (AEU) 64:344–350
7. Hayatleh K, Tammam AA, Hart BL (2010) Open-loop output characteristics of a current feedback operational amplifier. Int J Electron Commun (AEU) 64:1196–1202
8. Wilson B (1989) Universal conveyor instrumentation amplifier. Electron Lett 25:470–471
9. Payne A, Toumazou C (1992) High frequency self-compensation of current-feedback devices. IEEE Int Symp Circ Syst 3:1376–1379
10. THS3001 420-MHz High-speed Current-feedback amplifier. Texas Instruments Incorporated September 2009
11. Deboo GJ (1967) A novel integrator results by grounding its capacitor. Electron Design 15
12. OA-31 Current feedback amplifiers. National Semiconductor Corporation November 1992
13. Lee JL, Liu SI (1999) Dual-input RC integrator and differentiator with tunable time constants using current feedback amplifiers. Electron Lett 35:1910–1911
14. Lee JL, Liu SI (2001) Integrator and differentiator with time constant multiplication using current feedback amplifier. Electron Lett 37:331–333
15. Maundy B, Gift SJG, Aronhime PB (2004) A novel differential high-frequency CFA integrator. IEEE Trans Circ Syst-II 51:289–293
16. Fabre A (1994) New formulations to describe translinear mixed cells accurately. IEE Proc Circ Devices Syst 141:167–173
17. EL2260, EL 2460: Dual/Quad 130 MHz current Feedback Amplifiers. Intersil American Inc. January 1995, Rev B
18. HFA1130 850 MHz, Output limiting, low distortion current feedback operational amplifier. Intersil American Inc. 2005
19. Drachler W (1995) Two stage current-feedback amplifier. Analog Dialogue 29:1–2
20. Harvey B (1993) Current feedback opamp limitations: a state-of-the-art review. IEEE Int Symp Circ Syst 2:1066–1069

Chapter 3
Simulation of Inductors and Other Types of Impedances Using CFOAs

3.1 Introduction

Simulation of inductors by active RC networks has been a very prominent and popular area of analog circuit research. Due to the well-known difficulties of realizing on-chip inductors of moderate to high values and high quality factors, simulated inductors have been the alternative choice for realizing inductor-based circuits in integrated circuit (IC) form. Simulated inductors are also useful in discrete designs in which case they can replace bulky passive inductors and offer the advantages of reduced size, reduced cost and complete elimination of undesirable mutual-couplings when several inductors are being used in a circuit. The traditional voltage mode op-amp (VOA)-based simulated inductors had been extensively investigated in the seventies to nineties, for instance see [1–19] and the references cited therein. The well-known Antoniou's Generalized Impedance Convertor (GIC)-based circuit [4] requiring two op-amps and five passive components is regarded to be the best choice available for simulating a lossless grounded inductance. Apart from simulated inductors, two other useful circuit elements known as frequency-dependent-negative-resistance (FDNR) [2]-an element having input impedance of type $Z(s) = 1/Ds^2$ and frequency-dependent-negative-conductance (FDNC)-an element having input impedance of type $Z(s) = Ms^2$ also find numerous applications in active filters and sinusoidal oscillator designs.

3.2 An Overview of Op-Amp-RC Circuits for Grounded and Floating Inductor Simulation and Their Limitations

The objective of this chapter is to present a survey of some prominent CFOA-based circuits for the simulation of inductors and other types of impedances evolved since 1992 till date.

R. Senani et al., *Current Feedback Operational Amplifiers and Their Applications*, Analog Circuits and Signal Processing, DOI 10.1007/978-1-4614-5188-4_3,

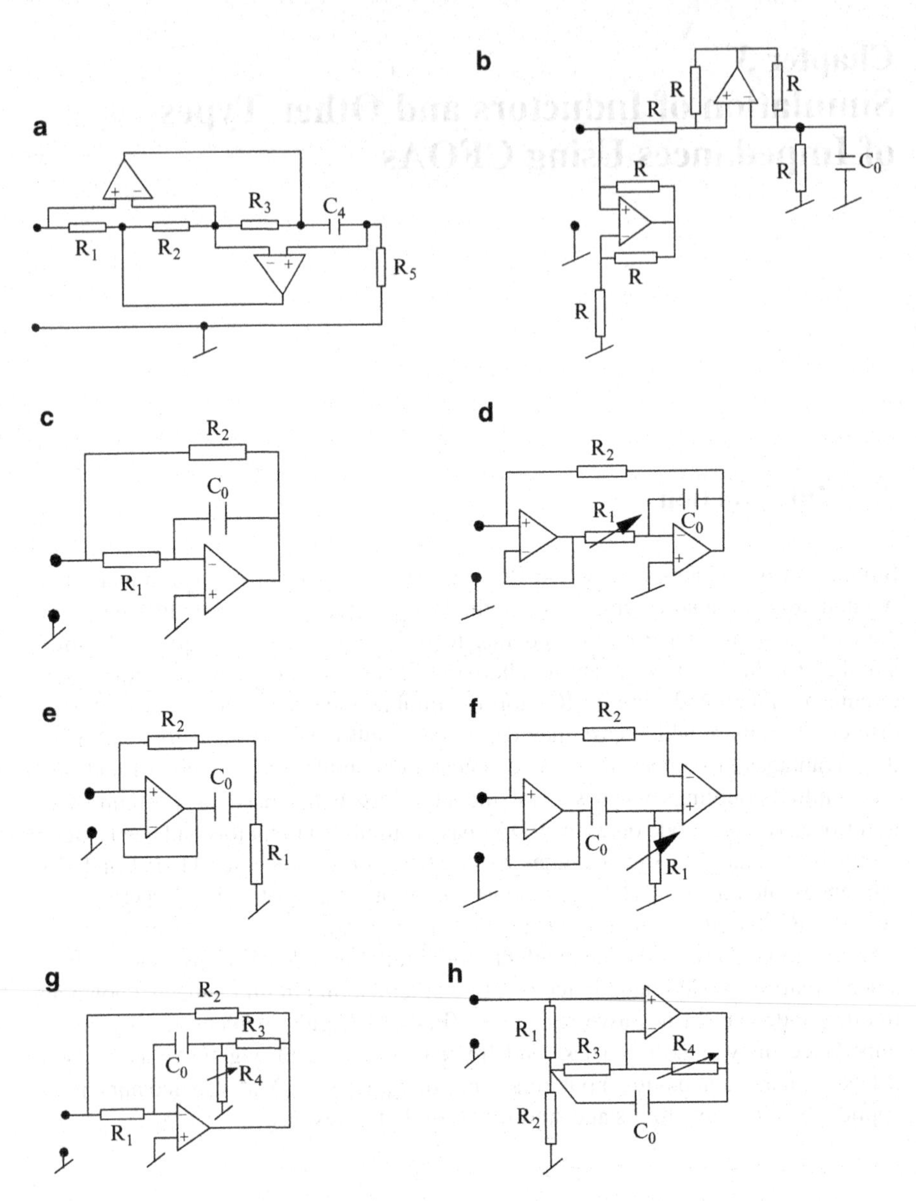

Fig. 3.1 Some well-known op-amp-RC circuits for grounded inductor simulation

However, in view of the large amount of work done on the simulation of grounded and floating inductors using VOAs, it is useful to take a quick review of some prominent VOA-based inductance simulators before discussing CFOA-based impedance simulation circuits and putting them in right perspective.

Some well-known op-amp-RC circuits for grounded inductance simulation from [1–9] are shown in Fig. 3.1.

The various circuits shown in Fig. 3.1 simulate lossless/lossy inductors of different kinds. The Antoniou's GIC-based circuit [2, 4] of Fig. 3.1a has input impedance given by $Z_{in}(s) = (sC_4R_1R_3R_5)/R_2$. On the other hand, the circuit of Fig. 3.1b from [2] also simulates a lossless grounded inductance of value $L = C_oR^2$. The circuits shown in Fig. 3.1c by Ford and Girling [7] and that of Fig. 3.1d due to Rao-Venkateshwaran [6] simulate lossy parallel RL type inductors with their input admittances given by

$$Y_{in}(s) = \left(\frac{1}{R_1} + \frac{1}{R_2}\right) + \frac{1}{sC_oR_1R_2}; \text{ and } Y_{in}(s) = \frac{1}{R_2} + \frac{1}{sC_oR_1R_2} \tag{3.1}$$

respectively. The circuit of Fig. 3.1e by Prescott [8] and that of Fig. 3.1f by Rao-Venkateshwaran [6] simulate lossy series RL with

$$Z_{in}(s) = (R_1 + R_2) + sC_oR_1R_2; \text{ and } Z_{in}(s) = R_2 + sC_oR_1R_2 \tag{3.2}$$

respectively.

The circuit of Fig. 3.1g due to Senani [9] simulates another type of parallel RL with the input admittance given by

$$Y_{in}(s) = \left[\frac{1}{R_1} + \frac{1}{R_2} + \frac{R_3}{R_1R_2} + \frac{(1 + \frac{R_3}{R_4})}{sC_0R_1R_2}\right] \tag{3.3}$$

while the circuit of Fig. 3.1h (also due to Senani [9]) simulates a grounded series RL impedance of value

$$Z_{in}(s) = \left[R_1 + R_2 + \frac{R_1R_2}{R_3} + sC_0R_1R_2(1 + \frac{R_4}{R_3})\right] \tag{3.4}$$

Note that in case of the circuits of Fig. 3.1g, h both, the inductance value is controllable through a single variable resistance R_4.

From the circuits given in Fig. 3.1, it may be noted that the op-amp based lossless simulated inductors require more than the minimum required number of passive components. If the capacitor needs to be grounded, as desirable for integrated circuit fabrication [20], then, apart from two op-amps, at least seven matched-resistors are needed as in the circuit of Fig. 3.1b. On the other hand, the passive component-matching can be avoided (as in the circuit of Fig. 3.1a) but the capacitor employed is still floating and the circuit still needs as many as four resistors. The circuits of Fig. 3.1c–f use a canonical number of passive components but require two op-amps if a variable inductance is needed. Detailed studies such as those of [5] have revealed that the grounded inductor circuits of the type shown in Fig. 3.1c–f, can be used satisfactorily (i.e., within permissible deviation in the inductance value and its quality factor) typically upto only a small fraction of the gain bandwidth product (GBP) of the op-amps employed (i.e. only about 10 kHz or so with an op-amp having GBP of 1 MHz).

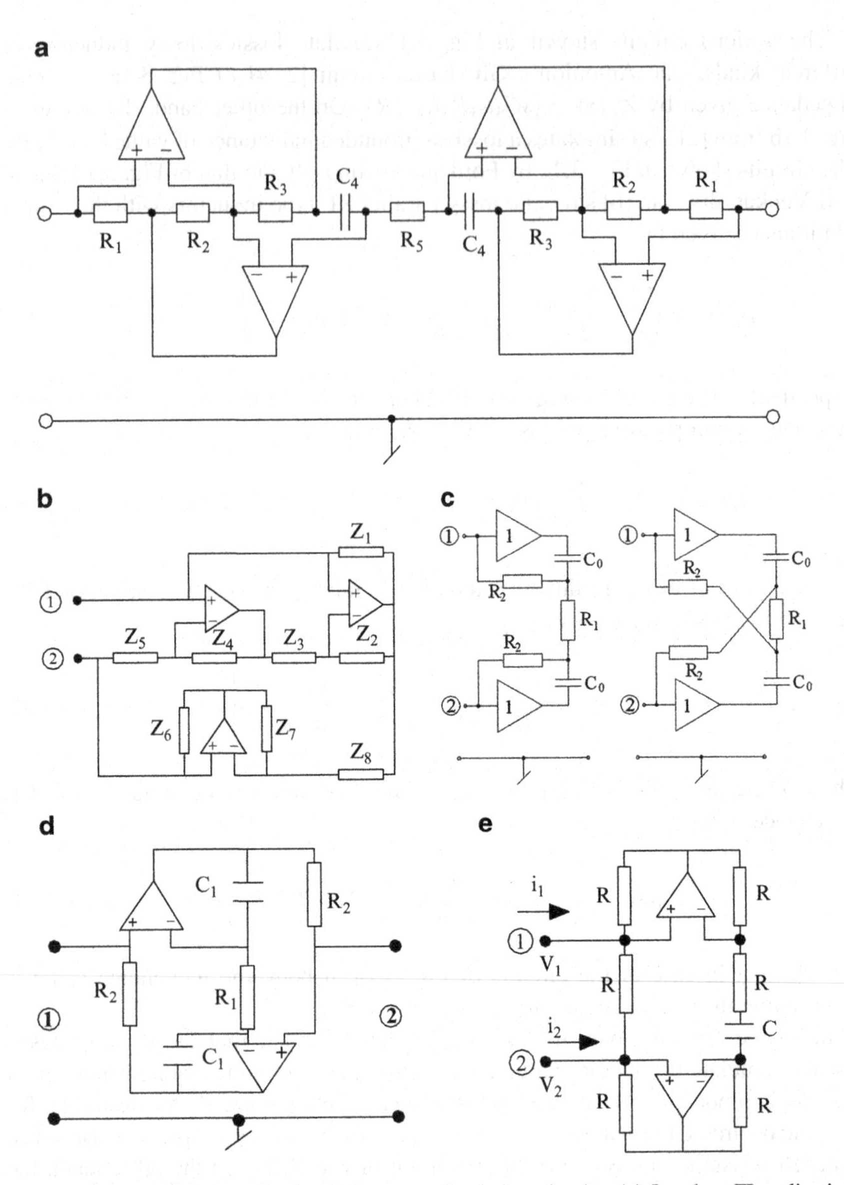

Fig. 3.2 Some well-known floating inductance simulation circuits. (**a**) Lossless FI realization based on Antoniou's GIC [11, 13, 14], (**b**) Lossless FI using only three op-amps-based upon Riordon gyrator [10, 12], (**c**) Lossy FI of Dutta Roy [15] and Wise [16]; Takagi and Fujii [43], (**d**) lossy FI by Sudo and Teramoto [17], (**e**) Lossy FI by Senani and Tewari [18]

Some prominent floating inductance/impedance simulation circuits using VOAs [10–19] are shown in Fig. 3.2.

The circuit of Fig. 3.2a simulates a lossless floating inductance having impedance value

$$Z_{1-2}(s) = sC_4R_1R_3R_5/R_2 \tag{3.5}$$

at the cost of employing four op-amps along with two capacitors and requiring matching of passive components in the two identical GIC networks. The circuit of Fig. 3.2b also requires one cancellation condition:

$$\frac{1}{Z_5} = \frac{Z_7}{Z_6Z_8} \tag{3.6}$$

and one matching condition

$$\frac{1}{Z_1} = \frac{Z_7}{Z_6Z_8} \tag{3.7}$$

to realize a lossless floating impedance of value:

$$Z_{1-2} = \frac{Z_1Z_3Z_5}{Z_2Z_4} \tag{3.8}$$

but has the advantage of employing one less op-amp than the circuit of Fig. 3.2a. In inductance simulation mode, this circuits has the advantage of employing only a single capacitance in contrast to two capacitors in the circuit of Fig. 3.2a.

The circuits of Fig. 3.2c simulate a lossy series RL type inductor with

$$Z(s) = (R_1 + 2R_2) + 2sCR_1R_2 \tag{3.9}$$

and have the disadvantage of employing two capacitors but have the advantage of requiring only two op-amps as unity gain amplifiers and a small number of only three resistors with a simple, practically adjustable condition of floatation [15]. The circuit of Fig. 3.2d simulates a lossy parallel RL inductor with

$$Y_{1-2} = \frac{1}{R_2} + \frac{1}{sC_1R_2R_1} \tag{3.10}$$

This circuit although employs two matched capacitors and two matched resistors but has the advantage of realizing a single resistance tunable inductance by having variable resistance R_1.

Lastly, the circuit of Fig. 3.2e employs only two op-amps, only a single capacitor but needs four matched resistors to realize a floating lossy inductance of value

$$Z(s) = R + sCR^2 \tag{3.11}$$

From the above described circuits, we note that for lossless FI simulation, three to four op-amps along with a non-canonical number of passive elements are needed and that the circuits usually require component-matching conditions and/or cancellation constraints for realizing the lossless floating inductance. On the other hand, circuits capable of simulating lossy (series RL/parallel RL) floating inductors although can be realized with usually two VOAs, these also employ non-canonical number of passive components (as in the circuits of Fig. 3.2c–e) and need identical resistor values or cancellation constraints in case of single-capacitor simulations.

Apart from the above mentioned difficulties and demerits, it has been established [19] that the useful frequency range of the FIs of the type of Fig. 3.2c is, disappointingly, restricted to a very small fraction of the GBP of the op-amps employed.

Since the well-known and popular method of designing active filters based upon passive RLC prototypes require both grounded and floating impedances (FI), the problem of realizing synthetic *floating* impedances was widely investigated during the late 1960s to 1990s. Eventually, it was found that using op-amps, it is impossible to realize synthetic FIs *without requiring any component matching conditions.*

In the subsequent sections of this chapter, we would highlight some prominent CFOA-based circuit configurations for realizing synthetic impedances, in both grounded and floating forms. It will be shown that using CFOAs, not only new types of simulators have been possible (the type of which cannot be realized with VOAs) but also that the resulting circuits offer a number of advantageous features not possible with VOA-based impedance simulators. We will discuss a number of CFOA-based impedance simulation circuits which can operate over frequency ranges several orders higher than those possible for VOA-based impedance simulators.

3.3 Realization of Gyrator and Grounded Impedances Using CFOAs

Among the first few applications of the CFOAs, which appeared soon after the CFOA was noticed as an interesting building block for analog cricuit design, was the gyrator implementation proposed by Fabre in [21]. This circuit is shown in Fig. 3.3. With port 2 terminated into a capacitance, the circuit simulates a lossless grounded inductor of value $L_{eq} = C_0R_1R_2$ looking into port 1.

In fact, CFOA-based grounded impedance circuits can be synthesized systematically starting from first principles [22]. To this end, we recall that impedance converters can be realized from inpedance inverters and vice versa. Also, a positive impedance inverter can be realised with two voltage-controlled current sources (VCCS) of opposite polarity connected back-to-back. Employing the non-inverting and inverting VCCS each realized with a single CFOA and no more than a single impedance, two impedance converter/inverter cicuits are shown in Fig. 3.4. An interconnection of the two VCCS as shown in Fig. 3.4a makes an effective use of the on-chip buffer to create a 2-port which realizes an impedance inverter. With its port 2 terminated into an impedance Z_3, the impedance looking into port 1 is given by

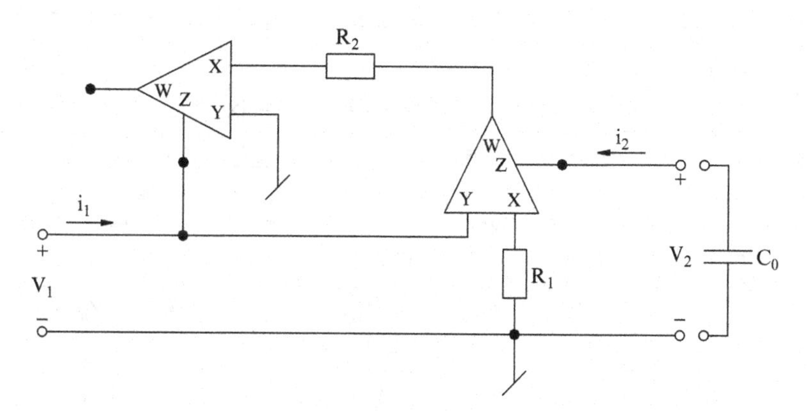

Fig. 3.3 A gyrator and inductance simulator proposed by Fabre (adapted from [21] ©1992 IEE)

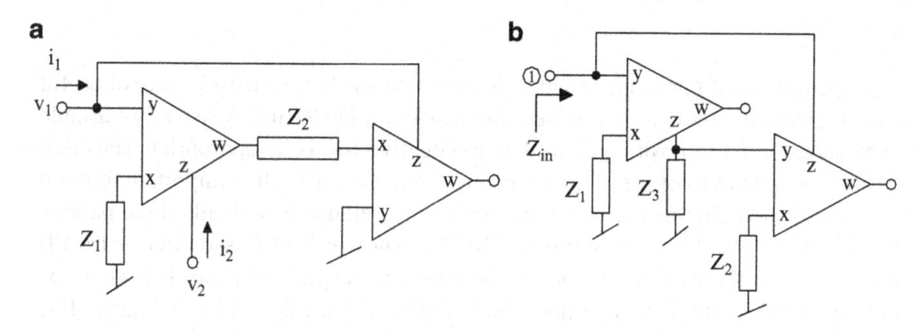

Fig. 3.4 Realization of grounded impedance converters/inverters. (**a**) Realization of GPII/GPIC, (**b**) Realization of GNII/GNIC (adapted from [22] © 1998 Walter de Gruyter GmbH & Co. KG, Germany)

$$Z_{in1}(s) = \frac{Z_1 Z_2}{Z_3} \tag{3.12}$$

This circuit is, hence, a generalized positive impedance inverter (GPII). The same circuit with port 2 terminated in Z_2, Z_1 deleted and the terminal thus created named as port 3, would function as a generalized positive impedance converter (GPIC).

The realization of the corresponding generalized negative impedance inverter (GNII) and generalized negative impedance converter (GNIC) elements is obtainable by a back-to-back interconnection of two inverting VCCSs or two non-inverting VCCSs. One of these two implementations of the GNII/GNIC is shown in Fig. 3.4b which realizes input impedance given by

$$Z_{in}(s) = -\frac{Z_1 Z_2}{Z_3} \tag{3.13}$$

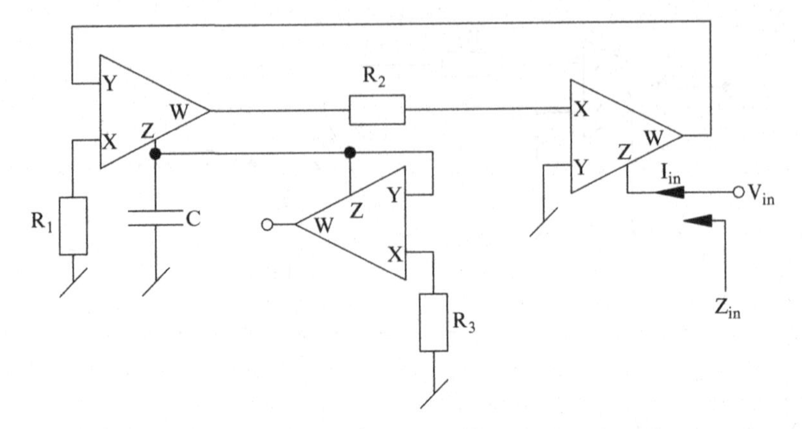

Fig. 3.5 Active-compensated lossless grounded inductance by Yuce and Minaei (adapted from [23] © 2009 Springer)

The circuit configurations of Fig. 3.4a, b can be used to realize a variety of useful circuit elements such as simulated inductance, FDNR and resistively-variable capacitance, both in positive as well as in negative forms by appropriate selection (resistive/capacitive) of the three impedances Z_1, Z_2 and Z_3. It is important to point out that no such circuits (i.e.using only two active elements and only three passive elements) are possible with traditional VOAs. Known VOA-based circuits for PII and NII typically require two VOAs and seven matched resistors [1] where as VOA-based GIC although does not need component-matching but still requires five impedances. By contrast, the CFOA-based circuits described above, apart from the capability of operating at relatively higher frequencies than VOA-based circuits, offer the following advantages: (1) use of a bare minimum (only three) of passive components (2) single-resistance tunability of the realized impedances in all the cases and (3) no component-matching requirements/realization constraints.

An interesting grounded lossless inductance circuit having reduced parasitic effect was suggested by Yuce and Minaei in [23] and is shown in Fig. 3.5. In this circuit, a third CFOA is employed in the mode of a current inversion type negative impedance converter (NIC) to reduce the Z-terminal parasitic resistance of the first CFOA although it slightly increases the total capacitance at the Z-terminal of the first CFOA.

3.4 Single-CFOA-Based Grounded Impedance Simulators

It is known that as compared to their lossless counterparts, lossy grounded inductors and FDNRs can be realized more economically typically requiring only a single op-amp along with three passive components. Similarly, it has been found that employing CFOAs too, lossy grounded inductors and FDNRs as well as grounded ideal negative inductance and negative capacitance elements can be realized with only

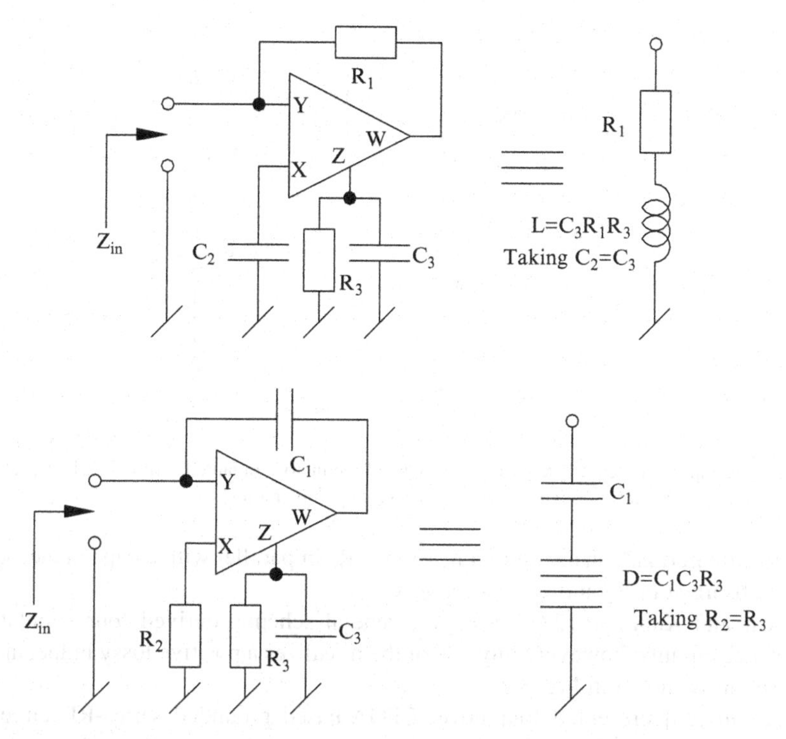

Fig. 3.6 Series-RL and series-CD simulators proposed by Liu and Hwang (adapted from [24] © 1994 IEE)

a single CFOA along with three/four passive components, with most of them offering the significant advantage of single-element-controllability of the realized inductance/capacitance/FDNR which is not possible in the mentioned VOA-based circuits.

A number of authors have proposed such single-CFOA-based grounded impedance simulators. In the following, we present some prominent and representative circuits from the references [24–30].

3.4.1 Lossy Grounded Inductors/FDNRs

In [24] Liu and Hwang presented a general single-CFOA circuit for grounded impedance simulation. Two interesting single CFOA-based circuits capable of simulating series-RL type lossy inductor and series-CD type lossy FDNR resulting from their general configuration, as special cases, are shown in Fig. 3.6.

In the first circuit, the inductance value is controllable through a variable resistance R_3 while in the second case, the FDNR value is adjustable through a variable capacitance C_3. Also, in both the circuits, the parasitic output impedance

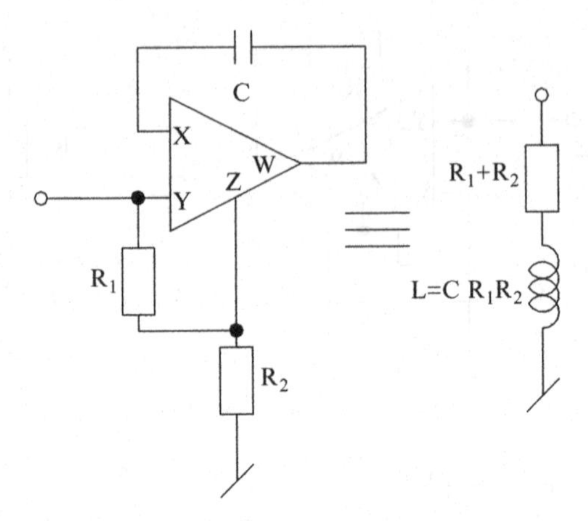

Fig. 3.7 Kacar and Kuntman's lossy grounded inductor (adapted from [28] © 2011 Radioengineering Society, Czech and Slovak Technical Universities)

looking into port Z (consisting of a resistance R_p in parallel with a capacitance C_p) can be absorbed in R_3 and C_3 respectively.

Kacar and Kuntman [28], through a general scheme, derived four simulated inductance circuits; however, only one of them realizes a positive lossy inductance. This circuit is shown in Fig. 3.7.

Yuce in [25] presented four novel CFOA-based grounded series-RL circuits which are shown here in Fig. 3.8. By a straight forward analysis, for all the circuits, the input impedance is found to be of type

$$Z_{in1}(s) = \frac{V_{in}}{I_{in}} = R_{eq} + sL_{eq} \tag{3.14a}$$

where

$$\left.\begin{array}{l} R_{eq} = R_1, L_{eq} = CR_1R_2(Circuits\ of\ Fig.3.8\text{a, b}) \\ R_{eq} = R_1, L_{eq} = 2CR_1R_2(Circuit\ of\ Fig.3.8\text{c}) \\ R_{eq} = R_1/2, L_{eq} = (CR_1R_2)/2(Circuit\ of\ Fig.3.8\text{d}) \end{array}\right\} \tag{3.14b}$$

Thus, an interesting feature of these circuits is that the realized inductance value can be controlled independently by resistance R_2 in all of them.

In [29], Abuelma'atti, through two single-CFOA-based generalized impedance simulation networks, derived a number of positive and negative impedances as special cases. A special case from [29], which simulates series RL impedance, is shown here in Fig. 3.9.

It is interesting to point out that the grounded series-CD impedances can be readily obtained from the grounded series-RL impedances, by the application of RC-CR transformation [1]. Two such exemplary grounded series-CD simulators, obtained from the circuits of Fig. 3.8c, b are shown in Fig. 3.10.

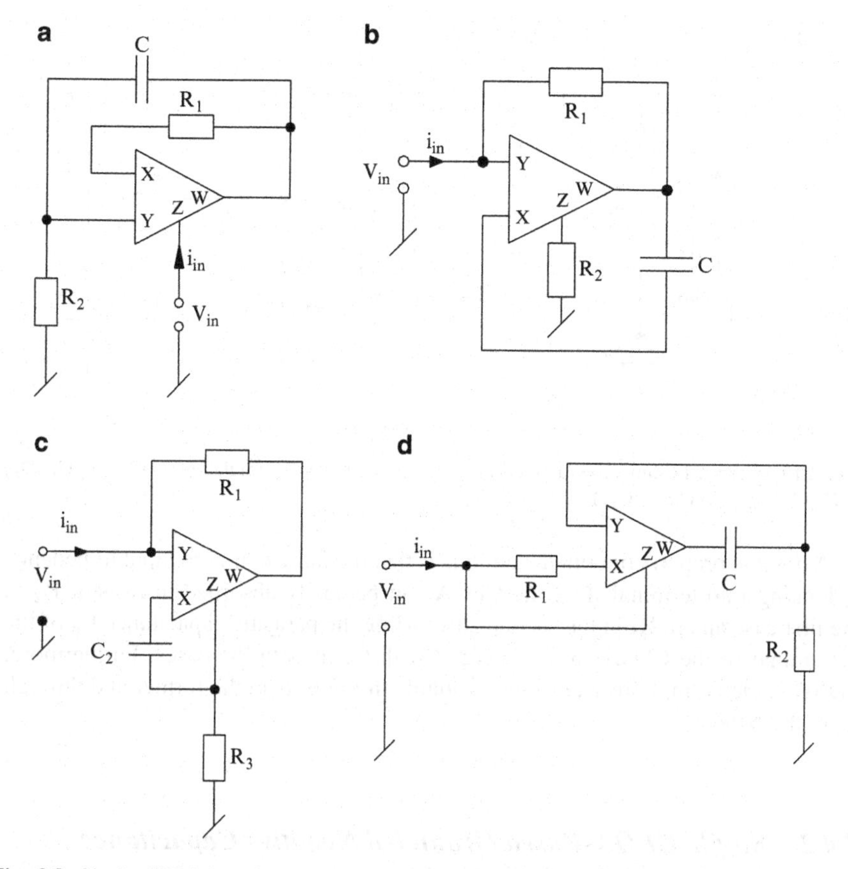

Fig. 3.8 Single-CFOA-based canonical lossy grounded inductors proposed by Yuce (adapted from [25] © 2009 Springer)

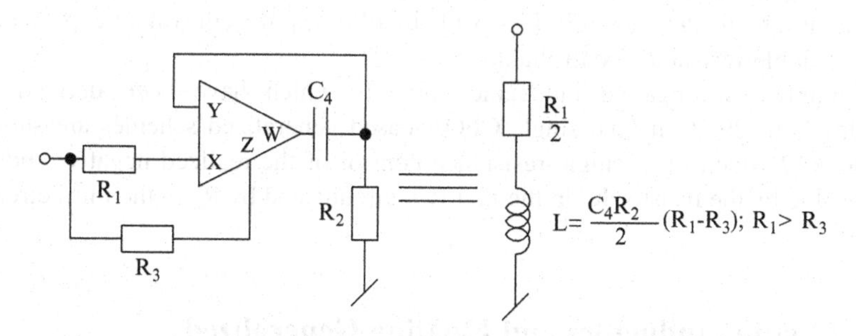

Fig. 3.9 An exemplary lossy inductance simulator proposed by Abuelma'atti (adapted from [29] © 2012 Springer)

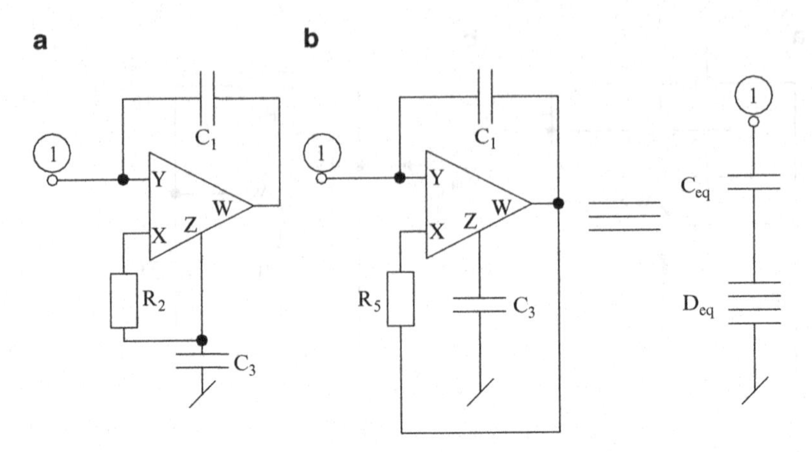

Fig. 3.10 Series CD simulators derived from Yuce's inductance simulators. (**a**) $C_{eq} = C_1$; $D_{eq} = (C_1C_3R_2)/2$, (**b**) $C_{eq} = C_1$; $D_{eq} = C_1C_3R_5$

A notable property of both the circuits is that the finite non-zero input impedance r_x looking into terminal X of the CFOA can be easily absorbed in resistor R_2 in the first case and in R_5 in the second case, while the parasitic capacitance C_p of the Z-terminal of the CFOA can be merged with C_3 in both the cases. Furthermore, FDNR value is single-resistance-controllable through R_2 in the former and through R_s in the latter.

3.4.2 *Single-CFOA-Based Grounded Negative Capacitance and Negative Inductance Simulators*

There have been a number of works [28–30] where single-CFOA-based grounded negative capacitance and grounded negative inductance have been simulated. Some prominent circuits are shown in Fig. 3.11. In all cases, the equivalent capacitance value is single-resistance-controllable.

A number of negative inductance circuits which have been derived by Abuelma'atti [29] from two single-CFOA-based generalized schemes are shown in Fig. 3.12 which offer single-resistance control of the realized negative inductance value by the resistor R_4 in the first two circuits and by R_2 in the third circuit.

3.5 Floating Inductors and Floating Generalized impedance Simulators Using CFOAs

Using traditional VOAs, no circuit is known to exist which can realize a lossless or lossy floating inductance (FI) *without requiring any component matching condition.* It was demonstrated in [31–34] for the first time that using a negative type CCII, it is

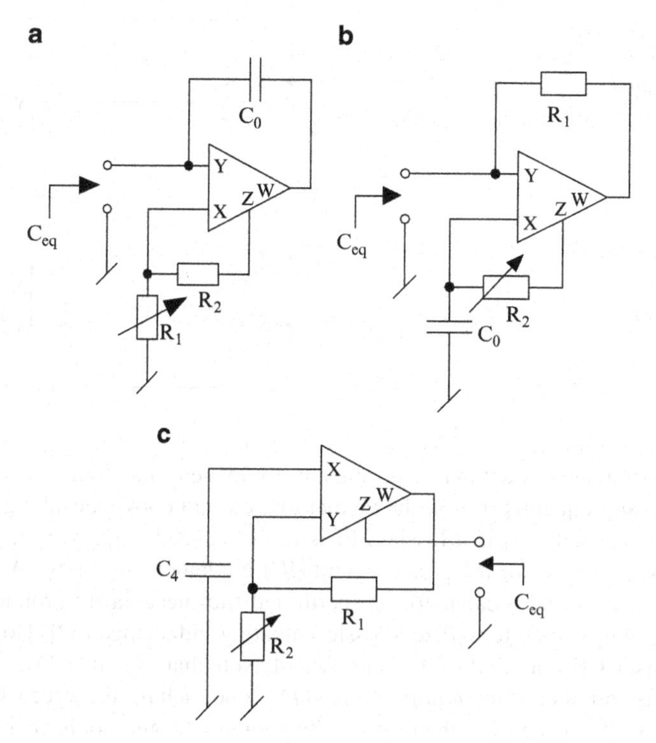

Fig. 3.11 Some grounded negative capacitance circuits (**a**), (**b**) $C_{eq} = -(C_0R_2/2R_1)$ (adapted from [30] © 2011 Springer; (**c**) $C_{eq} = -[C_4/(1 + R_1/R_2)]$ (adapted from [29], © 2011 Springer)

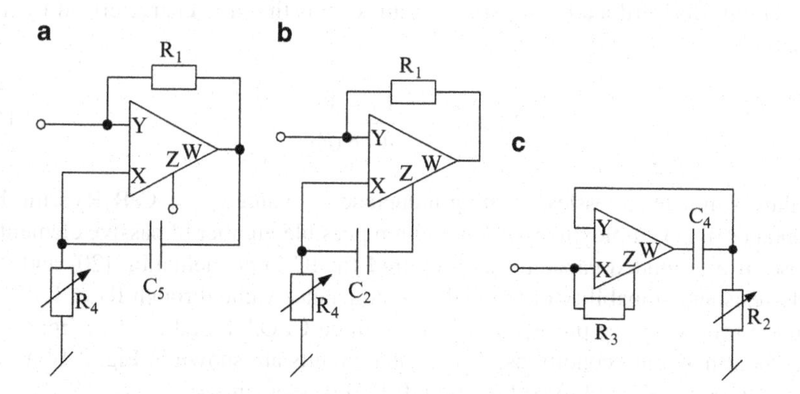

Fig. 3.12 Some grounded negative inductance simulation circuits proposed by Abuelma'atti. (**a**) $Z(s) = -sC_2R_1R_4$, (**b**) $Z(s) = -s\ C_5R_1R_4$, (**c**) $Z(s) = -s\ (C_4R_2R_3)/2$ (adapted from [29] © 2012 Springer)

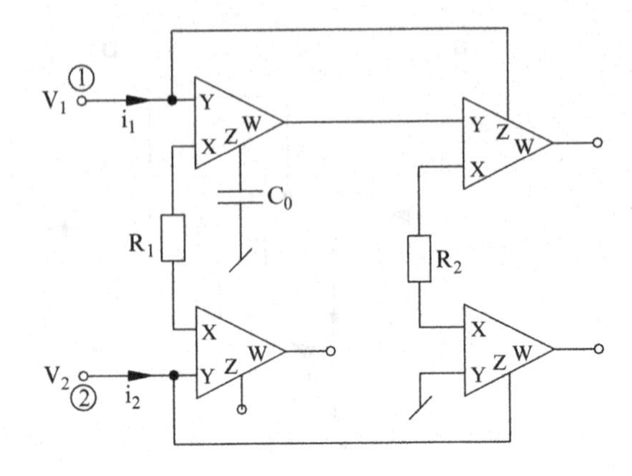

Fig. 3.13 CFOA-based lossless floating inductor employing a grounded capacitor

possible to simulate FIs without requiring any component-matching or cancellation constraints. Subsequently, it was shown that four current conveyors along with two resistors are needed [35] if a lossless FI is to be realized employing a grounded capacitor as preferred for integrated circuit (IC) implementation [20]. Although if only two resistors and a capacitor are permitted (not necessarily grounded) then three CCIIs will suffice to realize a lossless floating inductance [32]. However, in such circuits, CCII+ as well CCII− are needed. Note that, if realized with CFOAs, such circuits will, therefore, require four CFOAs (one each for the three CCII+s and two for the CCII−) and even then the resulting circuits would not have a grounded capacitor.

We now show that using four CFOAs, a lossless floating inductance can be realized using only two resistors and as desired, *a grounded capacitor*. Such a circuit, derivable from the four CC-based circuits of [35], is shown in Fig. 3.13.

A straight forward analysis of this circuit reveals that it is characterized by the equation

$$i_1 = -i_2 = \frac{V_1 - V_2}{sC_0R_1R_2} \tag{3.15}$$

and thus, simulates a lossless floating inductance of value $L_{eq} = C_0R_1R_2$ with the advantages of (1) employment of a minimum possible number of passive elements, (2) use of a grounded capacitor as preferred for IC implementation [20] and (3) single-resistance-tunability of the realized inductance value through R_1 or R_2.

In the following, we present a number of three-CFOA-based FIs free from any component-matching conditions. Three such circuits are shown in Fig. 3.14 all of which can be considered to be floating GPII/GPIC elements.

The first circuit [22] is obtained from the grounded impedance converter/inverter circuit of Fig. 3.4b (with its port 2 terminated into an impedance Z_3) by ungrounding the impedances Z_1 and Z_3, tying them together and connecting to the voltage V_2, inserting a third CFOA appropriately to make it possible to have the port 2 current i_2 of the overall circuit given by

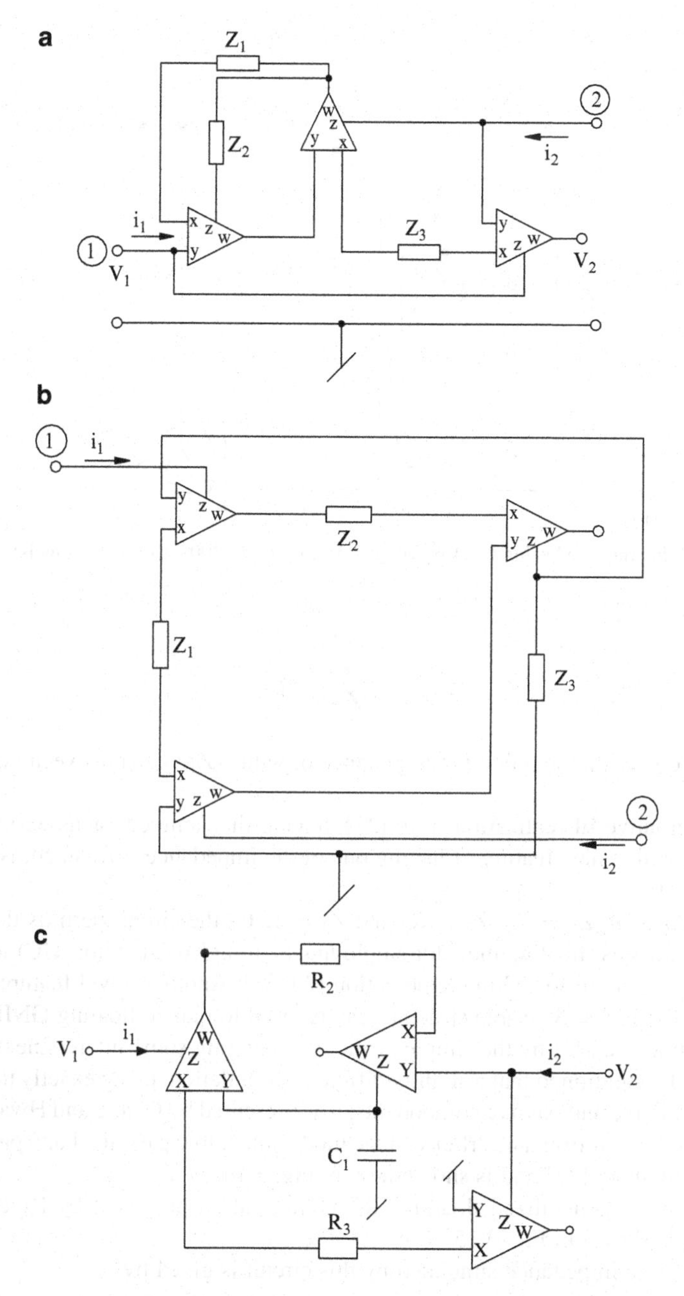

Fig. 3.14 Realization of floating impedance converters/inverters. (**a**) Floating GPII/GPIC configuration, (**b**) Configuration for realizing positive/negative generalized floating impedances, (**c**) A circuit for lossless FI simulation

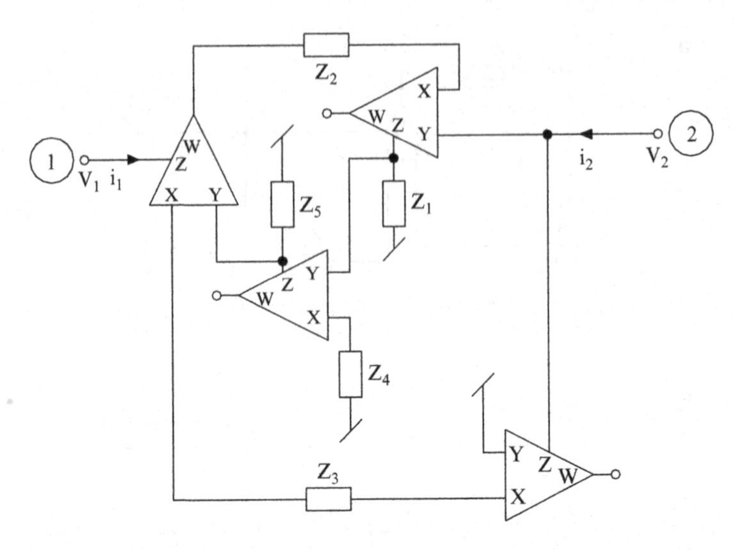

Fig. 3.15 A floating GIC using CFOAs (adapted from [38] © 2008 Taylor & Francis)

$$-i_2 = i_1 = \frac{Z_3}{Z_1 Z_2}(v_1 - v_2) \tag{3.16}$$

and therefore, realizing a floating impedance of value (Z_1Z_2/Z_3) between ports and 1 and 2.

An alternative FI realization from [22], having the same characterization as in (3.16) but with a new feature of having one of the impedances grounded, is shown in Fig. 3.14b.

Note that with $Z_1 = R_1$, $Z_2 = R_2$ and $Z_3 = 1/sC_3$ the circuit permits the realization of a lossless floating inductor employing a grounded capacitor (GC) which is an attractive feature for IC implementation [20, 35]. Another novel feature of this configuration is that the same structure can be used to realize floating GNII/GNIC elements too; see [22] by the simple artifice of changing some interconnections.

It must be mentioned that a slightly different formulation, using exactly the same number of active and passive components, was presented by Chang and Hwang [36] which was derived from an earlier voltage mode notch, low pass and band pass filter proposed by them [37] and is shown here in Fig. 3.14c.

A still more generalized floating GIC configuration proposed by Psychalinos et al. [38] is shown in Fig. 3.15.

The floating impedance simulated by this circuit is given by

$$Z_{1-2} = \frac{Z_2 Z_3 Z_4}{Z_1 Z_5} \tag{3.17}$$

From the above expression, it is easy to visualize that by selecting various impedances in the circuit appropriately, a variety of floating impedances can be simulated by this circuit. The following special cases are of practical interest.

1. *Floating FDNR* is obtained by choosing any two of the three impedances Z_2, Z_3 and Z_4 as capacitors. Taking Z_3 and Z_4 as capacitors while taking the remaining impedances as resistors leads to

$$Z_{1-2} = \frac{R_2}{s^2 R_1 C_3 C_4 R_5} \tag{3.18}$$

   It may be noted that the value of the simulated FDNR is controllable by a single grounded resistance R_5 and no component-matching is required.
2. *Floating inductance* is realizable with either of Z_1 or Z_5 selected as a capacitor. With Z_1 taken as a capacitance C_1 with all other impedances being resistors, the equivalent impedance is given by

$$Z_{1-2} = s\frac{C_1 R_2 R_3 R_4}{R_5} \tag{3.19}$$

3. *Floating Capacitance* is realizable when any one of Z_2, Z_3 or Z_4 is taken as a capacitance. For instance, taking Z_4 as a capacitor, the floating impedance realized is given by

$$Z_{1-2} = \frac{R_2 R_3}{s C_4 R_1 R_5} \tag{3.20}$$

3.6 Floating Inductance Circuits Employing Only Two CFOAs

We now show how lossless and lossy FIs can be simulated using only two CFOAs.

3.6.1 *Lossless/Lossy Floating Inductance Simulator*

In this section we present a circuit [39] which employs only two CFOAs along with only five passive components (namely two capacitors and three resistors) to realize a lossy/loss-less FI. This circuit is shown in Fig. 3.16. Assuming ideal characterization of the CFOAs, a straight forward analysis of the circuit reveals its y- matrix to be given by

$$[Y] = \left[\left(\frac{1}{R_1} - \frac{1}{R_2}\right) + \frac{1}{sC_1 R_1 R_2}\right]\begin{bmatrix} 1 & -1 \\ -1 & 1 \end{bmatrix} \tag{3.21}$$

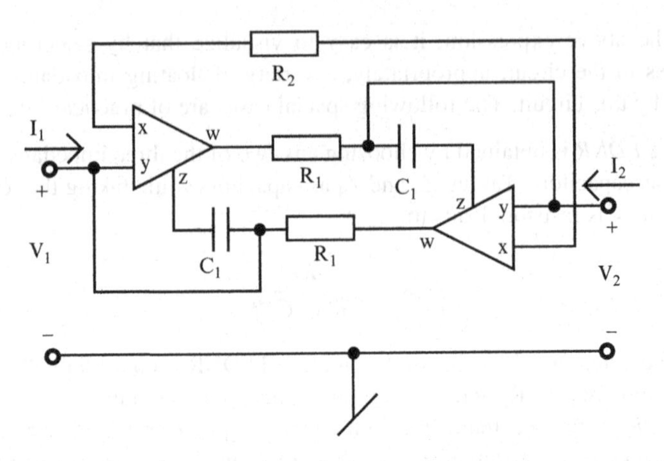

Fig. 3.16 A floating inductance configuration using only two CFOAs (adapted from [39] © 2012 Springer)

Thus, with $R_1 < R_2$, the circuit simulates floating parallel-RL admittance with equivalent resistance R_{eq} and equivalent inductance L_{eq} are given by

$$\frac{1}{R_{eq}} = \frac{1}{R_1} - \frac{1}{R_2}; \; L_{eq} = C_1 R_1 R_2 \tag{3.22}$$

On the other hand, with $R_1 = R_2 = R_0$, the circuit simulates a lossless FI with

$$L_{eq} = C_1 {R_0}^2. \tag{3.23}$$

With the parasitic impedances of the CFOAs accounted for, i.e. considering the finite input impedance looking into terminal-X as R_x and the output impedance looking into terminal-Z consisting of a parasitic resistance R_p in parallel with a parasitic capacitance C_p, it is found that in view of the symmetry of the circuit, the non-ideal y-parameters are such that $Y_{11}' = Y_{22}'$ and $Y_{12}' = Y_{21}'$. The values of these admittance parameters are found to be.

$$\begin{aligned} Y_{11}' = Y_{22}' = & \frac{1}{R_1} + \frac{sC_1}{(1 + sC_1 Z_p)} - \frac{sC_1 Z_p}{(1 + sC_1 Z_p)(R_2 + 2R_x)} \\ & + \frac{Z_p}{(1 + sC_1 Z_p)(R_1 R_2 + 2R_1 R_x)} \end{aligned} \tag{3.24}$$

$$\begin{aligned} & Y_{12}' = Y_{21}' \\ & = -\left[\frac{sC_1 Z_p}{R_1(1 + sC_1 Z_p)} - \frac{sC_1 Z_p}{(1 + sC_1 Z_p)(R_2 + 2R_x)} + \frac{Z_p}{(1 + sC_1 Z_p)(R_1 R_2 + 2R_1 R_x)}\right] \end{aligned} \tag{3.25}$$

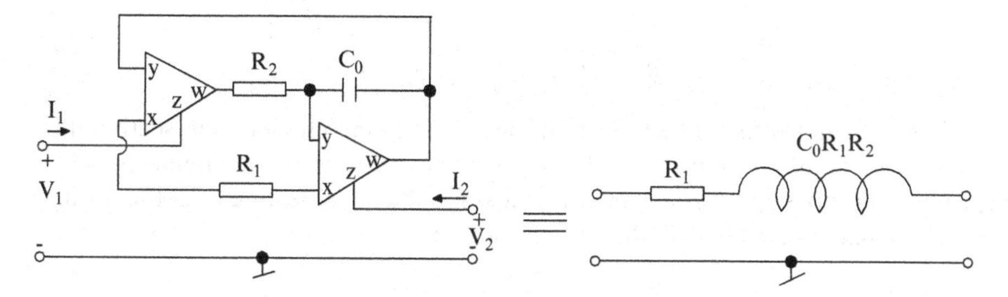

Fig. 3.17 A *new* series-RL type lossy floating inductance simulator

It may be seen that with $Z_p \to \infty$, $R_x \to 0$, the y-parameters in (3.24) and (3.25) reduce to those in (3.21).

From the above non-ideal expressions of the y-parameters of the circuit it may be easily visualized that the high frequency performance would be affected because of these parasitic impedances. The equivalent non-ideal inductive and resistive components resulting from all the four y-parameters of (3.24) and (3.25) would be frequency-dependent.

With $R_x = 50\,\Omega, R_p = 3\,\text{M}\Omega, C_p = 4.5\,\text{pF}$ and the circuit designed with $C_1 = 1\,\text{nF}$, $R_0 = 1\,\text{k}\Omega$ to realize a lossless FI of 1 mH, MATLAB frequency responses of $|Y_{11}| = |Y_{22}|, |Y_{12}| = |Y_{21}|$ have shown [39] that in the circuit of Fig. 3.16, the y-parameters remain intact (and hence, the circuit is useable) up to a frequency of around 10 MHz. This frequency range of the circuit has also been confirmed from a SPICE simulation of the circuit for the same component values using a macro model of AD844.

It is worth pointing out that in [40] Yuce and Minaei have described two FI circuits using the so-called modified CFOA (MCFOA). Each circuit therein employs two MCFOAs, two resistors and a single (grounded) capacitor. However, a MCFOA is not available commercially as an off-the-shelf integrated circuit. Furthermore, when an MCFOA is implemented with AD844-type CFOAs, as many as three CFOAs are needed for each MCFOA. Thus, each of the proposed FI circuits presented in [40] would require six CFOAs. Thus, the present circuit, although requires two identical capacitors and three resistors, it has the advantage of employing only two CFOAs.

3.6.2 A Lossy Floating Inductance Simulator

A possible circuit[1] for realizing floating series RL impedance is presented here in Fig. 3.17. By straight forward analysis, it is found that this circuit is characterized by

[1] R. Senani and D.R. Bhaskar, 'New floating lossy inductors, without component-matching, Employing only two CFOAs', May 2012 (unpublished).

$$\begin{bmatrix} i_1 \\ i_2 \end{bmatrix} = \frac{1}{(R_1 + sCR_1R_2)} \begin{bmatrix} 1 & -1 \\ -1 & 1 \end{bmatrix} \begin{bmatrix} v_1 \\ v_2 \end{bmatrix} \tag{3.26}$$

It may be noted that, like the circuit described in the previous sub-section this circuit also employs only two CFOAs but by contrast, uses a minimum possible number of (only three) passive elements and has the novel feature of not requiring any component-matching conditions whatsoever.

3.7 Applications of Simulated Impedances in Active Filter Designs

We now show how the circuits presented in this chapter can be used in the design of second order and higher order filters.

3.7.1 *Applications in the Design of Second Order Filters*

Those circuits which simulate lossless grounded inductance (for example, the two-CFOA-based circuits described earlier) and lossless floating inductance (such as the four/three CFOA-based circuits described earlier) can be used as direct replacements for grounded and floating inductors respectively in the passive RLC prototype second order filters. The resulting CFOA-based filters will possess the desirable property of employing grounded capacitors.

It is interesting to observe that in all the four circuits of Fig. 3.8 [25–27] if the only grounded element therein is ungrounded and a two port network is thus created, the resulting circuit will have the short circuit admittance matrix of the form

$$\begin{bmatrix} i_A \\ i_1 \end{bmatrix} = \begin{bmatrix} y_{11} & y_{12} \\ -\frac{1}{(R_{eq}+sL_{eq})} & +\frac{1}{(R_{eq}+sL_{eq})} \end{bmatrix} \begin{bmatrix} v_A \\ v_1 \end{bmatrix} \tag{3.27}$$

where the values of y_{11} and y_{12} vary from circuit to circuit but they are of no consequence because if a capacitor is connected from node 1 to ground, since the impedance looking into node 1 represents a series RL, a low pass filter would be realizable from all the circuits by connecting a capacitor across node 1 by applying input at node having voltage V_A, as shown in Fig. 3.18 [25–27].

Thus, all the four circuits shown in Fig. 3.18 realize second order low pass filter function having transfer function

$$\frac{V_o}{V_{in}} = \frac{1}{s^2L_{eq}C_1 + sC_1R_{eq} + 1} \tag{3.28}$$

where L_{eq} and R_{eq} for different circuits are same as given in (3.14b).

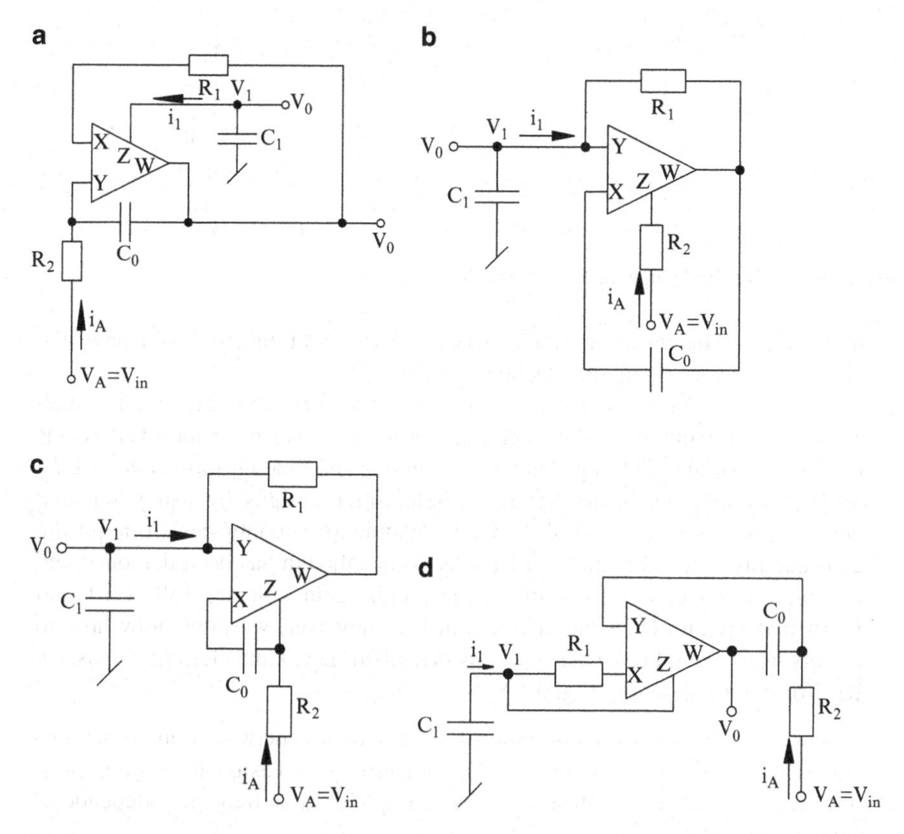

Fig. 3.18 Low pass filters based on series-RL type lossy inductance simulators proposed by Yuce (adapted from [27] © 2012 Springer)

Furthermore, two of these circuits, namely those of Fig. 3.18a, d, permit taking the output from the low-output-impedance terminal-W of the CFOA and thus, should be considered to be the best circuits of this set. Experimental results using AD 844 CFOA demonstrate [25] that the low pass filter of Fig. 3.18 can be readily used to realize LPF having $f_0 = 1.59$ MHz.

3.7.2 Application in the Design of Higher Order Filters

1. *Grounded-capacitor based designs:* A given RLC-prototype higher order passive filter can be readily converted into a CFOA-based active filter by simulating grounded and floating inductors by lossless grounded inductor circuit of Fig. 3.3 and any of the four-CFOA based or three-CFOA-based FI simulators of Fig. 3.13

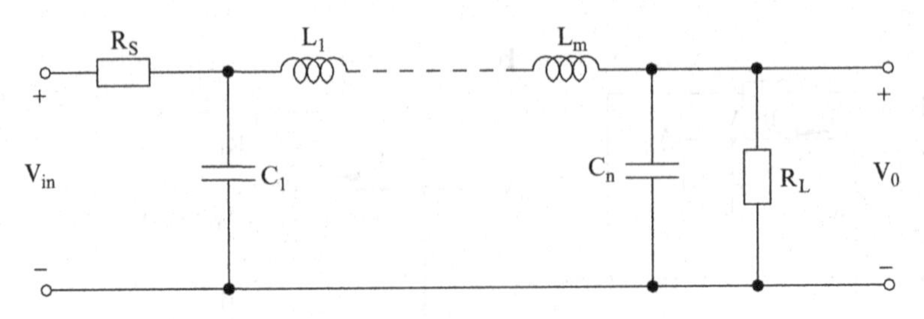

Fig. 3.19 Passive RLC prototype higher order low-pass ladder filter

or 3.14b, c. The resulting CFOA based higher order filters would have the advantage of employing all grounded capacitors.

2. *Designs with reduced number of CFOAs:* On the other hand, the circuits which simulate a lossy inductor (series-RL/parallel-RL impedance) or non-ideal FDNR (series-CD/parallel-CD impedance) in grounded and floating forms *can also be utilized as direct elements* in higher order filter designs by using Senani's network transformations [41, 42], thereby leading to economic designs requiring considerably reduced number of CFOAs. This although has been demonstrated for current-conveyor- based filters [41], unity-gain voltage- follower-based filters and op-amp-OTA-based filters in [42] however, we now show how to do this using the CFOA-based circuits described here. Consider now a passive RLC prototype shown in Fig. 3.19.

If we now apply Senani's transformation T-4 from the four network transformations proposed in [42] (T-2 was also proposed *independently* by Takagi and Fujii in [43]) on this ladder, each impedance is to be multiplied by a frequency-dependent-scaling-factor

$$F(s) = \left(\frac{1+s}{s}\right) \tag{3.29}$$

which transforms a resistor into series-RC impedance, an inductor into series-RL impedance and a capacitor into a series-CD impedance. The resulting transformed ladder turns out to be as shown in Fig. 3.20 which realizes exactly the same transfer function as the original RLC ladder of Fig. 3.19.

A CFOA-based circuit implementation can now be obtained by simulating the shunt CD-branches and series RL-branches by appropriate CFOA-based realizations. An exemplary implementation using the floating series-RL circuit of Fig. 3.17 and grounded series-CD impedance simulator of Fig. 3.10 is shown in Fig. 3.21.

It is, thus, clear that the above mentioned circuits of *lossy* floating series-RL impedance and *lossy* grounded CD-impedances can be directly used as elements in the transformed ladder network of type shown in Fig. 3.20, thereby leading to higher order CFOA-based filter designs with a reduced number of CFOAs, than those obtainable by simulating *lossless* floating inductors/FDNRs.

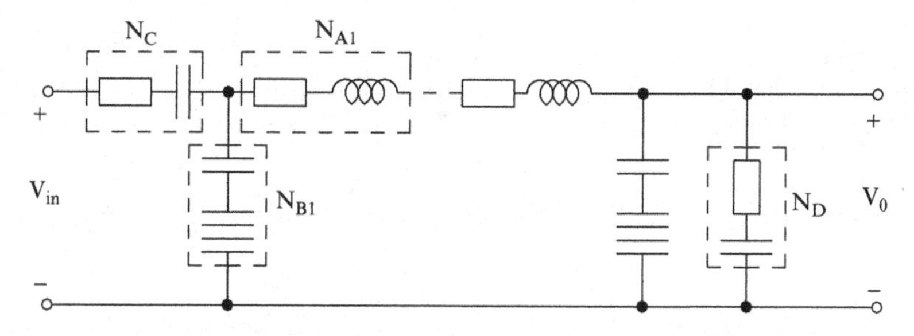

Fig. 3.20 Transformed version of the passive RLC ladder of Fig. 3.19 as per [42, 43]

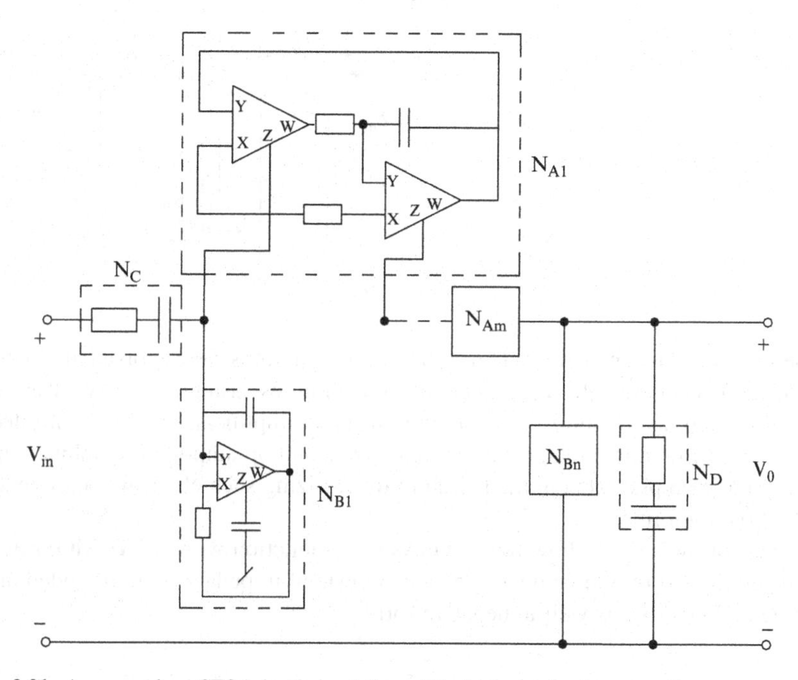

Fig. 3.21 An exemplary CFOA-implementation of the higher order low pass filter

3.8 Realization of Voltage-Controlled Impedances

Electronically-controlled impedances such as voltage-controlled-resistances (VCR) and voltage-controlled impedances (VCZ) find applications in automatic gain control, amplitude stabilization/control in oscillator circuits, design of analog multipliers/dividers, voltage-controlled filters and voltage-controlled oscillators etc. [44–57]. Op-amp-FET structures for realizing linearized positive/negative VCRs, providing wide dynamic range and low distortion, were first presented in [44, 45]

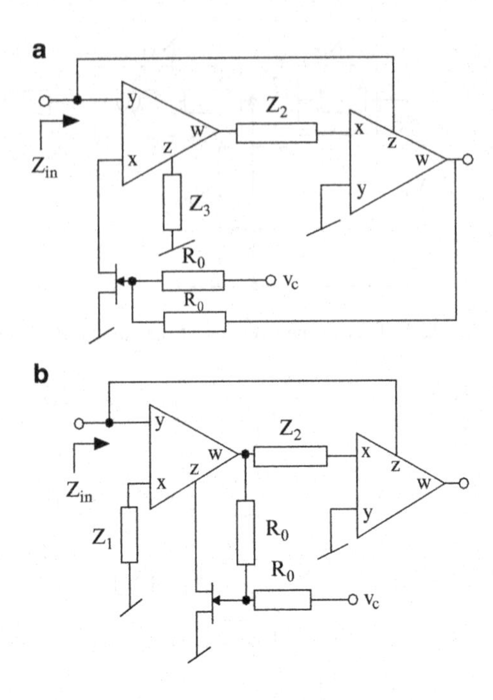

Fig. 3.22 Grounded voltage-controlled impedance configurations (adapted from [22] 1998 © Walter de Gruyter GmbH & Co. KG, Germany)

subsequent to which, generalized linear VCZ configurations were evolved in [46–51, 53, 54, 57]. Consequently a number of configurations, using a variety of active elements such as op-amps, operational-mirrored-amplifiers, current-controlled-conveyors, differential voltage current conveyors, and op-amp-OTA combinations, have so far been presented in the literature for realizing such elements in grounded and/or floating forms.

In the following, we show how CFOAs in conjunction with JFETs/MOSFETs, can be employed to realize novel voltage controlled impedances in grounded and floating and positive as well in negative forms.

3.8.1 Grounded Voltage Controlled Impedance Simulators

Consider now the VCZ structures shown in Fig. 3.22 [22] which are obtained by modifying the grounded impedance simulation circuits described earlier. Note that in both the circuits, one of the impedances has been replaced by a FET along with two equal valued resistors with their junction connected to the gate and the free ends connected to voltage V_c and the unused W terminal of a CFOA respectively thereby making the gate voltage as $(V_c + V_{DS})/2$ which results in the cancellation of square non-linearity of the FET thereby realizing a linear voltage controlled resistor (VCR). Assuming JFET to be confined to operate in the triode region, the drain current is given by

$$i_D = \frac{I_{DSS}}{V_p^2}\left[\left(v_{GS} - V_p\right)v_{DS} - \frac{v_{DS}^2}{2}\right] \tag{3.30}$$

Since

$$v_{GS} = \frac{1}{2}(v_c + v_1); \quad v_{DS} = v_1 \tag{3.31}$$

Substituting, (3.31) in (3.30) yields the modified input resistance realized by the FET circuit as

$$r_{DS} = \frac{v_1}{i_D} = \frac{2V_p^2}{I_{DSS}\left(v_C - 2V_p\right)} \tag{3.32}$$

Thus, the nonlinear term proportional to v_{DS}^2 in (3.30) is effectively cancelled, without requiring any additional active devices and as a consequence, the overall circuit realizes a linear voltage controlled (VC) input impedance (grounded) of value

$$Z_{in}(s) = \frac{Z_2}{Z_3} r_{DS} \tag{3.33}$$

This circuit can thus, realize linear VC-capacitance (VCC), VC-resistance (VCR) and VC inductance (VCL) by appropriate selection (resistive/capacitive) of impedances Z_2 and Z_3.

Alternatively, Z_{in} (s) = Z_1Z_2/r_{DS} is obtained from the circuit of Fig. 3.22a by replacing Z_3 in the same manner, as shown in Fig. 3.22b, which can be then used to realize linear VCR, VCC and VC-FDNR elements (the last one by selecting Z_1 and Z_2 both capacitors).

The same techniques applied to the circuit of Fig. 3.4b would yield structures providing $Z_{in}(s) = -Z_2\ r_{DS}/Z_3$ or $Z_{in}(s) = -(Z_1Z_2/r_{DS})$ respectively thus, facilitating realization of linear negative VCR, VCC and VCL in the former case and linear negative VCC, VCR and VC-FDNR in the latter case.

3.8.2 *Floating Voltage Controlled Impedance Simulators*

We now show a linear VC-floating impedance (VCFI) configuration which is obtained from the circuit of Fig. 3.14b by replacing Z_3 by a FET and using the W- terminal of the relevant CFOA for the non-linearity cancellation circuitry. This circuit is shown in Fig. 3.23 and realizes an FI of value $Z_{1-2} = Z_1Z_2/r_{DS}$. From this expression it is readily seen that this circuit can realize linear VCR, VCC and VC-FDNR elements in *floating* form. A novel feature of this circuit is that from the same circuit one can realize a VCR, a VCC and VCFDNR elements in floating forms with negative values also by changing the connections [a_1–a_2, b_1–b_2] to [a_1–b_2, a_2–b_1].

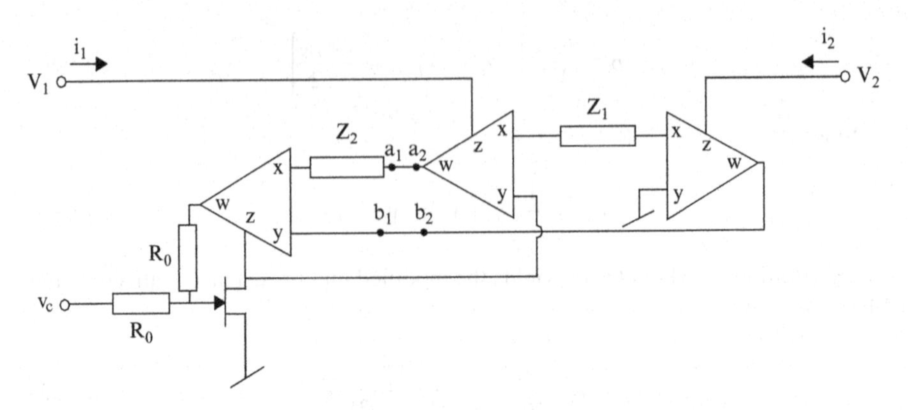

Fig. 3.23 Floating Voltage-controlled impedance configuration (adapted from [22] © 1998 Walter de Gruyter GmbH & Co. KG, Germany)

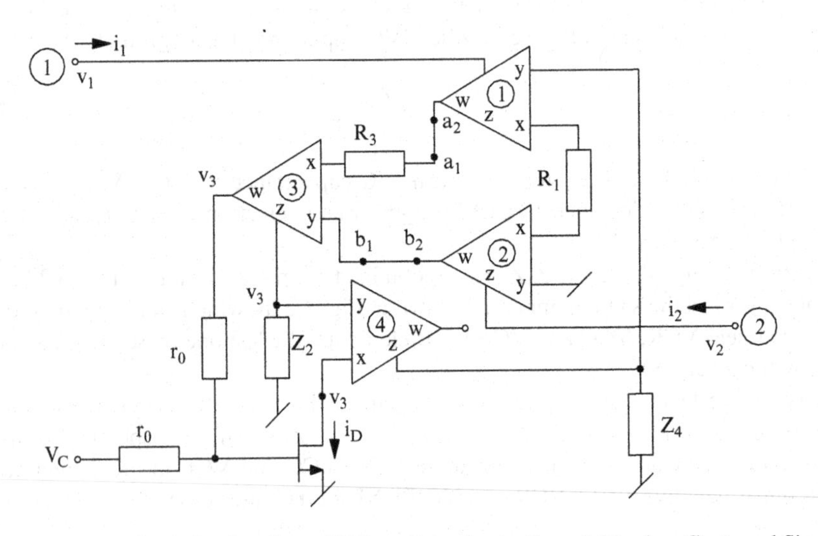

Fig. 3.24 Generalized, floating, linear VCZ configuration by Senani, Bhaskar, Gupta and Singh (adapted from [57] © 2008 John Wiley and Sons. Ltd)

While the circuit of Fig. 3.23 is capable of realizing floating VCR, voltage-controlled-capacitance (VCC) and VC-FDNR elements, this circuit, however, cannot realize floating VCL and VC-FDNC (frequency-dependent-negative-conductance characterized by $Z(s) = Ms^2$) elements.

A novel configuration which is capable of realizing linear VCR, VCL and VC-FDNC elements in positive as well as negative, floating as well as grounded - all possible forms, from the same topology, under appropriate conditions, is shown in Fig. 3.24.

From a straight forward analysis of the circuit of Fig. 3.24, the equivalent floating impedance realized by the circuit between terminals (1) and (2) is given by:

$$Z_{1-2} = \left(\frac{R_1R_3}{Z_2Z_4}\right) r_{DS} \tag{3.34}$$

from which linear, floating, positive, VCR, VCL and VC-FDNC elements can be realized from the circuit by the following choice (resistive/capacitive) of impedances Z_2 and Z_4:

1. VCR: $Z_2 = R_2$ and $Z_4 = R_4$
2. VCL: either $Z_2 = 1/sC_2$ or $Z_4 = 1/sC_4$
3. VC-FDNC: $Z_2 = 1/sC_2$ and $Z_4 = 1/sC_4$.

It is interesting to mention that the various negative-valued elements corresponding to the equivalent impedance given in (3.34) can be obtained by the simple artifice of connecting a_1-b_2 and a_2-b_1 in the circuit of Fig. 3.24, thereby leading to floating negative impedance given by

$$Z_{1-2} = -\left(\frac{R_1R_3}{Z_2Z_4}\right) r_{DS} \tag{3.35}$$

Furthermore, the grounded forms of all the above-mentioned floating impedances can be obtained by grounding either port-1 or port-2. However, in the circuit of Fig. 3.24, with port-2 grounded, CFOA2 becomes redundant (y-terminal of CFOA-3 and R_1 can be connected to ground directly) and as a consequence, the circuit can be simplified to have only three CFOAs while still being capable of realizing a grounded impedance.

$$Z_{in} = \pm\left(\frac{R_1R_3}{Z_2Z_4}\right) r_{DS} \tag{3.36}$$

It is important to keep in mind that in order to reduce the effect of various parasitic impedances of the CFOAs (i.e. finite input resistance R_x (typically, 50–100 Ω) at port x and parasitic impedance Z_p at port z which contains a parasitic resistance R_p (typically, 3 MΩ) in parallel with a parasitic capacitance C_p (typically, 4.5 pF)), in all the cases, the external circuit impedances are to be chosen such that they are larger than R_x but smaller than the magnitude of Z_p (over the frequency range of interest).

Floating VCR: The experimentally observed v-i characteristics using AD844 type CFOAs biased with ±15 V DC in conjunction with BFW11 JFETs (with $R_1 = R_3 = r_0 = 10$ kΩ and $R_2 = R_4 = 1$ kΩ) shows (Fig. 3.25) the linear range of the resulting circuit to be nearly ±6 V DC which is about two orders of magnitude larger than that of a conventional FET-VCR.

Floating negative VCR: The v–i characteristic of the negative VCR realized from the circuit (for the same component values as in positive floating VCR) is shown in Fig. 3.26, which shows a linear range of the same order (± 6.0 V DC) as that of the positive VCR (Fig. 3.25).

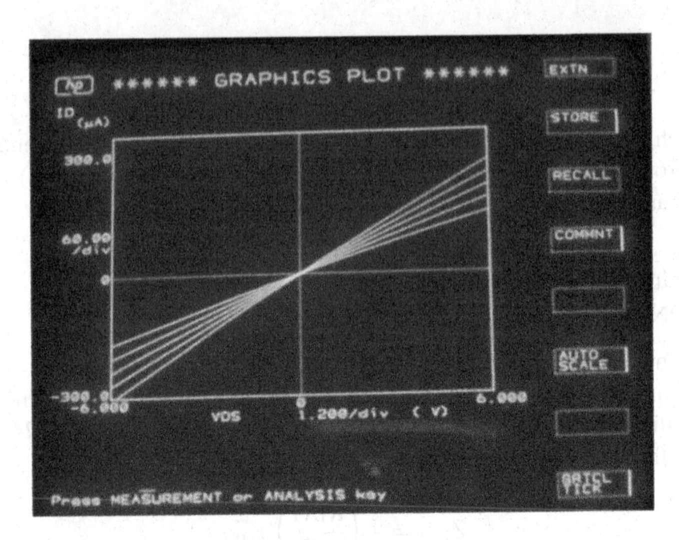

Fig. 3.25 *v–i* characteristics of the floating VCR

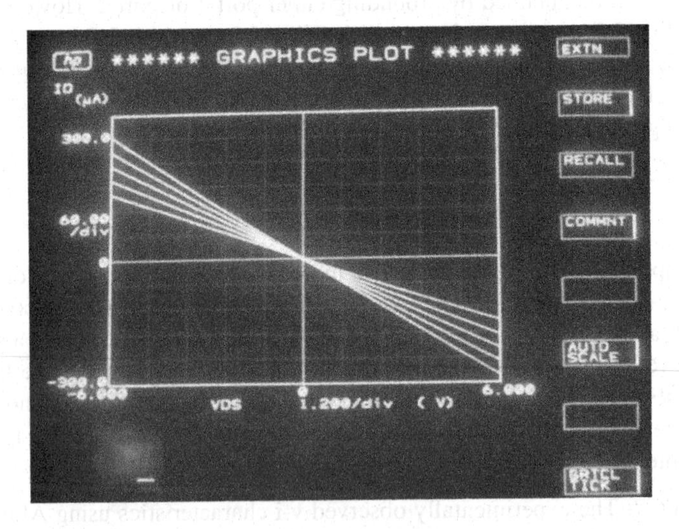

Fig. 3.26 *v–i* characteristics of the floating negative VCR

For the details about the workability of this circuit in realizing VC-L, VC-FDNR and V-FDNC elements and applications thereof, the reader is referred to [57].

Lastly, it may be mentioned that if R_1, R_3 are replaced by general impedances Z_1 and Z_3, the equivalent floating impedance realized becomes $\pm\left(\frac{Z_1Z_3}{Z_2Z_4}\right) r_{DS}$ and in this mode, the circuit[2] can be treated as a floating positive/negative generalized-impedance-converter.

In view of the above novelty, the circuit of Fig. 3.24 can, therefore, be considered to be a *universal linear voltage controlled floating impedance circuit.*

3.9 Concluding Remarks

In this chapter we presented a variety of CFOA-based circuits for simulating inductors and other kinds of impedances such as FDNR and FDNC, in grounded/floating and positive/negative forms. It was shown that CFOA-based grounded impedance as well as floating impedance circuits require a minimum possible number of passive components without requiring any component-matching conditions. It was also shown that most of the economic grounded simulators realizable with only a single CFOA and three passive components provided an attractive advantage of single resistance control of the inductance value which cannot be attained by any single VOA-three passive component circuits such as those in [5, 7, 8] known in literature. Applications of some of these circuits in the design of second order and higher order filters were exemplified.

It was also demonstrated that CFOAs provide novel solutions to the realization of linear voltage-controlled impedances in grounded/floating, positive/negative all forms, while maintaining the same advantageous features.

Although a new floating series-RL impedance employing only two CFOAs has been presented here; there could be a family of such circuits which is still waiting to be discovered! It is believed that the circuits presented in this chapter provide a good repertoire of impedance simulation circuits which may be employed in the new designs of filters and oscillators and may also find interesting other applications.

Finally, it may be pointed out that negative capacitance elements appear to have an application in the area of high frequency oscillator design where the frequency of oscillation can be scaled up by having in the expression for frequency a capacitive difference term in the denominator such that by making this difference term small, the frequency generated by the circuit can be scaled up. On the other hand, a negative inductance, together with negative FDNR and negative FDNC elements, are still elements of academic curiosity and have yet to find practical applications. This constitutes an interesting area for research.

[2] This circuit was first reported in R. Senani, 'Novel linear voltage controlled floating-impedance configuration', ELL/96/53450, dated 25th November 1996 (unpublished) and has been subsequently published later in [57].

References

1. Mitra SK (1969) Analysis and synthesis of linear active networks. Wiley, New York. pp 468–469
2. Bruton LT (1980) RC-active circuits theory and design. Prentice-Hall, Inc., Englewood Cliffs, NJ, pp 145–180
3. Newcomb RW (1968) Active integrated circuit synthesis. Prentice-Hall, Inc., Englewood Cliffs, New Jersey, p 151
4. Antoniou A (1969) Realisation of gyrators using operational amplifiers, and their use in RC-active-network synthesis. Proc IEE 116:1838–1850
5. Ahmed MT, Dutta Roy SC (1976) Evaluating the frequency limitations of operational amplifier RC networks: the dominant-pole technique and its application to inductance simulation. J Inst Electron Telecom Eng 22:703–708
6. Rao KR, Venkateshwaran S (1970) Synthesis of inductors and gyrators with voltage- controlled voltage sources. Electron Lett 6:29–30
7. Ford RL, Girling FEJ (1966) Active filters and oscillators using simulated inductance. Electron Lett 2:52
8. Presstcott AJ (1966) Loss—compensated active gyrator using differential—input operational amplifiers. Electron Lett 2:283–284
9. Senani R (1982) Author's reply in Comments on new canonic active RC realizations of grounded and floating inductors. Proc IEEE 70:101–103
10. Von Grunigen DC, Ramseier D, Moschytz GS (1986) Simulation of floating impedances for low-frequency active filter design. Proc IEEE 74:366–367
11. Reddy MA (1976) Some new operational-amplifier circuits for the realization of lossless floating inductance. IEEE Trans Circ Syst 23:171–173
12. Riordan RHS (1967) Simulated inductors using differential amplifiers. Electron Lett 3:50–51
13. Senani R (1989) Three op-amp floating immittance simulators: A retrospection. IEEE Trans Circ Syst 36:1463–1465
14. Rathore TS, Singhi BM (1980) Active RC synthesis of floating immittances. Int J Circ Theor Appl 8:184–188
15. Dutta Roy SC (1974) A circuit for floating inductance simulation. Proc IEEE 64:521–523
16. Wise DR (1974) Active simulation of floating lossy inductances. Proc IEE 121:85–87
17. Sudo S, Teramoto M (1977) Constitution of floating inductance using operational amplifiers. IEICE Trans E60-E:185–186
18. Senani R, Tiwari RN (1978) New Canonic active RC realizations of grounded and floating inductors. Proc IEEE 66:803–804
19. Ahmed MT, Dutta Roy SC (1977) A critical study of some non-ideal floating inductance simulators. Arch Elektron Uevertragungstch (AEU) 31:182–188
20. Bhushan M, Newcomb R (1967) Grounding of capacitors in integrated circuits. Electron Lett 3:148–149
21. Fabre A (1992) Gyrator implementation from commercially available transimpedance operational amplifiers. Electron Lett 28:263–264
22. Senani R (1998) Realization of a class of analog signal processing/signal generation circuits: novel configurations using current feedback op-amps. Frequenz 52:196–206
23. Yuce E, Minaei S (2009) On the realization of simulated inductors with reduced parasitic impedance effects. Circ Syst Sign Process 28:451–465
24. Liu SI, Hwang YS (1994) Realization of R-L and C-D impedances using current feedback amplifier and its applications. Electron Lett 30:380–381
25. Yuce E (2009) Novel lossless and lossy grounded inductor simulators consisting of a canonical number of components. Analog Integr Circ Sign Process 59:77–82
26. Abuelma'atti MT (2011) Comment on "Novel lossless and lossy grounded inductor simulator consisting of canonical number of components. Analog Integr Circ Sign Process 68:139–141

27. Yuce E (2012) Reply to comment on "Novel lossless and lossy grounded inductor simulator consisting of canonical number of components". Analog Integr Circ Sign Process 72:505–507
28. Kacar F, Kuntman H (2011) CFOA-based lossless and lossy inductance simulators. Radioengineering 20:627–631
29. Abuelma'atti MT (2012) New grounded immittance function simulators using single current feedback operational amplifier. Analog Integr Circ Sign Process 71:95–100
30. Lahiri A, Gupta M (2011) Realizations of grounded negative capacitance using CFOAs. Circ Syst Sign Process 30:143–155
31. Senani R (1979) Novel active RC circuit for floating—inductor simulation. Electron Lett 15:679–680
32. Senani R (1980) New tunable synthetic floating inductors. Electron Lett 16:382–383
33. Senani R (1984) Floating ideal FDNR using only two current conveyors. Electron Lett 20(5):205–206
34. Senani R (1986) On the realization of floating active elements. IEEE Trans Circ Syst 33:323–324
35. Senani R (1982) Novel lossless synthetic floating inductor employing a grounded capacitor. Electron Lett 18:413–414; also see *Erratum, ibid*, August 1982 issue
36. Chang CM, Hwang CS (1995) Comment on voltage-mode notch, low pass and band pass filter using current-feedback amplifiers. Electron Lett 31:246
37. Chang CM, Hwang CS, Tu SH (1994) Voltage-mode notch, lowpass and band pass filter using current-feedback amplifiers. Electron Lett 30:2022–2023
38. Psychalinos C, Pal K, Vlassis S (2008) A floating generalized impedance converter with current feedback operational amplifiers. Int J Electron Commun (AEU) 62:81–85
39. Senani R, Bhaskar DR (2012) New lossy/loss-less synthetic floating inductance configuration realized with only two CFOAs. Analog Integr Circ Sign Process 73:981–987
40. Yuce E, Minaei S (2008) A modified CFOA and its applications to simulated inductors, capacitance multipliers, and analog filters. IEEE Trans Circ Syst-I 55:266–275
41. Senani R (1985) Novel higher-order active filter design using current conveyors. Electron Lett 21:1055–1057
42. Senani R (1987) Network transformations for incorporating nonideal simulated immittances in the design of active filters and oscillators. Proc IEE Circ Devices Syst 134:158–166
43. Takagi S, Fujii N (1985) High-frequency active RC simulation of impedance-scaled LC filters using voltage followers. Proc IEEE Int Symp Circ Syst Kyoto, Jpn:295–298
44. Nay K, Budak A (1983) A voltage-controlled resistance with wide dynamic range and low distortion. IEEE Trans Circ Syst 30:770–772
45. Nay KW, Budak A (1985) A variable negative resistance. IEEE Trans Circ Syst 32:1193–1194
46. Senani R, Bhaskar DR (1991) Realization of voltage-controlled impedances. IEEE Trans Circ Syst 38:1081–1086, also see ibid, 1991: 39: 162
47. Senani R, Bhaskar DR (1992) A simple configuration for realizing voltage-controlled impedances. IEEE Trans Circ Syst 39:52–59
48. Senani R, Bhaskar DR (1994) Versatile voltage-controlled impedance configuration. IEE Proc Circ Devices Syst 141:414–416
49. Senani R (1995) Universal linear voltage-controlled impedance configuration. IEE Proc Circ Devices Syst 142:208
50. Ndjountche T (1996) Linear voltage-controlled impedance architecture. Electron Lett 32:1528–1529
51. Leuciuc A, Goras L (1998) New general immittance converter JFET voltage-controlled impedances and their applications to controlled biquads synthesis. IEEE Trans Circ Syst 45:678–682
52. Senani R (1994) Realisation of linear voltage-controlled resistance in floating form. Electron Lett 30:1909–1911
53. Senani R (1995) Floating GNIC/GNII configuration realized with only a single OMA. Electron Lett 31:423–425

54. Ndjountche T, Unbehauen R, Luo FL (1999) Electronically tunable generalized impedance converter structures. Int J Circ Theor Appl 27:517–522
55. Maundy B, Gift S, Aronhime P (2008) Practical voltage/current controlled grounded resistor with wide dynamic range extension. IET Circ Devices Syst 2:201–206
56. Senani R, Bhaskar DR (2008) Comment on practical voltage/current controlled grounded resistor with wide dynamic range extension. IEE Circ Devices Syst 2:465–466, also see ibid, 2: 467
57. Senani R, Bhaskar DR, Gupta SS, Singh VK (2009) A configuration for realizing floating, linear, voltage-controlled resistance, inductance and FDNC elements. Int J Cir Theor Appl 37:709–719

Chapter 4
Design of Filters Using CFOAs

4.1 Introduction

In the area of analog circuit design, considerable attention has been devoted to the realization of the so-called *universal biquad filters* using a variety of active elements such as the classical op-amps, OTAs, various forms of current conveyors and a host of other building blocks of relatively more recent origin such as OTRAs, CFOAs, CDBAs, CDTAs and CFTAs etc. The term universal biquad filter, strictly speaking, is supposed to mean circuits which are capable of realizing, from the same topology, all the five basic filtering functions namely, Low pass (LP), band pass (BP), high pass (HP), band stop (BS also referred to as band reject, band elimination or notch filter) and all pass (AP). However, quite often, some authors also use the term *biquad* loosely, to refer to configurations which can realize three (LP, BP and HP) or even two functions only.

Several researchers and practicing engineers have often wondered and even questioned the utility of a circuit which simultaneously realizes three responses (usually LP, BP, and HP) from the same circuit arguing that at a given time, after all, the circuit would be used for realizing only a single type of filter and thus, there may not be any great utility of simultaneously realizing all the three responses from the same circuit. It is, therefore, also argued that a circuit realizing three simultaneous responses is not necessarily better than the one which realizes only a single type of filter response.

In the above context, we would like to point out that simultaneous realization of several filter functions from the same topology, particularly if the realized responses are LP, BP and HP, indeed finds many applications such as in phase locked loops, FM stereo demodulators, touch-tone telephone tone decoders and crossover networks used in a three-wave high-fidelity loud speaker; see [1]. Moreover, a *universal* biquad, if available as a standard integrated circuit, gives the versatility and flexibility of designing any second order or higher order filter using such biquad filters as standard building blocks. It is worth mentioning that several such universal filters using op-amps are, indeed, commercially available as ICs, for

R. Senani et al., *Current Feedback Operational Amplifiers and Their Applications*, Analog Circuits and Signal Processing, DOI 10.1007/978-1-4614-5188-4_4,

example, UAF-42 from Texas Instruments, AF-151 from National Semiconductors and MF10 from National Semiconductors, to name a few.

Since the advent of CFOAs in the area of analog circuit design, there have been numerous investigations and proposals for realizing universal filters using CFOAs as building blocks. The objective of this chapter is to highlight some prominent filter circuit configurations employing CFOAs. Besides this, some work has also been done on realizing MOSFET-C biquads and higher order filters using CFOAs and hence, some prominent results in these two areas would also be highlighted.

4.2 The Five Generic Filter Types, Their Frequency Responses and Parameters

Before moving further, it is useful to outline the second order transfer functions of the five standard filtering functions to establish the notations employed and to understand the basic terminology and the parameters which shall be often used in the discussion of various circuits in the subsequent sections of this chapter. These five basic filter functions are as follows:

Low Pass:

$$\mathrm{T(s)} = \frac{H_0\omega_0^2}{s^2 + \frac{\omega_0}{Q_0}s + \omega_0^2} \tag{4.1}$$

High Pass:

$$\mathrm{T(s)} = \frac{H_0 s^2}{s^2 + \frac{\omega_0}{Q_0}s + \omega_0^2} \tag{4.2}$$

Band pass:

$$\mathrm{T(s)} = \frac{H_0\left(\frac{\omega_0}{Q_0}\right)s}{s^2 + \frac{\omega_0}{Q_0}s + \omega_0^2} \tag{4.3}$$

Band stop:

$$\mathrm{T(s)} = \frac{H_0\left(s^2 + \omega_0^2\right)}{s^2 + \frac{\omega_0}{Q_0}s + \omega_0^2} \tag{4.4}$$

All pass:

$$\mathrm{T(s)} = \frac{H_0\left(s^2 - \frac{\omega_0}{Q_0}s + \omega_0^2\right)}{s^2 + \frac{\omega_0}{Q_0}s + \omega_0^2} \tag{4.5}$$

It may be noted that in all cases, H_0 represents the maximum gain or the gain factor whereas ω_0 represents the central frequency (sometimes also referred as resonant frequency) in case of BP and BS responses. In case of BP and BS responses, $\frac{\omega_0}{Q_0}$ represents the bandwidth. Lastly, Q_0 represents the quality factor which is normally taken as $\frac{1}{\sqrt{2}}$ in case of LP and HP filters to attain maximally-flat response in the pass band.

4.3 Voltage-Mode/Current-Mode Biquads Using CFOAs

The CFOA-based biquad circuits can be classified in two main categories[1]: variable-topology type biquads and fixed topology type biquads. The latter can be further subdivided into two main categories : voltage mode (VM) and current mode (CM) each of which can be further divided into the following categories: single input single output (SISO) type, single input multiple output (SIMO), multiple input single output (MISO) type. Finally, there are universal mixed-mode biquads which can realize both VM/CM and in an extended case, even transimpedance type and transadmittance type biquads. In the following sections, we present prominent circuits in each type chosen from a vast amount of literature [2–52] existing on the topic. In the literature, various authors have proposed a number of topologies employing one to five CFOAs exhibiting different characteristic features. In general, single CFOA-based biquad circuits are not capable of realizing ideally infinite input impedance in case of voltage-mode (VM) filter realizations and ideally zero input impedance in case of current-mode (CM) filter realizations. On the other hand, there are number of two-CFOA-based VM topologies which possess interesting properties, a number of these have been included. A number of three-CFOA-based circuits have also been proposed out of which we have included only those which employ both grounded capacitors as preferred for IC implementation. Thus, we have endeavored to include only the most prominent universal biquad circuits in the following sections.

4.3.1 *Dual Function VM Biquads*

Two single input dual output biquads introduced by Soliman in [2] derived from RLC filters, are shown in Fig. 4.1.

A routine circuit analysis (assuming ideal CFOAs) of the circuit of Fig. 4.1a yields the following transfer functions:

[1] In the category of SIMO-type CM biquad, surprisingly, no configuration based on CFOAs is known to have been published in the technical literature till the time of writing this chapter.

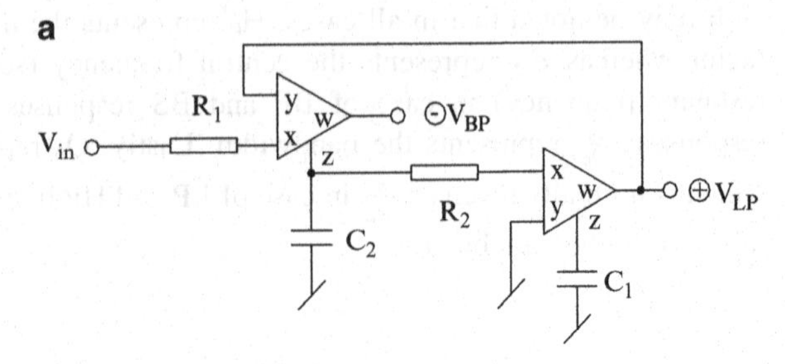

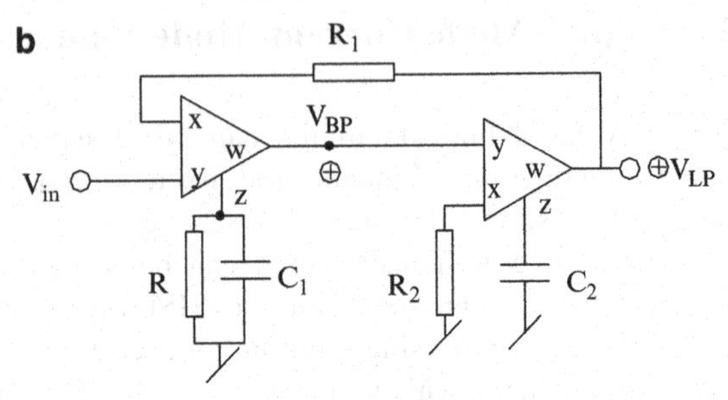

Fig. 4.1 Dual function circuits proposed by Soliman (**a**) non-inverting low pass and inverting band pass filters, (**b**) non-inverting low pass and non-inverting band pass filters (adapted from [2] © 1998 Taylor & Francis)

$$T(s)|_{LP} = \frac{\frac{1}{C_1C_2R_1R_2}}{s^2 + \frac{1}{C_2R_2} + \frac{1}{C_1C_2R_1R_2}}; \; T(s)|_{BP} = \frac{-s\frac{1}{C_2R_1}}{s^2 + \frac{1}{C_2R_2} + \frac{1}{C_1C_2R_1R_2}} \tag{4.6}$$

Similarly, the transfer functions for the circuit of Fig. 4.1b are given by

$$T(s)|_{LP} = \frac{\frac{1}{C_1C_2R_1R_2}}{s^2 + \frac{1}{C_1R} + \frac{1}{C_1C_2R_1R_2}}; \; T(s)|_{BP} = \frac{s\frac{1}{C_1R_1}}{s^2 + \frac{1}{C_1R} + \frac{1}{C_1C_2R_1R_2}} \tag{4.7}$$

Both the circuits of Fig. 4.1 employ two CFOAs and two GCs and have the attractive feature in that the parasitic input resistance r_x of port-X and the parasitic output capacitance C_p of port-Z of both the CFOAs can be absorbed in the external passive elements. On the other hand, the circuit of Fig. 4.1a uses only two resistors and does not provide infinite input impedance while the circuit of Fig. 4.1b does provide an infinite input impedance though it employs three resistors. SPICE simulation results given in [2] demonstrate that filters having ω_0 of the order of 1 Mrad/s and Q_0 of the order of 10 using AD 844 macromodel show excellent performance.

4.3.2 *Single Input Multiple Output (SIMO) Type VM Biquads*

Indeed, the state variable Kerwin-Huelsman-Newcomb (KHN) [53] biquad, popularly known as KHN-biquad, which originally employed classical VOAs, has been not only the first but also one of the most prominent active filter arrangements due

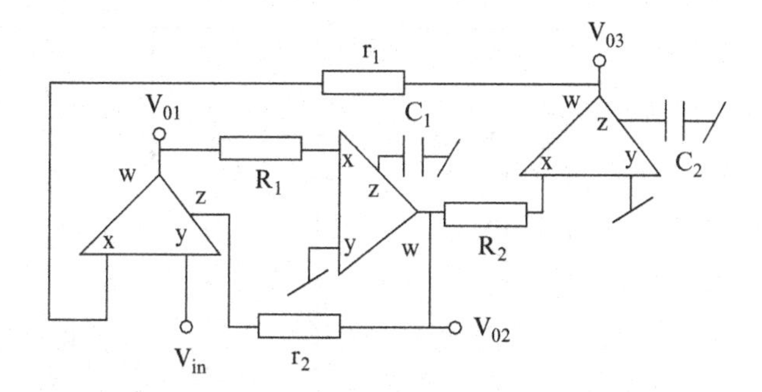

Fig. 4.2 Senani's low-component-count KHN-equivalent biquad (adapted from [3] © 1998 Walter de Gruyter GmbH & Co. KG)

to its novel feature of simultaneously providing LP, BP and HP filter responses from the basic three-VOA-structure. With a fourth amplifier added (configured as a summer) this configuration also makes it possible to realize BS/notch and AP functions, subject to fulfillment of appropriate conditions. As a consequence, the technical literature is flooded with a large number of KHN-equivalent biquads using various building blocks such as OTAs, CCII, CCIII, and numerous others. In [3], Senani derived a minimum-component CFOA version of the KHN-biquad which is shown in Fig. 4.2. The circuit realizes a LP response at V_{01}, a BP response at V_{02} and a HP response at V_{03} with the relevant parameters of the realized filters given by

$$\omega_0 = \sqrt{\frac{r_2}{r_1 C_1 C_2 R_1 R_2}} \quad (\text{LP, BP, HP}) \tag{4.8}$$

$$Q_0 = \sqrt{\frac{C_1 R_1 r_2}{C_2 R_2 r_1}} \quad (\text{LP, HP}) \tag{4.9}$$

$$\text{Bandwidth (BW)} = \left(\frac{\omega_0}{Q_0}\right) = \frac{1}{R_1 C_1} \quad (\text{BP}) \tag{4.10}$$

$$\text{H}_0 = 1(\text{for noninverting LP at } \text{V}_{01}) \tag{4.11}$$

and

$$\text{H}_0 = \frac{r_2}{r_1} \; (\text{for non-inverting BP at } \text{V}_{02} \text{ and noninverting HP at } \text{V}_{03}) \tag{4.12}$$

Note that as compared to VOA-based KHN biquad, this circuit has the advantage of providing ideally infinite input impedance and employing a reduced number of resistors (only four as against six in the original circuit).

Yet another three-CFOA biquad, which realizes LP, BP and BS functions was proposed by Bhaskar [4] and is shown in Fig. 4.3. The circuit is, in fact, a single resistance controlled oscillator (SRCO) in the form shown in Fig. 4.3, but becomes

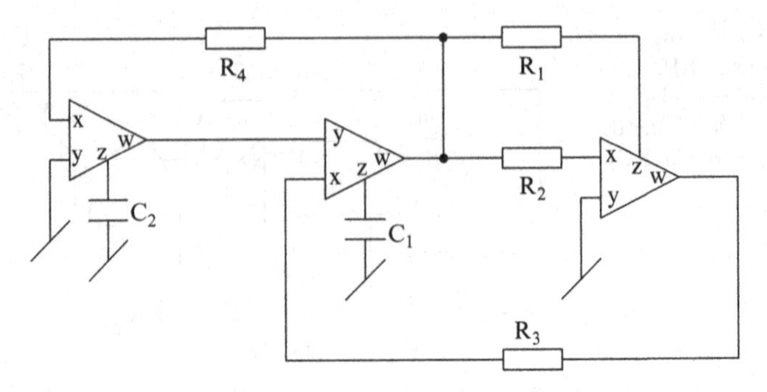

Fig. 4.3 Bhaskar's SRCO-cum-Multifunction biquad (adapted from [4] © 2003 Walter de Gruyter GmbH & Co. KG)

a multifunction biquad filter when resistor R_2 is disconnected from the junction of R_1, R_4 and the w-output of the middle CFOA and an input V_{in} is applied at the free end of the resistor R_2. In this case, the circuit realizes a BP at the w-output of the middle CFOA, inverting LP at w-output of first CFOA and inverting BS at the w-terminal of the third CFOA.

The characterizing parameters of the various filter responses are given by

$$\omega_0 = \sqrt{\frac{1}{C_1 C_2 R_3 R_4}} \tag{4.13}$$

$$Q_0 = \sqrt{\frac{C_1 R_3}{C_2 R_4}} \tag{4.14}$$

$$\mathrm{BW} = \frac{1}{R_3 C_1} \text{ (BP and BS) and } |\mathrm{H_0}| = \frac{R_1}{R_2} \tag{4.15}$$

In case of BP and BS, BW can be adjusted by R_3 after which ω_0 can be adjusted through R_4 and finally, gain H_0 can be adjusted by R_1 or R_2.

Another three-CFOA-based biquad proposed by Chang et al. [5] is shown in Fig. 4.4, which realizes BS at V_{01}, LP at V_{02} and BP at V_{03}. The relevant filter parameters of this circuit are given by

$$\omega_0 = \sqrt{\frac{1}{C_1 C_2 R_2 R_3}};\ Q_0 = \frac{1}{R_1}\sqrt{\frac{C_1 R_2 R_3}{C_2}} \tag{4.16}$$

$$\mathrm{BW} = \frac{R_1}{C_1 R_2 R_3} \text{(BP and BS)} \tag{4.17}$$

$$\mathrm{H_{BS}} = \mathrm{H_{LP}} = 1 \text{ and } |\mathrm{H_0}|_{\mathrm{BP}} = \frac{R_3}{R_1} \tag{4.18}$$

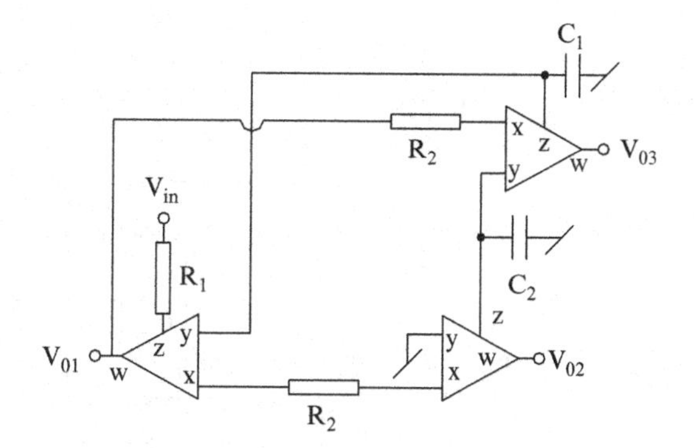

Fig. 4.4 Voltage-mode BS, LP and BP filter configuration (adapted from [5] © 1998 IEE)

While the CFOA-version of the KHN biquad of Fig. 4.2 has ideally infinite input Impedance it does not provide any independent/orthogonal tunability of the various parameters of the realizable filters. On the other hand, the circuits of Figs. 4.3 and 4.4 although do provide the required tunability of the parameters but do not offer infinite input impedance.

SPICE simulation of the biquad of Fig. 4.3 [4] and experimental results of the circuit of Fig. 4.4 [5] demonstrate that using AD844 type CFOAs, the circuits can be satisfactorily used to design filters having f_0 of the order of 100 kHz.

We now present another state-variable biquad circuit proposed by Singh and Senani [6] which offers infinite input impedance, possesses the feature of the tunability of the parameters and in addition, makes it possible to apply passive compensation for the degradation of the high frequency response of the filter in case of HP (at V_{o1}) response. This circuit which realizes BP at V_{o2} and LP at V_{o3} is shown in Fig. 4.5. The relevant parameters of this circuit are given by

$$\omega_0 = \sqrt{\frac{R_2}{C_1 C_2 R_3 R_4 R_6}};\ Q_0 = R_5 \sqrt{\frac{C_1 R_3}{C_2 R_2 R_4 R_6}};\ \mathrm{BW} = \frac{R_2}{C_1 R_3 R_5} \tag{4.19}$$

$$\mathrm{H_{LP}} = \frac{R_6}{R_1};\ \mathrm{H_{BP}} = \frac{R_5}{R_1} \quad \text{and} \quad \mathrm{H_{HP}} = \frac{R_2}{R_1} \tag{4.20}$$

It may be observed that after fixing ω_0, the quality factor Q_0 or the BW can be adjusted by R_5 and finally, H_0 can be adjusted by R_1.

Consider now the effect of the various parasitics, namely, the finite input resistance r_x at port- X and compensation pin parasitics ($R_p \,||\, 1/sC_p$) at port-Z. Since resistors R_1, R_3 and R_4 are connected at port-X of CFOAs, r_x of the respective CFOA can be easily accommodated in these resistors. Similarly, the Z-port parasitic capacitances of CFOA-II and CFOA-III can be easily accommodated in C_1 and C_2 respectively. A non-ideal analysis of the integrator made from CFOA-II gives the transfer function

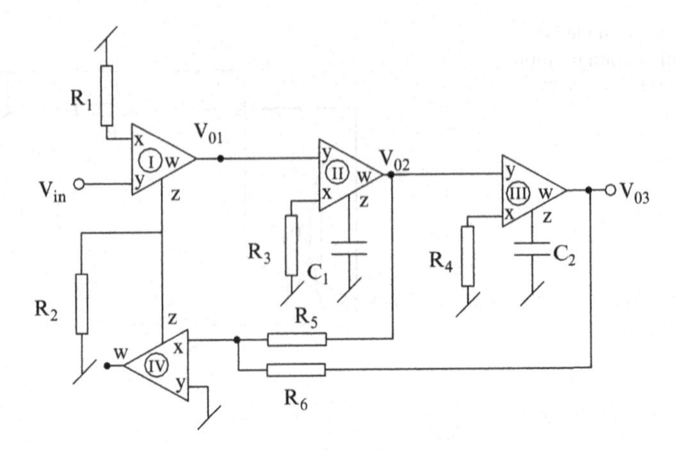

Fig. 4.5 The CFOA-based state-variable biquad proposed by Singh and Senani (adapted from [6] © 2005 IEICE)

$$\mathrm{T(s)} = \frac{1}{\left[s(C_1 + C_{p2}) + \frac{1}{R_{p2}}\right](R_3 + r_{x2})} \approx \frac{1}{s(C_1 + C_{p2})(R_3 + r_{x2})}; \tag{4.21}$$

provided $\omega(C_1 + C_{p2}) \geq \frac{1}{R_{p2}}$.

With $C_1 = 1$ nF, $C_{p3} = 4.5$ pF and $R_{p3} = 3$ MΩ, the above constraints implies f >> 53 Hz, which does not appear to be very restrictive.

It has been shown in [6] that the effect of Z-port parasitics of CFOA-I and CFOA-IV constituting the summer, can be accomplished by shunting resistors R_1, R_5 and R_6 by small external capacitors C_{c1}, C_{c5} and C_{c6} as shown in Fig. 4.6.

An analysis of this circuit reveals that the output of CFOA-I is now given by

$$\begin{aligned} V_{o3} &= \left\{ \frac{V_{in}(sC_{c1}R_1 + 1)}{R_1} - \frac{V_{o2}(sC_{c5}R_5 + 1)}{R_5} - \frac{V_{o1}(sC_{c6}R_6 + 1)}{R_6} \right\} \\ &* \left\{ \frac{R_{p1} \| R_{p2} \| R_2}{s(C_{p1} + C_{p2})(R_{p1} \| R_{p2} \| R_2 + 1)} \right\} \end{aligned} \tag{4.22}$$

Thus, if we select C_{c1}, C_{c5} and C_{c6} such that

$$C_{c1}R_1 = C_{c5}R_5 = C_{c6}R_6 = (C_{p1} + C_{p2})(R_{p1} \| R_{p2} \| R_2) \tag{4.23}$$

then (4.22) reduces to

$$V_{o3} = \frac{R_2}{R_1}V_{in} - \frac{R_2}{R_5}V_{o2} - \frac{R_2}{R_6}V_{o1}; \text{ assuming } (R_{p1} \| R_{p2} \| R_2) \cong R_2 \tag{4.24}$$

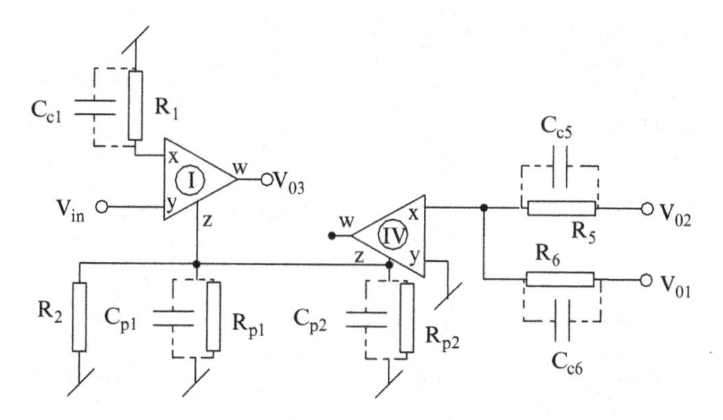

Fig. 4.6 A simple passive-compensation of the summer (adapted from [6] © 2005 IEICE)

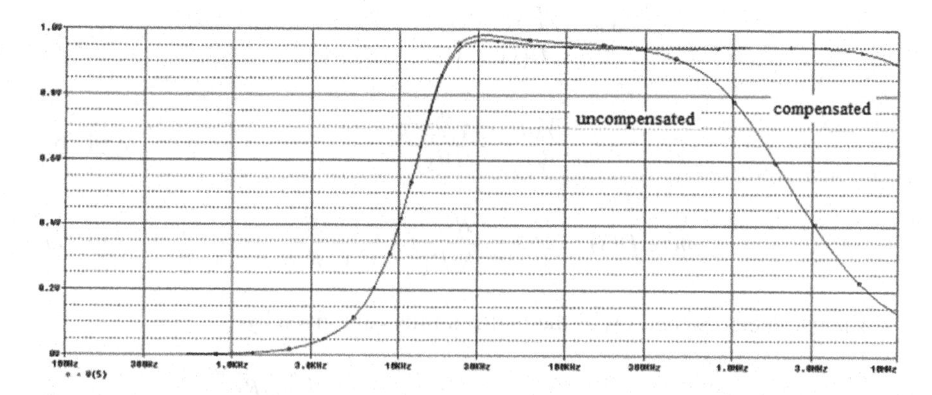

Fig. 4.7 Frequency response of the HP filter realized from the biquad of Fig. 4.5 with and without compensation (adapted from [6] © 2005 IEICE)

which is the ideal value of V_{o3} and thus, perfect compensation for the Z-port parasitics of CFOA-I and CFOA-IV would be achieved subject to the satisfaction of the conditions given in equation (4.23). The SPICE simulations of the uncompensated and compensated designs as contained in [6] are shown in Fig. 4.7 from where it is seen that the passive compensation is able to extend the operation frequency range by almost one decade.

A very interesting circuit shown in Fig. 4.8 was advanced by Soliman in [7] which although uses five CFOAs but has the following novel features (1) realizing three basic filter functions namely LP, BP and HP at V_{03}, V_{02} and V_{01} respectively (2) providing ideally infinite input impedance and zero output impedance at all the three outputs and (3) providing controllability of ω_0 and Q_0 (or bandwidth) and (4) using all grounded passive elements thereby making the circuit attractive for IC implementation.

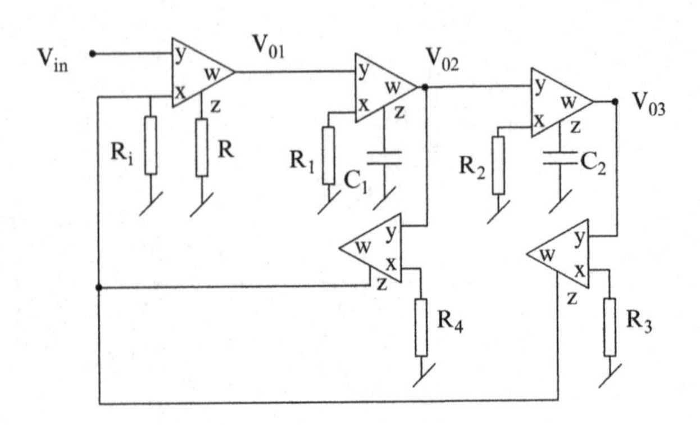

Fig. 4.8 Soliman's non-inverting HP-BP-LP filter using all grounded passive elements (adapted from [7] © 1996 Springer)

The various filter functions realized by the circuit of Fig. 4.8 are given by

$$\frac{V_{01}}{V_i}=\frac{\frac{R}{R_i}s^2}{D(s)}\quad \frac{V_{02}}{V_i}=\frac{\frac{R}{C_1R_1R_i}s}{D(s)} \tag{4.25}$$

$$\frac{V_{03}}{V_i}=\frac{\frac{R}{C_1C_2R_1R_2R_i}}{D(s)} \tag{4.26}$$

$$where\ D(s)=s^2+\frac{R}{C_1R_1R_4}s+\frac{R}{C_1C_2R_1R_2R_3} \tag{4.27}$$

The parameter ω_0 and Q_0 of the realized filters are given by

$$\omega_0=\frac{R}{\sqrt{C_1C_2R_1R_2R_3}}\quad \text{and}\quad Q_0=R_4\sqrt{\frac{C_1R_1}{C_2R_2R_3R}} \tag{4.28}$$

It may be noted from equation (4.28) that ω_0 and $\frac{\omega_0}{Q_0}$ can be adjusted independently; the former, by R_2 or R_3 and subsequently, the latter by R_4.

Two universal VM biquad configurations, each employing five CFOAs and eight passive elements and possessing the unique feature of providing all the five standard filters at five different output terminals were introduced by Abuelma'atti and Alzaher in [8, 9]. Here, we present one of these circuits which has the advantage of offering ideally infinite input impedance and the tunability of the various filter parameters. The circuit requires only a single matching condition in case of AP (at V_{o5}) response ($R_1 = R_8$). This configuration which realizes BS at V_{o1}, LP at V_{o2}, BP at V_{o3} and HP at V_{o4} is shown in Fig. 4.9.

The various characteristic parameters of this configuration are given by

$$\omega_0=\sqrt{\frac{1}{C_4C_6R_3R_5}};\ \text{BW}=\frac{R_2}{C_6R_1R_5} \tag{4.29}$$

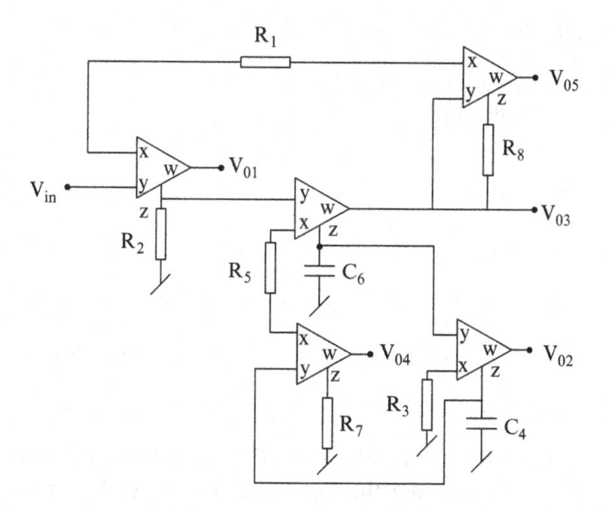

Fig. 4.9 Universal filter structure introduced by Abuelma'atti and Al-zaher (adapted from [8] © 1998 Springer)

$$H_{LP} = \frac{R_2}{R_1} = H_{BS}, \quad H_{BP} = 1 = H_{AP} \quad \text{and} \quad H_{HP} = \frac{R_2R_7}{R_1R_5} \tag{4.30}$$

Hence, this circuit enjoys grounded-resistance-controllability of ω_0 by R_3 while realizing all the five standard functions from five different low-output-impedance output terminals. Experimental results of the circuit realized with AD844 type CFOAs [8] demonstrate that the circuit works well in all the five modes with f_0 of the order of 100 kHz realizable quite well.

4.3.3 Multiple Input Single Output (MISO) Type VM Biquads

A variety of circuits are available in literature which realize MISO-type VM biquads using one to three CFOAs. In the following, we have included some prominent configurations from among those available in [10–15].

We first consider an interesting MISO-type VM biquad using a single CFOA proposed by Horng et al. [10], as shown in Fig. 4.10.

Assuming ideal CFOA, the output voltage V_0 of the circuit, in terms of various input voltages is given by

$$V_0 = \frac{s^2V_3 + s\left(\frac{1}{R_1C_2}V_4 + \frac{1}{C_1R_2}V_1 - \frac{1}{C_2R_3}V_2\right) + \left(\frac{1}{C_1C_2R_1R_2}\right)V_1}{s^2 + s\left(\frac{1}{R_1C_2} + \frac{1}{C_1R_2} - \frac{1}{C_2R_3}\right) + \frac{1}{R_1R_2C_1C_2}} \tag{4.31}$$

From (4.31), the various filter responses can be obtained by proper selection of inputs as follows

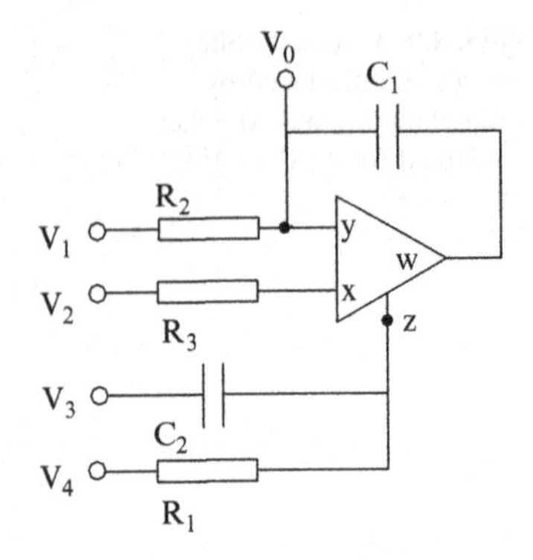

Fig. 4.10 VM universal biquadratic filter by Horng et al. (adapted from [10] © 2002 IEICE)

(1) LP: if $V_4 = V_3 = 0$, $C_1 = C_2$, $R_2 = R_3$ and $V_1 = V_2 = V_{in}$ (2) BP: if $V_1 = V_3 = 0$ and the input signal is either V_2 or V4 (3) HP: if $V_1 = V_2 = V_4 = 0$ and $V_3 = V_{in}$ (4) BS: if $V_4 = 0$, $C_1 = C_2$, $R_2 = R_3$ and $V_1 = V_2 = V_3 = V_{in}$ (5) AP: if $V_1 = V_2 = V_3 = -V_4 = V_{in}$ and $R_2 = R_3$ with $C_1 = C_2$

The filter parameters ω_0 and Q_0 are given by

$$\omega_0 = \sqrt{\frac{1}{C_1 C_2 R_1 R_2}} \quad \text{and} \quad Q_0 = \frac{R_3\sqrt{C_1 C_2 R_1 R_2}}{C_1 R_2 R_3 + C_2 R_1 R_3 - C_1 R_1 R_2} \tag{4.32}$$

This circuit does not provide a low impedance output node, hence would require a voltage follower (i.e., another CFOA) to be able to connect load impedance without altering the transfer function. This difficulty is, however, overcome in some of the available two-CFOAs-based biquads. There are a number of MISO-type universal filters available in literature employing two CFOAs [11–14].

A MISO-type universal filter proposed by Abuelma'atti and Al-Shahrani in [11] which, indeed, offers ideally infinite input impedance though only in one case, is shown in Fig. 4.11.

Assuming ideal CFOAs, the output voltage V_0 of the circuit in terms of the input voltages is given as

$$V_0 = \frac{s^2 V_3 - s\dfrac{1}{R_1 C_2} V_2 + \dfrac{1}{R_1 R_3 C_1 C_2} V_1}{s^2 + s\dfrac{1}{C_2 R_2} + \dfrac{1}{R_1 R_3 C_1 C_2}} \tag{4.33}$$

From (4.33), the various filter responses can be obtained by proper selection of inputs as follows: (1) LP: if $V_2 = V_3 = 0$ and $V_1 = V_{in}$ (2) BP: if $V_1 = V_3 = 0$ and $V_2 = V_{in}$ (3) HP: if $V_1 = V_2 = 0$ and $V_3 = V_{in}$ (4) BS: if $V_2 = 0$ and $V_1 = V_3 = V_{in}$ (5) AP: if $V_1 = V_2 = V_3 = V_{in}$ and $R_1 = R_2$

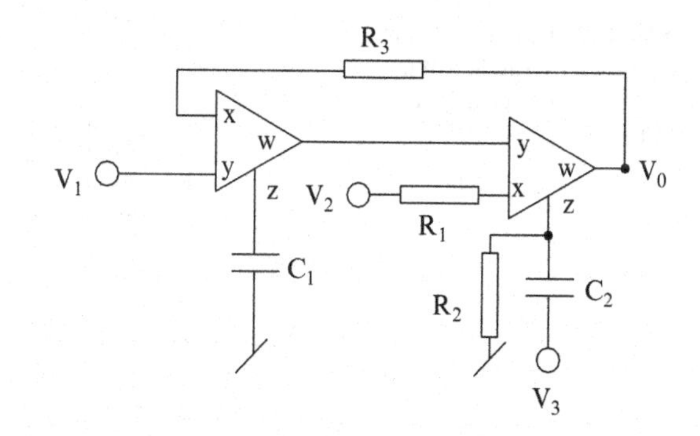

Fig. 4.11 Low component universal filter proposed by Abuelma'atti and Al-Shahrani (adapted from [11] © 1996 Taylor & Francis)

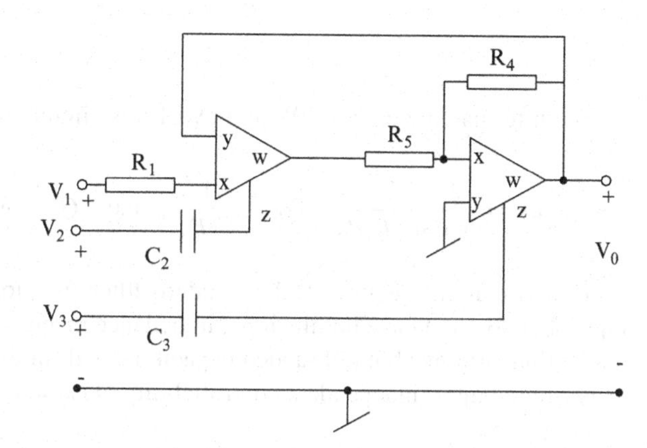

Fig. 4.12 Universal filter proposed by Abuelma'atti and Al-Shahrani (adapted from [12] © 1997 Hindawi Publishing Corporation)

The various filter parameters are given by

$$\omega_0 = \sqrt{\frac{1}{C_1C_2R_1R_3}};\ Q_0 = R_2\sqrt{\frac{C_2}{C_1R_1R_3}} \quad \text{and} \quad \text{BW} = \frac{1}{C_2R_2} \tag{4.34}$$

Thus, although this circuit does offer independent tunability of ω_0 and BW independently, the former by R_1 and/or R_3 and the latter by R_2, it does not have ideally infinite input impedance in case of realizing BP, BS and AP filters. Neither the circuit retains both capacitors grounded in all the cases.

The same authors in [12] presented yet another biquad using exactly the same number of active and passive components which is shown in Fig. 4.12.

A straight forward analysis of this circuit gives

$$V_0 = \frac{s^2V_3 - s\dfrac{1}{C_3R_5}V_2 + \dfrac{1}{C_2C_3R_1R_5}V_1}{s^2 + s\dfrac{1}{C_3R_4} + \dfrac{1}{C_2C_3R_1R_5}} \tag{4.35}$$

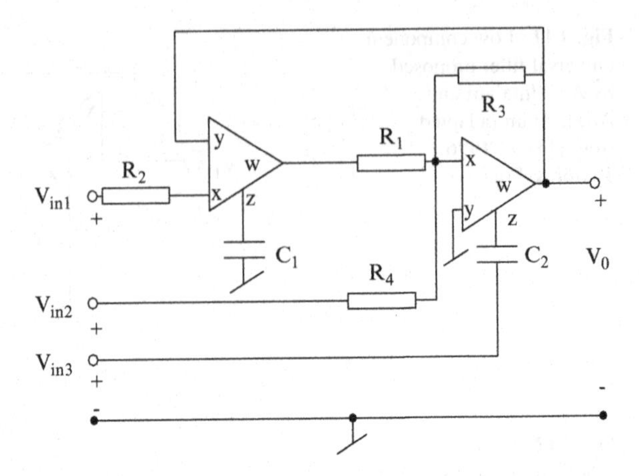

Fig. 4.13 Universal filter proposed by Liu and Wu (adapted from [13] © 1995 IEEE)

The filter parameters ω_0, BW and Q_0 of this circuit are given by

$$\omega_0 = \sqrt{\frac{1}{C_2C_3R_1R_5}},\ \mathrm{BW} = \frac{1}{C_3R_4} \quad \text{and} \quad Q_0 = R_4\sqrt{\frac{C3}{C_2R_1R_5}} \tag{4.36}$$

This circuit can realize all the standard filter functions by proper selection of inputs but does not have infinite input impedance in any of the modes. However, LP realization employs both grounded capacitors as desirable for IC implementation. This circuit enjoys independent controllability of ω_0 and BW and orthogonal tuning of ω_0 and Q_0.

Experimental results based on AD 844 CFOAs show [12] that the circuit can be used to realize BP and Notch filters with f_0 of the order of 1 MHz quite satisfactorily.

Figure 4.13 shows yet another biquad proposed by Liu and Wu [13] using two CFOAs and six passive elements which realizes all the five standard filter responses.

The expression for the output voltage in terms of its input voltages for this circuit is given by

$$V_0 = \frac{s^2V_{in3} - s\frac{1}{C_2R_4}V_{in2} + \frac{1}{C_1C_2R_1R_2}V_{in1}}{s^2 + s\frac{1}{C_2R_3} + \frac{1}{C_1C_2R_1R_2}} \tag{4.37}$$

From (4.37), it is clear that the various filter responses can be obtained by proper selection of inputs.

The filter parameters ω_0, BW and Q_0 of this biquad are given by

$$\omega_0 = \sqrt{\frac{1}{C_1C_2R_1R_2}},\ \mathrm{BW} = \frac{1}{C_2R_3} \quad \text{and} \quad Q_0 = R_3\sqrt{\frac{C_2}{C_1R_1R_2}} \tag{4.38}$$

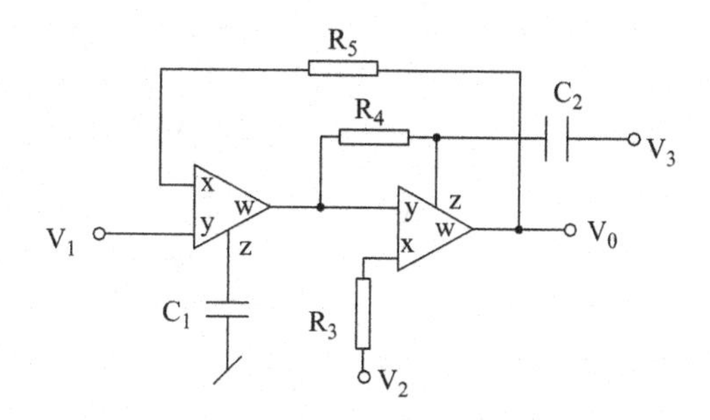

Fig. 4.14 MISO-type biquad deduced from a Mason graph by Wu et al. (adapted from [14] © 1994 Taylor & Francis)

Thus, Q_0 is independently tunable through resistor R_3. However, none of the filters have ideally infinite input impedance.

Another universal biquad employing two CFOAs, two capacitors and three resistors, deduced from a Mason graph, was presented by Wu et al. in [14] which is shown in Fig. 4.14.

The expression for V_0 in terms of its input voltages for this circuit is given by

$$V_0 = \frac{s^2 V_3 - s\frac{1}{C_2 R_3} V_2 + \frac{R_3 + R_4}{C_1 C_2 R_3 R_4 R_5} V_1}{s^2 + s\frac{1}{C_2 R_4} + \frac{R_3 + R_4}{C_1 C_2 R3 R_4 R_5}} \tag{4.39}$$

The various filter parameters ω_0, BW and Q_0 of this biquad are given by

$$\omega_0 = \sqrt{\frac{R_3 + R_4}{C_1 C_2 R_3 R_4 R_5}}, \ \mathrm{BW} = \frac{1}{C_2 R_4} \quad \text{and} \quad Q_0 = R_4 \sqrt{\frac{C_2 (R_3 + R_4)}{C_1 R_3 R_4 R_5}} \tag{4.40}$$

A high input impedance biquad using two CFOAs proposed by Liu [15] is shown in Fig. 4.15. Although this circuit does not belong to VM MISO-type category, however, it provides infinite input impedance to realize the three filter functions.

The various filter responses can be obtained by proper selection of admittances y_1, y_2, y_3 and y_4 as follows:

LP at node Vo1: (1) $y_1 = 1/R_1$, $Y_2 = 1/R_2$, $y_3 = (sC_3 + 1/R_3)$ and $y_4 = 1/R_4$ (2) $y_1 = 1/R_1$, $Y_2 = 1/R_2$, $y_4 = (sC_4 + 1/R_4)$ and $y_3 = 1/R_3$

BP at node V_{o2}: (1) $y_1 = sC_1$, $y_2 = (sC_2 + 1/R_2)$, $y_3 = 1/R_3$ and $y_4 = 1/R_4$ (2) $y_1 = 1/R_1$, $y_2 = (sC_2 + 1/R_2)$, $y_3 = sC_3$ and $y_4 = sC_4$

HP at node v_{o2}: (1) $y_1 = sC_1$, $y_2 = sC_2$, $y_3 = (sC_3 + 1/R_3)$ and $y_4 = 1/R_4$ (2) $y_1 = sC_1$, $y_2 = sC_2$, $y_3 = sC_3$ and $y_4 = (sC_4 + 1/R_4)$

The circuit provides orthogonal control of ω_0 and Q_0 by grounded resistors or capacitors which makes it suitable for easy conversion into voltage tuned filter by replacing grounded resistor by a JFET.

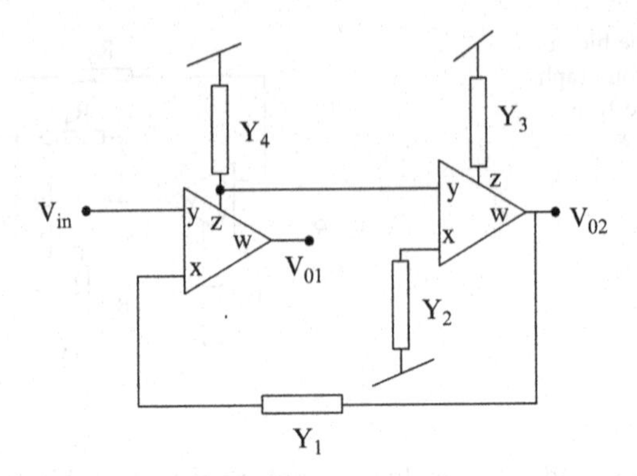

Fig. 4.15 High input impedance biquad introduced by Liu (adapted from [15] © 1995 IEE)

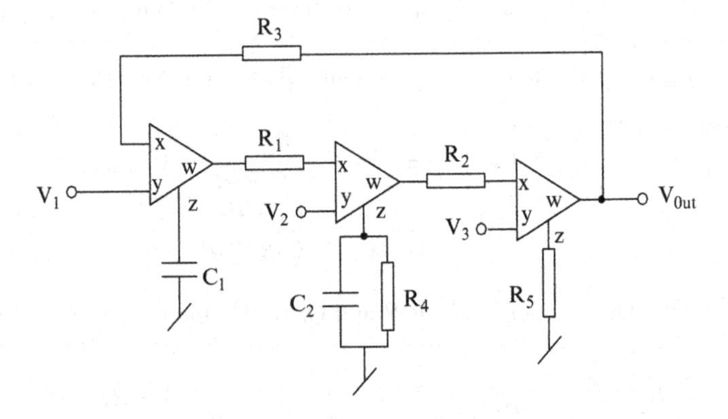

Fig. 4.16 Topaloglu-Sagbas-Anday configuration (adapted from [16] © 2012 Elsevier)

The biquad presented by Topaloglu et al. in [16] (shown in Fig. 4.16) is a three-input single-output second-order universal filter which realizes all the five basic filtering functions, by selecting different input signal combinations using three CFOAs, five resistors and two GCs as preferred for integrated circuit implementation. The circuit offers the attractive feature of providing ideally-infinite input impedance at all the three input terminals.

The output voltage V_{out} of this circuit, in terms of the various input signals, is given by

$$V_{out} = \frac{V_3\left\{\left(\frac{R_5}{R_2}\right)s^2 + s\left(\frac{R_5}{C_2R_2R_4}\right)\right\} - V_2 s\left(\frac{R_5}{C_2R_1R_2}\right) + V_1\left(\frac{R_5}{C_1C_2R_1R_2R_3}\right)}{s^2 + s\left(\frac{1}{R_4C_2}\right) + \frac{R_5}{C_1C_2R_1R_2R_3}} \tag{4.41}$$

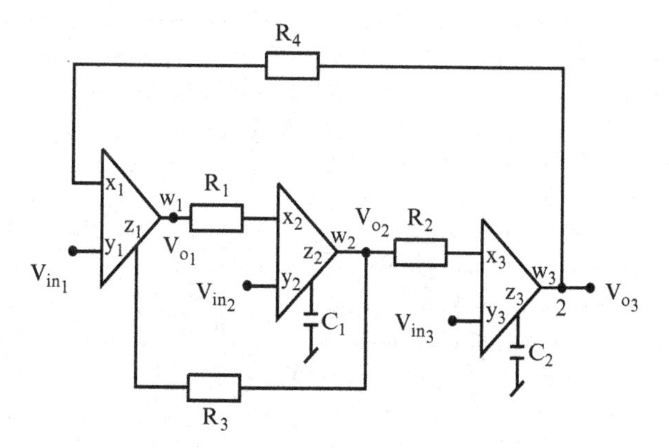

Fig. 4.17 An alternative three-input-three-output universal biquad

The various filter responses can be realized as follows: (1) HP: $V_1 = 0$, $V_2 = V_3 = V_{in}$ and $R_1 = R_4$ (2) BP (inverting): $V_1 = V_3 = 0$, $V_2 = V_{in}$ (3) LP: $V_2 = V_3 = 0$, $V_1 = V_{in}$ (4) BS: $V_1 = V_2 = V_3 = V_{in}$, $R_1 = R_4$ and $R_2 = R_5$ (5) AP: $V_1 = V_2 = V_3 = V_{in}$, $2R_1 = R_4$ and $R_2 = R_5$

Note that this circuit although employs three CFOAs but offers (1) ideally infinite input impedance in all the cases (2) employment of both grounded capacitors and (3) tunability of the parameters ω_0 and $\left(\frac{\omega_0}{Q_0}\right)$ but needs realization constraints in case of HP, BS, and AP responses. SPICE simulations have demonstrated [16] the workability of this structure in realizing corner/centre frequencies between 75 and 80 kHz using AD844-type CFOAs.

An alternative MISO-type VM biquad configuration is, derivable from the KHN-equivalent biquad of [3] shown in Fig. 4.2 of this chapter by un-grounding the y-input terminals of $CFOA_1$ and $CFOA_2$ and applying two additional inputs on the terminals thus created. The resulting circuit, which uses the same number of CFOAs and GCs as in the circuit of Fig. 4.16 *but requires one less resistor*, is shown here in Fig. 4.17.

By straight forward analysis, the output voltage V_{o1} of the circuit of Fig. 4.17 is given by

$$V_{o1} = \frac{s^2\left(\frac{R_3}{R_4}\right)V_{in1} + \left\{s\left(\frac{1}{R_1C_1}\right) + \frac{R_3}{C_1C_2R_1R_2R_4}\right\}V_{in2} - \left\{s\left(\frac{R_3}{C_2R_2R_4}\right)\right\}V_{in3}}{s^2 + s\left(\frac{1}{R_1C_1}\right) + \frac{R_3}{C_1C_2R_1R_2R_4}} \tag{4.42}$$

Note that with $V_{in2} = 0 = V_{in3}$, $V_{in1} = V_{in}$, the circuit becomes exactly same as the circuit of Fig. 4.2 and realizes a LP at V_{o3}, BP at V_{o2} and HP at V_{o1} without requiring any realization constraints.

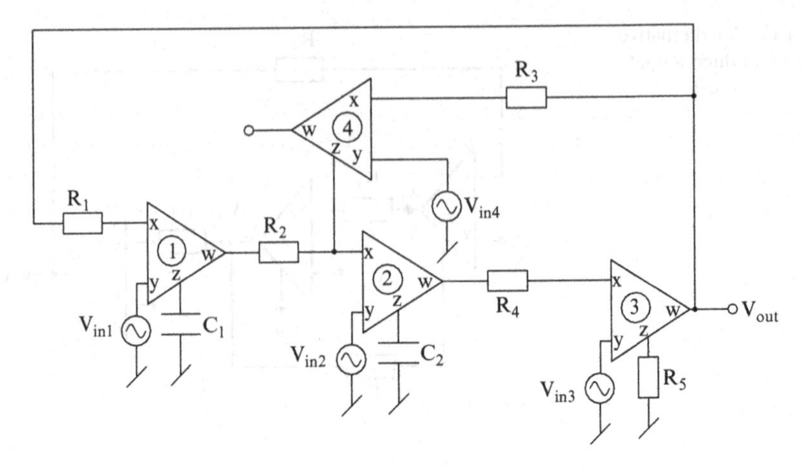

Fig. 4.18 A multiple-input single output type universal biquad proposed by Nikoloudis and Psychalinos (adapted from [17] © 2010 Springer)

However in the modified form of Fig. 4.17, the additional filter functions BR and AP are realizable as follows:

BR: $V_{in2} = V_{in3} = V_{in1} = V_{in}$, taking output at V_{o1} with C_1 $R_1 = C_2$ R_2 and $R_3 = R_4$

AP: $V_{in2} = V_{in3} = V_{in1} = V_{in}$, taking output at V_{o1} with C_1 $R_1 = 2C_2$ R_2 and $R_3 = R_4$

A careful comparison of this circuit with that of Fig. 4.16 reveals the following: (1) The circuit of Fig. 4.17 employs *one less resistor* than the circuit of Fig. 4.16 (2) the circuit of Fig. 4.16 requires realization conditions in case of HP, BS[2] and AP whereas the circuit of Fig. 4.17 *requires realization conditions only in BS and AP*. Furthermore, like the circuit of Fig. 4.16, this circuit also provides ideally infinite input impedance in all the cases.

Lastly, it may be pointed out that this version has not been described earlier in [3] or elsewhere.

We now present a MISO-type universal biquad proposed by Nikoloudis and Psychalinos [17], which is shown in Fig. 4.18.

The expression for the output voltage, in terms of the input voltages for this circuit, is given by

$$v_{out} = \frac{s^2\frac{R_5}{R_4}v_{in3} - s\frac{R_5}{R_4}\left(\frac{1}{R_2C_2}v_{in2} - \frac{1}{R_3C_2}v_{in4}\right) + \frac{R_5}{R_1R_2R_4C_1C_2}v_{in1}}{s^2 + \frac{R_5}{R_3R_4C_2}s + \frac{R_5}{R_1R_2R_4C_1C_2}} \tag{4.43}$$

[2] A re-analysis of the circuit of Fig. 4.16 reveals that this circuit needs two realization constraints $G_1 = G_4$ and $G_2 = G_5$, in case of BS filter realization also, which appear to have been missed in [16] inadvertently.

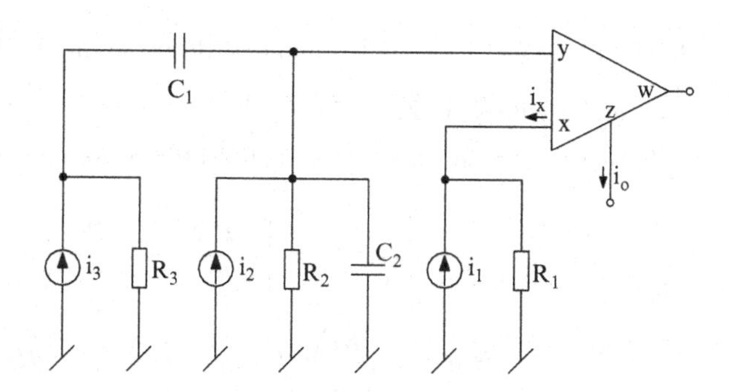

Fig. 4.19 A MISO-type CM universal biquad introduced by Sharma and Senani (adapted from [21] © 2004 Taylor & Francis)

The various filter functions obtained are as follows: LP: if $v_{in2} = v_{in3} = v_{in4} = 0$ and $v_{in1} = v_{in}$, HP: if $v_{in1} = v_{in2} = v_{in4} = 0$ and $v_{in3} = v_{in}$, BP: if $v_{in1} = v_{in2} = vin3 = 0$ and $v_{in4} = v_{in}$ (non-inverting) or: $v_{in1} = v_{in3} = v_{in4} = 0$ and $v_{in2} = v_{in}$ (inverting), BS: if $v_{in1} = v_{in3} = v_{in}$, $v_{in2} = v_{in4} = 0$, AP: if $v_{in1} = v_{in2} = v_{in3} = v_{in}$ and $v_{in4} = 0$. In addition, $R_2 = R_3$ and $R_4 = R_5$.

It is interesting to note that this circuit, except having two realization constraints in case of AP, has all the desirable properties namely, (1) infinite input impedance in all the cases, (2) employment of both grounded capacitors, (3) independent single resistance-controllability of ω_0 and $\frac{\omega_0}{Q_0}$ and (4) realisability of all the five standard filter responses. SPICE simulations show [17] that the configuration successfully realizes filters with f_0 in the vicinity of 100 kHz or so.

4.3.4 MISO-Type Universal Current-Mode (CM) Biquads

A MISO-type of universal CM biquad can be realized with only a single CFOA and a number of such circuits have been advanced by various researchers, for instance, see [18–22]. In the following, we describe one such circuit which provides a number of advantageous features as compared to other alternatives. This circuit from [21] is shown here in Fig. 4.19.

By straight forward analysis, the network function of interest for this circuit is found to be

$$i_o = \frac{i_3\left[\frac{s}{R_1C_2}\right] + i_2\left[\frac{s}{R_1C_2} + \frac{1}{R_1R_3C_2C_1}\right] - i_1\left[s^2 + s\left\{\frac{1}{R_3C_1} + \frac{1}{C_2}\left(\frac{1}{R_2} + \frac{1}{R_3}\right)\right\} + \frac{1}{R_2R_3C_1C_2}\right]}{\left[s^2 + s\left\{\frac{1}{R_3C_1} + \frac{1}{C_2}\left(\frac{1}{R_2} + \frac{1}{R_3}\right)\right\} + \frac{1}{R_2R_3C_1C_2}\right]} \tag{4.44}$$

From (4.44), the various filter functions can be realized as follows:

(1) LP: by setting $i_1 = 0$ and choosing $i_2 = -i_3 = i_{in}$ (2) BP: by setting $i_1 = i_2 = 0$, and choosing $i_3 = i_{in}$; (3) HP: by choosing $i_1 = i_2 = i_3 = i_{in}$, along with $R_1 = R_2$ and $\frac{R_3}{R_2} = \left(1 + \frac{C_2}{C_1}\right)$ (4) notch: by setting $i_2 = 0$ and choosing $i_1 = i_3 = i_{in}$,

along with $\frac{R_3}{R_1} = \left(1 + \frac{C_2}{C_1} + \frac{R_3}{R_2}\right)$ and (5) AP: by setting $i_2 = 0$ and choosing $i_1 = i_3 = i_{in}$, along with $\frac{R_3}{R_1} = 2\left(1 + \frac{C_2}{C_1} + \frac{R_3}{R_2}\right)$

The various parameters of the realized filters are now given by

$$\omega_O = \frac{1}{\sqrt{R_2 R_3 C_2 C_1}}; \; Q_O = \left[\sqrt{\frac{R_2 C_2}{R_3 C_1}} + \sqrt{\frac{R_2 C_1}{R_3 C_2}} + \sqrt{\frac{R_3 C_1}{R_2 C_2}}\right]^{-1} \tag{4.45}$$

$$H_O|_{LP} = \left(\frac{R_2}{R_1}\right), \; H_O|_{HP} = H_O|_{notch} = H_O|_{AP} = 1 \tag{4.46}$$

and

$$H_O\Big|_{BP} = \left[\frac{R_1 C_2}{R_3 C_1} + \frac{R_1}{R_3} + \frac{R_1}{R_2}\right]^{-1} \tag{4.47}$$

It is seen that LP and BP responses do not have realization constraints in terms of component values. As a consequence, from the expression for Q_O it is found that the upper bound on Q_O for these cases is $Q_{Omax} = 1/(2\sqrt{2})$. In other cases, due to the realization conditions involving all component values, it may be difficult to achieve this value of Q_O.

The low values of Q_O, however, does not hamper the usability of the proposed circuit in practical applications. One such application is in the realization of constant-Q graphic equalizers [54], where band-pass and band-reject filters with Q range varying from 0.1 to 1 are needed for attaining wide-band characteristics. As another example, simple LP filters realizable with only one active element may also be used as low-cost anti-aliasing/band-limiting filters [55]. Simulations demonstrate that like other circuits, f_O of the order of 100 kHz is realizable easily using AD844 type CFOAs.

An inspection of the expressions for the various parameters of the filters, in conjunction with the relevant realization conditions, reveals that in all cases, constant-H_O, constant-Q_O and variable-ω_O realizations can be achieved by simultaneous variation of all the three resistors *i.e.* $R_1 = R_2 = R_3 = R$. If these are replaced by equal-valued MOSFET-based voltage-control-resistances (VCR), (such as the one proposed by Banu-Tsividis [56]) driven by a common control voltage V_c, electronic tunability of ω_O is achievable.

4.3.5 Dual-Mode Universal Biquads Using a Single CFOA

In this section, we first present two single CFOA-based MISO-type CM/VM multifunction biquad configurations [19] which are shown in Figs. 4.20 and 4.21 respectively.

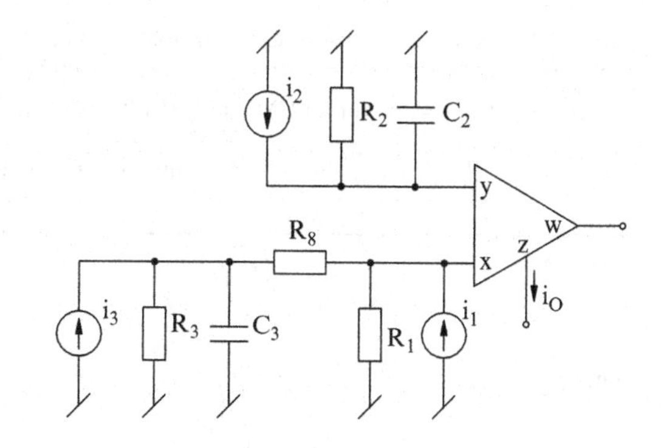

Fig. 4.20 MISO-type CM Multifunction biquad proposed by Sharma and Senani (adapted from [19] © 2003 Elsevier)

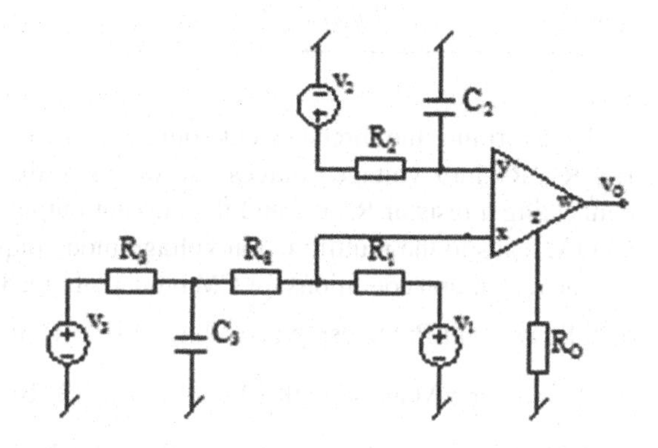

Fig. 4.21 MISO-type VM Multifunction biquad introduced by Sharma and Senani (adapted from [19] © 2003 Elsevier)

A straight forward analysis of the circuit of Fig. 4.20 reveals that the relation between the three input currents i_1, i_2, i_3 and the output current i_o is given by

$$i_0 = \frac{k_2 i_2 - k_3 i_3 - k_1 i_1}{D(s)} \tag{4.48}$$

where

$$k_2 = \left[\frac{1}{C_2C_3R_3R_8} + \frac{1}{C_2C_3R_1}\left(\frac{1}{R_3} + \frac{1}{R_8}\right) + s\left\{\frac{1}{C_2}\left(\frac{1}{R_1} + \frac{1}{R_8}\right)\right\}\right]$$

$$k_3 = \left[\frac{1}{C_2C_3R_2R_8} + \frac{s}{C_3R_8}\right]$$

$$k_1 = \left[s^2 + s\left\{\frac{1}{C_3}\left(\frac{1}{R_3} + \frac{1}{R_8}\right) + \frac{1}{C_2R_2}\right\} + \frac{1}{C_2C_3R_2}\left(\frac{1}{R_3} + \frac{1}{R_8}\right)\right]$$

$$D(s) = \left[s^2 + s\left\{\frac{1}{C_3}\left(\frac{1}{R_3} + \frac{1}{R_8}\right) + \frac{1}{C_2R_2}\right\} + \frac{1}{C_2C_3R_2}\left(\frac{1}{R_3} + \frac{1}{R_8}\right)\right]$$

Table 4.1 Parameters of the filters realizable from the circuits of Figs. 4.20 and 4.21

ω_o and ω_o/Q_o are same for all CM and VM filters and are given by $\omega_o = \sqrt{\frac{1}{C_2C_3R_2}\left(\frac{1}{R_3}+\frac{1}{R_8}\right)}$ and $BW = \frac{\omega_o}{Q_o} = \left\{\frac{1}{C_3}\left(\frac{1}{R_3}+\frac{1}{R_8}\right)+\frac{1}{C_2R_2}\right\}$ The values of H_O are as follows:

Filter type	H_O (current-mode)	H_O (voltage-mode)
Low-pass	$H_{oI}\vert_{LP} = \frac{\left(1+\frac{R_1}{R_3}+\frac{R_8}{R_3}\right)-\frac{R_1}{R_2}}{\frac{R_1}{R_2}\left(1+\frac{R_8}{R_3}\right)}$	$H_{oV}\vert_{LP} = \left(\frac{R_o}{R_2}\right)H_{oI}\vert_{LP}$
Band-pass	$H_{oI}\vert_{BP} = \frac{\left(1+\frac{R_8}{R_1}\right)-\frac{C_2}{C_3}}{\frac{C_2}{C_3}\left(1+\frac{R_8}{R_3}\right)+\frac{R_8}{R_2}}$	$H_{oV}\vert_{BP} = \left(\frac{R_o}{R_2}\right)H_{oI}\vert_{BP}$
High-pass	$H_{oI}\vert_{HP} = 1$	$H_{oV}\vert_{HP} = \left(\frac{R_o}{R_1}\right)H_{oI}\vert_{HP}$
Notch	$H_{oI}\vert_{Notch} = 1$	$H_{oV}\vert_{Notch} = \left(\frac{R_o}{R_1}\right)H_{oI}\vert_{Notch}$

Transforming the three current sources i_1, i_2, i_3 along with parallel resistances R_1, R_2, R_3 into voltage sources v_1, v_2, v_3 with series resistances R_1, R_2, R_3, connecting a resistor R_o at z and then taking output V_o from the w-terminal of the CFOA, leads to the multifunction voltage mode biquad shown in Fig. 4.21.

The realization conditions for CM biquad of Fig. 4.20/VM biquad of Fig. 4.21 are as follows: (1) LP: choosing $i_1 = 0$, $i_2 = i_3 = i_{in}$ for CM and $v_1 = 0$, $v_2 = v_{in}$, $v_3 = \left(\frac{R_3}{R_2}\right)v_{in}$ for VM along with $\left(1+\frac{R_8}{R_1}\right) = \frac{C_2}{C_3}$ (2) BP choosing $i_1 = 0$, $i_3 = i_2 = i_{in}$ for CM and $v_1 = 0$, $v_2 = v_{in}$, $v_3 = \left(\frac{R_3}{R_2}\right)v_{in}$ for VM along with $\left(1+\frac{R_1}{R_3}+\frac{R_8}{R_3}\right) = \frac{R_1}{R_2}$ (3) HP: taking $i_1 = i_2 = i_3 = i_{in,}$ for CM and $v_1 = v_{in}$, $v_2 = \left(\frac{R_2}{R_1}\right)v_1$, $v_3 = \left(\frac{R_3}{R_1}\right)v_1$ for VM along with $\frac{R_1}{R_2}\left(2+\frac{R_8}{R_3}\right) = \left(1+\frac{R_1}{R_3}+\frac{R_8}{R_3}\right)$ and $\frac{C_2}{C_3}\left(2+\frac{R_8}{R_3}\right) = \left(1+\frac{R_8}{R_1}\right)-\frac{R_8}{R_2}$ (4) Notch taking $i_1 = i_2 = i_3 = i_{in}$ for CM and $v_1 = v_{in}$, $v_2 = \left(\frac{R_2}{R_1}\right)v_1$, $v_3 = \left(\frac{R_3}{R_1}\right)v_1$ for VM along with $\left(1+\frac{R_3}{R_1}+\frac{R_8}{R_1}\right) = \frac{R_3}{R_2}$ and $\frac{C_2}{C_3}\left(2+\frac{R_8}{R_3}\right) = \left(1+\frac{R_8}{R_1}\right)-\frac{R_8}{R_2}$

Table 4.1 shows the values of the various parameters of the realized filters.

It may be seen that the realisability conditions for the various responses in voltage mode are analogous to those of current mode case (with i_j replaced by v_j, $j = 1$–3), which are also given in Table 4.1. Note that the expression for ω_o and ω_o/Q_o remain exactly the same in both modes, however in voltage-mode version, the gains H_o in all the cases become controllable through the resistor R_o. Circuit makes it possible to realize filters having f_0 in the vicinity of 1 MHz [19] using AD 844 type CFOAs.

A limitation of the circuits of Figs. 4.20 and 4.21 is the non-availability of independent tunability of ω_o and ω_o/Q_o thereby restricting the realizations to low Q_o values. However, this does not hamper the usability of the proposed circuits in practical applications particularly those outlined in Sect. 4.3.4. Lastly, although no tunability is available for H_o and BW (or Q_O), ω_O can be tuned electronically if all the four resistors are replaced by appropriately-valued floating CMOS voltage-controlled-resistors (VCR) circuits. Results in [19] show that in a specific case, tunability of f_0 over a range of more than a decade (500 Hz–20 kHz) is attainable with this method.

4.3.6 Mixed-Mode Universal Biquads

The term 'mixed-mode' in the context of universal biquads is normally used for circuits which can be realise two or more of the voltage-ratio, current-ratio, trans-impedance and transadmittance mode filters, from the same configuration under appropriate conditions.

We now present a configuration which realizes all the five standard filter responses in both CM and VM [3] and thus, can be called a universal VM/CM biquad. The starting point of the development is the passive LCR filter of Fig. 4.22a, which is basically a VM BP filter. Applying the source transformation on this circuit, the circuit becomes as shown in Fig. 4.22b. If we denote the currents in C_0, L_0 and R_0 as I_{C0}, I_{L0} and I_{R0} respectively, the three current transfer functions can be obtained as

$$\frac{I_{L0}}{I_{IN}} = \frac{\frac{1}{L_0C_0}}{s^2 + \frac{1}{C_0R_0}s + \frac{1}{L_0C_0}} \tag{4.49}$$

$$\frac{I_{R0}}{I_{IN}} = \frac{s\frac{1}{R_0C_0}}{s^2 + \frac{1}{C_0R_0}s + \frac{1}{L_0C_0}} \tag{4.50}$$

and

$$\frac{I_{C0}}{I_{IN}} = \frac{s^2}{s^2 + \frac{1}{C_0R_0}s + \frac{1}{L_0C_0}} \tag{4.51}$$

From (4.49)–(4.51) it is, thus, obvious that the circuit of Fig. 4.22b is an excellent vehicle for making a CM biquad, by simulating the inductor actively and by sensing the currents I_{C0}, I_{L0}, and I_{R0} and making them available at high output impedance nodes, by using appropriate circuitry. A CFOA-based circuit which implements these mechanisms is shown in Fig. 4.22c. The various characterizing equations of this circuit are given by

$$I_{L0} = \frac{\frac{1}{C_0C_1R_1R_2}}{D(s)}\left(I_{IN} - \frac{V_{IN}}{R_s}\right); \; I_{C0} = -\frac{s^2}{D(s)}\left(I_{IN} - \frac{V_{IN}}{R_s}\right) \tag{4.52}$$

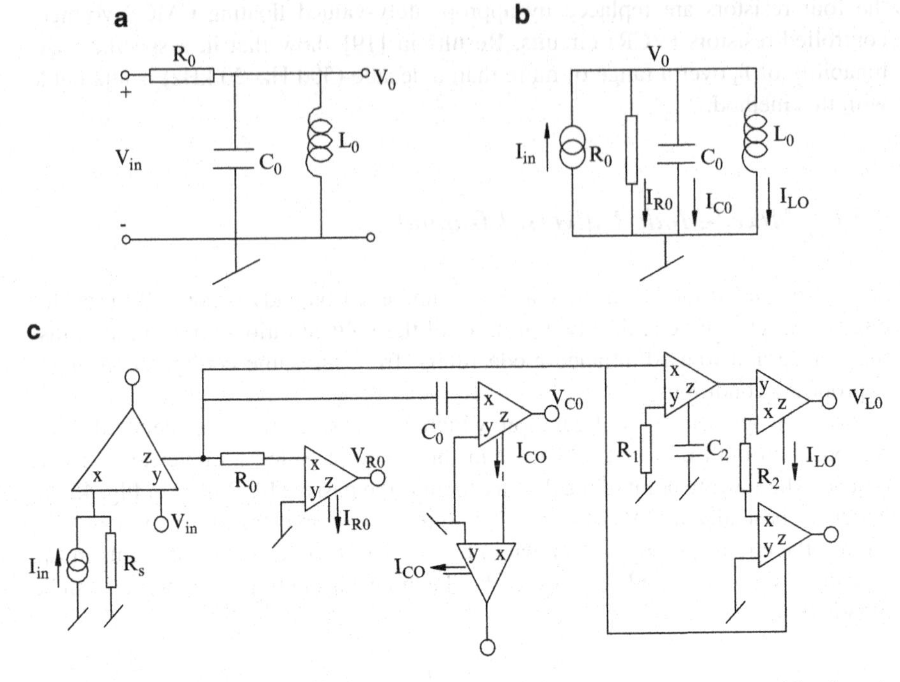

Fig. 4.22 Voltage-mode (VM) current-mode (CM) universal biquad filter proposed by Senani (adapted from [3]© 1998 Water de Gruyter GmbH & Co. KG)

and

$$I_{RO} = -\frac{\frac{s}{C_0 R_0}}{D(s)}\left(I_{IN} - \frac{V_{IN}}{R_s}\right) \tag{4.53}$$

where

$$D(s) = s^2 + \left(\frac{1}{R_0 C_0}\right)s + \left(\frac{1}{R_1 R_2 C_0 C_1}\right) \tag{4.54}$$

The various filter parameters are given by

$$\omega_0 = \sqrt{\frac{1}{C_0 C_1 R_1 R_2}} \text{ for LP, BP, and HP, BW} = \frac{1}{R_0 C_0} \text{ for BP,}$$

$$Q_0 = R_0\sqrt{\frac{C_0}{C_1 R_1 R_2}} \text{for LP, HP and } H_0 = 1 \text{ for LP, BP and HP} \tag{4.55}$$

From the above equations, it is clear that Q_0 is tunable by varying R_0 in case of LP and HP, and for BP, the bandwidth is adjustable by R_0 whereas the centre frequency is tunable with R_1 (or R_2).

To realize BS filter, we create $-i_{CO}$ by using an additional CFOA (shown by dotted lines) as an inverting current follower and then add $-i_{CO}$ and i_{LO} which yields

$$I_{o1} = \frac{s^2 + \frac{1}{C_0 C_1 R_1 R_3}}{D(s)} \left(I_{IN} - \frac{V_{IN}}{R_s} \right) \tag{4.56}$$

For realizing the AP function, the three current outputs $-I_{CO}$, I_{RO} and I_{LO} need to be added (i.e., $I_{o2} = I_{RO} - I_{CO} + I_{LO}$) and requires the node thus created to be treated as the output terminal. This results in

$$I_{o2} = \frac{s^2 - s\frac{1}{R_0 C_0} + \frac{1}{C_0 C_1 R_1 R_3}}{D(s)} \left(I_{IN} - \frac{V_{IN}}{R_s} \right) \tag{4.57}$$

The structure can be converted into a VM biquad by terminating the various output currents in to load resistors and then taking the outputs from the z-terminals of the respective CFOAs. For example, with the z-terminal of CFAOs terminated into a load R_L and voltage output taken from the w-terminal of CFOAs, we obtain (with $I_{IN} = 0$) a non-inverting BP function having the transfer function

$$\frac{V_{BP}}{V_{IN}} = \frac{\frac{R_L}{R_s}\left(\frac{1}{C_0 R_0}\right)s}{D(s)} \tag{4.58}$$

Thus, the same kind of transfer functions are realizable in VM too, except that each carries a negative sign and, H_0 can be adjusted by means of the various load resistors mentioned above. Experimental results using AD 844 type CFOAs show the workability of the circuit with f_0 around 159 kHz achievable quite satisfactorily [3].

We now present a novel mixed-mode universal biquad configuration from [23] which is shown in Fig. 4.23.

This circuit employs two inverting integrators and two specially devised summers, each realized by a single CFOA. The various transfer functions realizable from this configuration in its various modes of operation are as follows.

(i) *Voltage-mode universal biquad filter*: Assuming ideal CFOAs, a routine analysis of the circuit yields the following five transfer functions

$$\frac{V_{o3}}{V_{in}} = \frac{H_0 \omega_0{}^2}{D(s)} = \frac{-\left(\frac{r_2}{r_1}\right)\frac{r_3}{r_2 R_1 C_1 R_2 C_2}}{s^2 + \frac{1}{R_1 C_1}s + \frac{r_3}{r_2 R_1 C_1 R_2 C_2}} \tag{4.59}$$

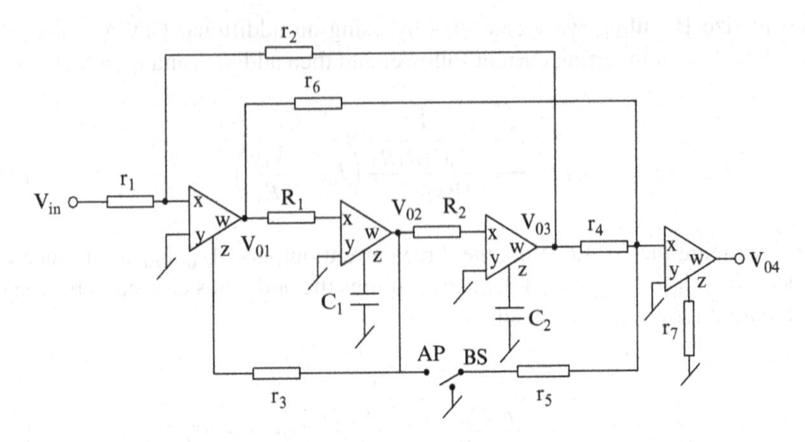

Fig. 4.23 CFOA-based mixed mode universal biquad (adapted from [23] © 2005 IEICE)

$$\frac{V_{o2}}{V_{in}} = \frac{H_0\left(\frac{\omega_0}{Q_0}\right)s}{D(s)} = \frac{\left(\frac{r_3}{r_1}\right)\frac{r_3}{R_1C_1}s}{s^2 + \frac{1}{R_1C_1}s + \frac{r_3}{r_2R_1C_1R_2C_2}} \tag{4.60}$$

$$\frac{V_{o1}}{V_{in}} = \frac{H_0 s^2}{D(s)} = \frac{-\left(\frac{r_3}{r_1}\right)s^2}{s^2 + \frac{1}{R_1C_1}s + \frac{r_3}{r_2R_1C_1R_2C_2}} \tag{4.61}$$

$$\frac{V_{o4}}{V_{in}} = \frac{H_0(s^2 + {\omega_0}^2)}{D(s)} = \frac{\left(\frac{r_3}{r_1}\right)\left(s^2 + \frac{r_3}{r_2R_1C_1R_2C_2}\right)}{s^2 + \frac{1}{R_1C_1}s + \frac{r_3}{r_2R_1C_1R_2C_2}} \tag{4.62}$$

and

$$\frac{V_{o4}}{V_{in}} = \frac{H_0\left(s^2 - \left(\frac{\omega_0}{Q_0}\right)s + {\omega_0}^2\right)}{D(s)} = \frac{\left(\frac{r_3}{r_1}\right)\left(s^2 - \frac{1}{R_1C_1}s + \frac{r_3}{r_2R_1C_1R_2C_2}\right)}{s^2 + \frac{1}{R_1C_1}s + \frac{r_3}{r_2R_1C_1R_2C_2}} \tag{4.63}$$

with the switch at position 'AP' and choosing $r_2 = r_3$; $r_4 = r_5 = r_6 = r_7$.

Thus, the circuit realizes a LP filter at V_{o3}, BP response at V_{o2}, a HP response at V_{o1}, and BR and AP responses at V_{o4} under appropriate conditions.

(ii) *Current-mode universal biquad filter*: With r_1 and r_7 deleted, the circuit can be converted into an universal current-mode biquad with ideally zero input impedance and ideally infinite output impedance. With an input current I_{in} injected into input terminal ‘m’ (x-terminal of the first CFOA) and output current I_{out} taken out from the node ‘n’ (z-terminal of the last CFOA) the circuit can realize all the five filter responses in current mode. The general transfer function for this single-input-single-output universal current-mode filter is given by

$$\frac{I_{out}}{I_{in}} = \frac{r_3\left\{\frac{s^2}{r_6} - \left(\frac{1}{C_1R_1r_5}\right)s + \frac{1}{C_1C_2R_1R_2r_4}\right\}}{s^2 + \frac{1}{C_1R_1}s + \frac{r_3}{C_1C_2R_1R_2r_2}} \tag{4.64}$$

The circuit realizes a LP with r_5 and r_6 open circuited; a BP with r_6 and r_4 open circuited; a HP with r_5 and r_4 open circuited; a BR with r_5 open circuited (along with $r_2 = r_4 = r_6 = r_0$ (say) thereby yielding $H_0 = r_3/r_0$) and finally, an AP with $r_2 = r_3 = r_4 = r_5 = r_6$ yielding $H_0 = 1$.The gains for LP, BP, and HP filters are r_3/r_4, r_3/r_5 and r_3/r_6 respectively.

In the CM biquad, LP and HP responses have only H_0 controllable (through r_4 and r_6 respectively); in BR and AP, H_0 is not tunable, however, BW and ω_0 can be independently adjusted (through R_1 and R_2) respectively and finally, in BP realization, BW, ω_0 and H_0, all are independently controllable (through R_1, R_2 and r_5 respectively).

(iii) *Trans-admittance universal biquad filter:*
In this mode, we retain the input resistor r_1 but take the output I_{out} from z-terminal of the last CFOA. The various responses realized and their features are similar to those of case (ii).

(iv) *Trans-impedance universal biquad filter:*
In this case, with r_1 deleted, the input will be a current I_{in}, however, the output voltages will be V_{o1}, V_{o2}, V_{o3} and V_{o4}. The realisability conditions, parameters of filters and their features are similar to those of case (i).

Thus, the configuration of Fig. 4.23 is an universal *mixed-mode* biquad capable of realizing all the five standard responses in all the four possible modes.

The hardware implementation of the circuit using AD 844 type CFOAs has been demonstrated [23] to work well in realizing filters with corner/centre frequencies of the order of 100 kHz.

4.4 Active-R Multifunction VM Biquads

The active–R biquads utilizing the CFOA-pole have been shown to be superior alternatives to the active-R circuits designed using the compensation-poles of the traditional voltage-mode op-amps [24]. In this section, we present two CFOA-pole-based active-R biquads [24, 25] which overcome the limitations of the op-amp-based active-R biquads such as strong temperature-dependence of filter center frequency and the limited dynamic range (due to the finite slew rate of the VOAs).

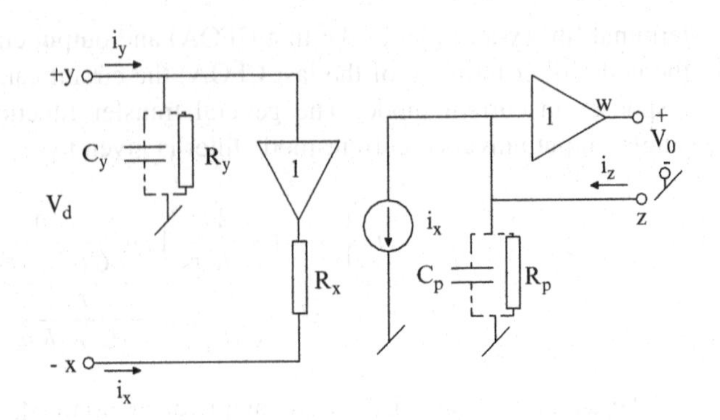

Fig. 4.24 Non-ideal equivalent circuit of the CFOA including the various parasitics

The non-ideal equivalent circuit of the CFOA is shown in Fig. 4.24, where R_x is the input resistance at x-port, $R_p \| 1/sC_p$ is the parasitic impedance at the z-port and $R_y \| 1/sC_y$ represents the parasitic impedance at y-port.

An analysis of the circuit of Fig. 4.24 yields

$$\frac{V_o(s)}{V_d(s)} = \frac{1}{R_x C_p\left(s + \frac{1}{R_p C_p}\right)} \tag{4.65}$$

Thus, $s = -1/C_pR_p$ represents the pole of the CFOA.

Figure 4.25 shows an active-R biquad employing two CFOAs and five resistors proposed by Toumazou, Payne and Pookaiyaudom.

Assuming CFOAs to be characterized by the non-ideal model of Fig. 4.24, a straightforward analysis of the circuit of Fig. 4.25 shows that this circuit would give BP response at V_{o1} and LP response at V_{02} with the relevant filter parameters given by

$$\omega_0 = \sqrt{\frac{1}{kR_2R_4C_{z1}C_{z2}}} \text{ and } Q_0 = R_3\sqrt{\frac{C_{z1}}{kC_{z2}R_2R_4}} \tag{4.66}$$

where $k = \dfrac{R_a}{R_a + R_b}$

From (4.66) it is clear that Q_0 of this biquad can be controlled independently through R_3. The circuit was shown [24] capable of realizing f_0 of the order of 1 MHz using AD 846 type dual CFOAs.

Yet another active-R CFOA-based biquad introduced by Singh and Senani in [25] is shown in Fig. 4.26.

A routine circuit analysis (taking $R'_1 = R_{p1}$ which is required to ensure that BP response contains only the first order term of s in the numerator) analysis shows the realisability of the LP filter at V_{01} and BP filter at V_{02}. The various filter parameters of this circuit are given by

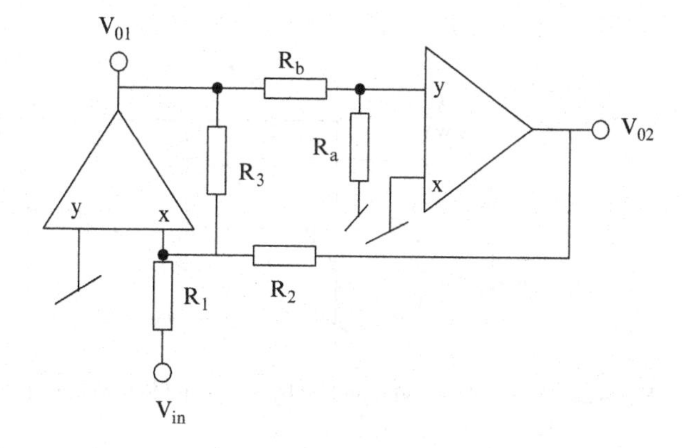

Fig. 4.25 Active-R VM biquad proposed by Toumazou et al. (adapted from [24] ©1995 IEEE)

For BP:

$$BW = \frac{1}{C_{p23}}\left(\frac{1}{R_{p1}} + \frac{1}{R'_4}\right) \tag{4.67}$$

$$\omega_0 = \sqrt{\frac{\left(1 + \frac{R'_2}{R_{p1}}\right)}{C_{p1}C_{p23}R_{p1}R'_2}} \tag{4.68}$$

$$H_{BP} = \frac{R_{p1}}{R'_3}\left(1 + \frac{R_{p1}}{R'_4}\right)^{-1} \tag{4.69}$$

where $C_{p23} = C_{p2} + C_{p3}$; $\frac{1}{R'_4} = \frac{1}{R_4} + \frac{1}{R_{p2}} + \frac{1}{R_{p3}}$; $R'_2 = R_2 + R_{x2}$ $R'_3 = R_3 + R_{x3}$

For LP: $Q_0 = \sqrt{\frac{C_{p23}}{C_{p1}}\left(\frac{1 + \frac{R_{p1}}{R'_2}}{1 + \frac{R_{p1}}{R'_4}}\right)}$;

$$H_{LP} = \left(\frac{R'_2}{R'_3}\right)\left(1 + \frac{R'_2}{R_{p1}}\right)^{-1} \quad \text{while } \omega_0 \text{ remains same.} \tag{4.70}$$

A comparison of these two biquads reveals that the biquad shown in Fig. 4.26 has the following novel features : (1) high input impedance is available (2) Y-port parasitics of all the three CFOAs become ineffective (3) R_{xi} can be accommodated in R_i; i = 1–3 and (4) tunability of parameters is available (in case of BP filter, having set BW by R_4, ω_0 and H_{BP} can be independently adjusted by R_2 and R_3, respectively and in case of LP response, having fixed ω_0 by R_2, Q_0 and H_{LP} can then be adjusted independently through R_4 and R_3 respectively). The hardware realizations have demonstrated that this circuit can also realise filters with f_0 of the order of 1 MHz.

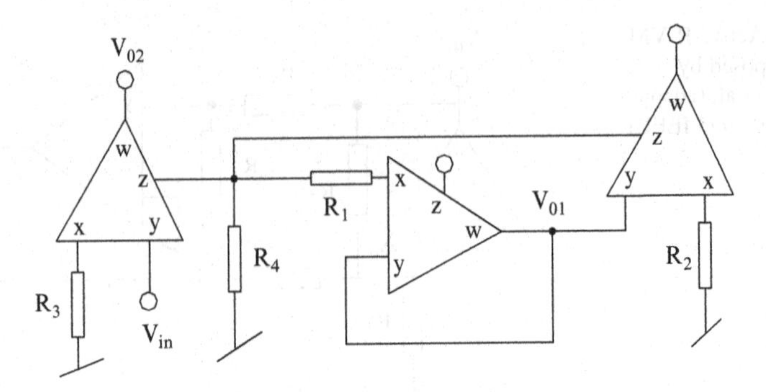

Fig. 4.26 Active-R biquad proposed by Singh and Senani (adapted from [25] © 2001 IEEE)

4.5 Inverse Active Filters Using CFOAs

In communication, control and instrumentation systems, there are applications where inverse filters are required to correct the distortion of the signal caused by the signal processors or transmission system. This correction can be done by using an inverse filter which is required to have the frequency response as reciprocal of the frequency response of the system which has caused the distortion. Although several techniques are known to design such inverse digital filters however, in the domain of analog signal processing, very few attempts have been made in the past to synthesize inverse continuous time analog filters, for instance, see [50–52] and the references cited therein.

In most of the earlier works (see references cited in [50, 51]) quite often four-terminal floating nullors (FTFNs) have been used as building blocks to realize inverse active filter. However, so far, FTFNs are not available as off the shelf ICs whereas CFOAs are. In view of the commercial availability of CFOAs as off-the-shelf ICs coupled with their popularity, recently, a number of attempts have been made to realize inverse filters using CFOAs [50–52]. In the following, we show some exemplary realizations for inverse low pass, inverse band pass, inverse high pass and inverse band reject filters from [50, 51]. These circuits are shown in Fig. 4.27.

The transfer functions and the parameters of the filters realized by these circuits are as follows:
Circuit of Fig. 4.27a:

$$\frac{V_0}{V_{in}} = -\frac{1}{\dfrac{\left(\dfrac{R_1}{R_0}\right)s^2}{s^2 + \dfrac{s}{C_1R_2} + \dfrac{1}{C_1C_2R_2R_3}}} \tag{4.71}$$

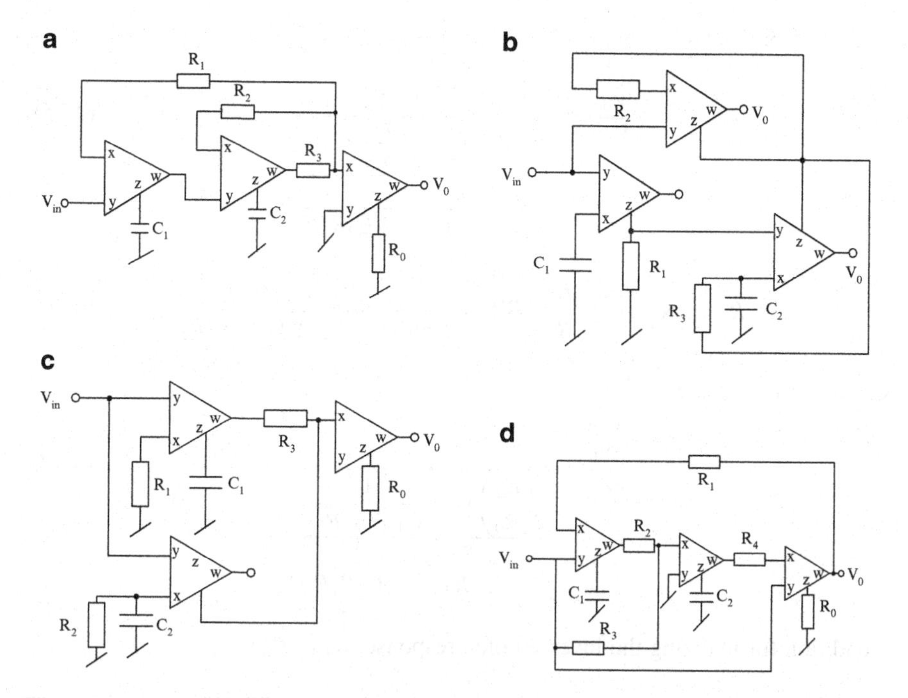

Fig 4.27 Inverse active filters proposed by Gupta et al. (**a**) Inverse HP filter (adapted from [50] ©2009 Springer), (**b**) Inverse low pass filter, (**c**) Inverse band pass filter, (**d**) Inverse band reject filter (adapted from [51] 2009 Taylor & Francis)

$$R_{in} = \infty;\ \mathrm{H}_0 = \frac{R_1}{R_0};\ BW = \frac{1}{C_1R_2};\ \omega_0 = \sqrt{\frac{1}{C_1C_2R_2R_3}} \tag{4.72}$$

Circuit of Fig. 4.27b:

$$\frac{V_0}{V_{in}} = \frac{1}{\dfrac{\left(1+\dfrac{R_2}{R_3}\right)\dfrac{2}{C_1C_2R_1R_2}}{s^2+\dfrac{2}{C_2R_3}s+\dfrac{2}{C_1C_2R_1R_2}}} \tag{4.73}$$

$$R_{in} = \infty;\ H_0 = \left(1+\frac{R_2}{R_3}\right);\ BW = \frac{2}{C_2R_3};\ \omega_0 = \sqrt{\frac{2}{C_1C_2R_1R_2}} \tag{4.74}$$

Circuit of Fig. 4.27c:

$$\frac{V_0}{V_{in}} = -\frac{1}{\frac{\left(\frac{R_2}{R_0}\right)\frac{s}{C_2R_2}}{s^2 + \frac{s}{C_2R_2} + \frac{1}{C_1C_2R_1R_3}}} \tag{4.75}$$

$$R_{in} = \infty; \; H_0 = \frac{R_2}{R_0}; \; BW = \frac{1}{C_2R_2}; \; \omega_0 = \sqrt{\frac{1}{C_1C_2R_1R_3}} \tag{4.76}$$

Circuit of Fig. 4.27d:

$$\frac{V_0}{V_{in}} = \frac{1}{\frac{\left(\frac{R_4}{R_0}\right)s^2 + \frac{1}{C_1C_2R_1R_2}}{s^2 + \frac{s}{C_2R_3} + \frac{1}{C_1C_2R_1R_2}}} \tag{4.77}$$

Condition for realizing the inverse notch response: $R_4 = R_0$

$$R_{in} = R_3; \; H_0 = 1; \; BW = \frac{1}{C_2R_3}; \; \omega_0 = \sqrt{\frac{1}{C_1C_2R_1R_2}} \tag{4.78}$$

The Inverse filters shown in Fig. 4.27 can be readily implemented to have f_0 of the order of 159 kHz [51]. Recently, a generalized CFOA-based configuration for realizing inverse filters has also been presented [52] from which, using different selection of various circuit elements, all the four types of inverse filters can be realized as special cases.

4.6 MOSFET-C Filters Employing CFOAs

MOSFET-C filters were evolved as fully-integratable continuous-time alternatives to the clock-frequency-tunable switched-capacitor filters and digital filters to be implemented in VLSI technology. Traditionally, MOSFET-C filters were derived from classical op-amp based active-RC filters using dual complementary output type op-amps, MOS capacitors and MOSFETs as the basic elements [57–60]. Apart from compatibility with CMOS VLSI techniques, an interesting property of the MOSFET-C filters is that the various parameters of the realized filters, namely ω_0 Q_0 and H_0 can be electronically-tuned through external voltages applied at the gate terminals of the appropriate pairs of MOSFETs replacing a given resistor in the parent active-RC filter from where a corresponding MOSFET-C filter is evolved. Both, balanced-input balanced-output types as well as single-input single-output type MOSFET-C circuits have been evolved so far. These circuits fall into the category of 'externally linear internally non-linear' (ELIN) class of networks and the use of the technique so developed was not limited to only filters but was extendable to other functional circuits such as oscillators, voltage-controlled amplifiers, automatic gain control circuits and others. Such filters have also been proposed subsequently using other building blocks such as CCII, CCCS, and OTRAs for instance, see [61–70] and the references cited therein. In this section, we discuss a number of MOSFET-C filter configurations based upon the use of CFOAs as active building blocks.

4.6.1 MOSFET-C Fully Differential Integrators

Two interesting methods of realizing MOS-C lossy and lossless integrators with in-built mechanism for cancellation of the square nonlinearity of the MOSFETs are shown in Fig. 4.28a, b. These circuits were proposed by Mahmoud and Soliman [71]. Assuming MOSFETs to be operating in triode region having equal threshold voltages (V_{TH}), by a straight forward analysis, using the equation of drain current as

$$I_D = \mu_s C_{ox}\left(\frac{W}{L}\right)\left[(V_{GS} - V_{TH}) - \frac{V_{DS}}{2}\right]V_{DS} \tag{4.79}$$

the transfer function of the circuit of Fig. 4.28a is given by

$$\frac{V_0}{V_i} = \frac{1}{sC_1R_1} \tag{4.80}$$

where $R_1 = \frac{1}{2K_1(V_{G1} - V_{TH})}$ is the equivalent resistance of the MOS transistor M_1 and $K_1 = \mu_n Cox\left(\frac{W_1}{L_1}\right)$

Similarly, the transfer function of the circuit of Fig. 4.28b is given by

$$\frac{V_0}{V_i} = \frac{\frac{1}{R_1C_1}}{s + \frac{1}{R_2C_1}} \tag{4.81}$$

where $R_2 = \frac{1}{2K_2(V_{G2} - V_{TH})}$ is the equivalent resistance of MOS transistor M_2

Two alternative circuits were also proposed by the same authors in [72] which are shown in Fig. 4.29.

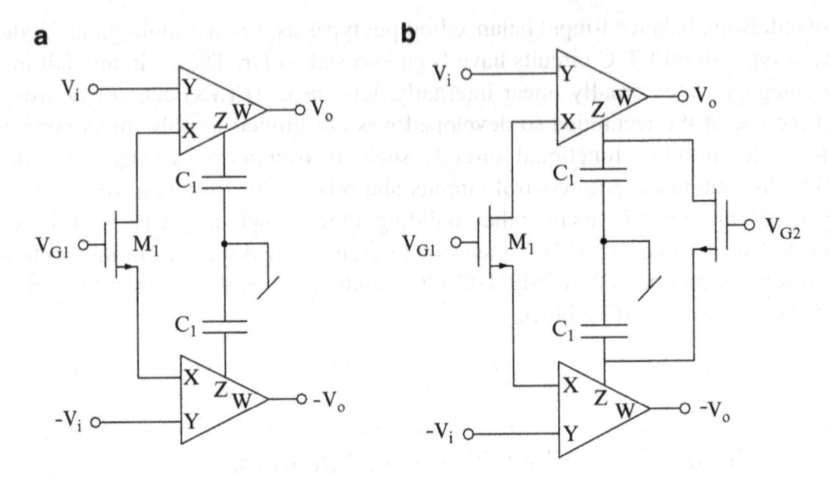

Fig. 4.28 MOS-C integrators proposed by Mahmoud and Soliman (**a**) lossless integrator, (**b**) lossy integrator (adapted from [71] © 1998 Taylor & Francis)

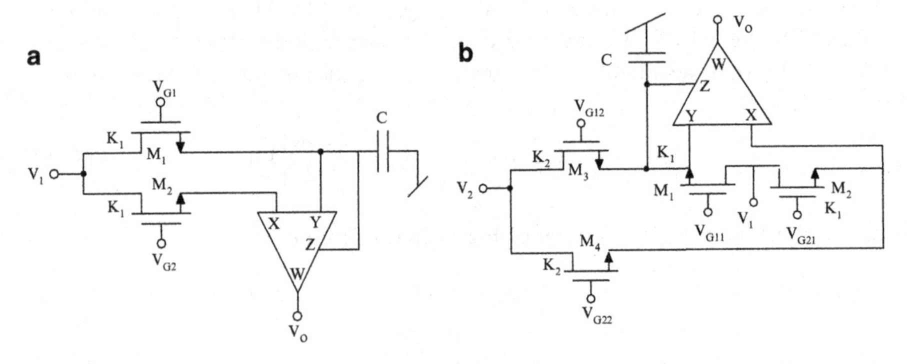

Fig. 4.29 Alternative MOS-C integrators proposed by Mahmoud and Soliman (**a**) lossless integrator, (**b**) generalized integrator (adapted from [72] © 1999 IEEE)

An analysis of the circuit of the Fig. 4.29a (for matched transistors (M_1, M_2) and triode region operation) gives

$$V_o = \frac{\frac{1}{R}}{s + \frac{1}{R}} V_1 \tag{4.82}$$

where

$$R = \frac{1}{K(V_{G1} - V_{G2})} \tag{4.83}$$

is the equivalent resistance of matched transistors for $(V_{Gi} - V_{TH}) >$ max (V_1, V_o); for i = 1, 2

and where $K_1 = \mu_n C_{ox}\left(\frac{W_1}{L_1}\right)$ is the transconductor parameter of each transistor M1 and M2.

On the other hand, an analysis of the circuit of Fig. 4.29b (for matched transistor pairs (M_1, M_2) and (M_3, M_4)) yields its output voltage as

$$V_o = \frac{\frac{1}{R_1}V_1 + \frac{1}{R_2}V_2}{s + \frac{1}{R_1} + \frac{1}{R_2}} \tag{4.84}$$

where

$$R_1 = \frac{1}{K_1(V_{G11} - V_{G12})} \tag{4.85}$$

is the equivalent resistance of matched transistors (M_1, M_2) for $(V_{G1i} - V_{TH}) > \max(V_1, V_O)$; for $i = 1, 2$ and

$$R_2 = \frac{1}{K_2(V_{G21} - V_{G22})} \tag{4.86}$$

is the equivalent resistance of matched transistors (M_3, M_4) for $(V_{G2i} - V_{TH}) > \max(V_2, V_O)$; for $i = 1, 2$.

4.6.2 *MOSFET-C Fully Differential Biquads*

A MOSFET-C biquad filter proposed by Mahmoud and Soliman [71] is shown in Fig. 4.30.

Assuming MOSFETs to be operating in triode region, by a straight forward analysis of the circuit, the two transfer functions realizable by this circuit are given by

$$\frac{V_{BP}}{V_i} = \frac{\frac{s}{R_1C_1}}{D(s)} \quad \text{and} \quad \frac{V_{LP}}{V_i} = \frac{\frac{1}{R_3R_4C_1C_2}}{D(s)} \tag{4.87}$$

where

$$D(s) = s^2 + \frac{1}{R_2C_1}s + \frac{1}{R_3R_4C_1C_2} \tag{4.88}$$

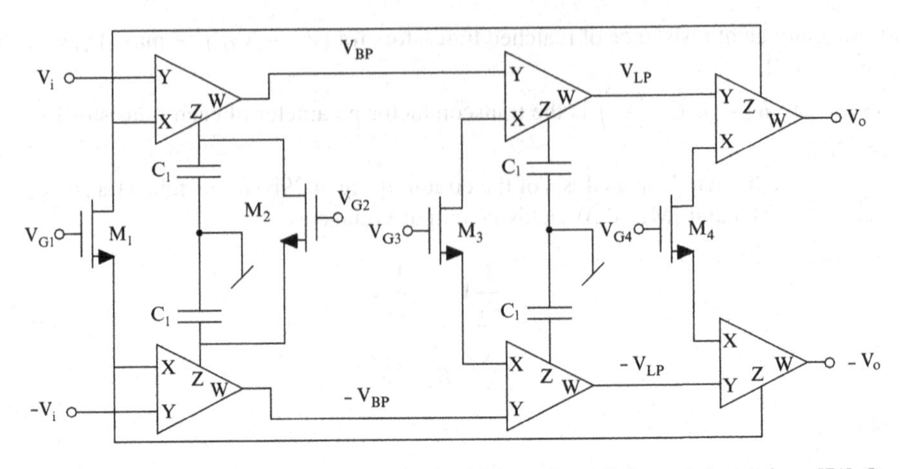

Fig. 4.30 MOSFET-C LP/BP filter proposed by Mahmoud and Soliman (adapted from [71] © 1998 Taylor & Francis)

$$\omega_0 = \sqrt{\frac{1}{R_3 R_4 C_1 C_2}},\ Q_0 = R_2 \sqrt{\frac{C_1}{R_3 R_4 C_2}} \tag{4.89}$$

and $R_i = 1/K_i(V_{Gi} - V_{TH})$ for $(i = 1, 2, 3, 4)$ is the equivalent resistance of i_{th} MOS transistor.

From the above, it is seen that for the realized filters, the parameter ω_0 can be controlled by V_{G3} and/or V_{G4} whereas Q_0 in case of LP and bandwidth ω_0/Q_0 in case of BP can be controlled by external voltage V_{G2}.

4.6.3 *MOSFET-C Single-Ended Biquad*

A single-ended MOSFET-C biquad was advanced in [72] which uses the lossy integrator of Fig. 4.29a and lossy summing integrator of Fig. 4.29b and is shown in Fig. 4.31.

Assuming that all transistors are operating in triode region, the two transfer functions realized by this circuit are given by

$$\frac{V_{BP}}{V_i} = \frac{-s\dfrac{1}{R_1 C_1}}{s^2 + s\dfrac{\left(\dfrac{1}{R_2} - \dfrac{1}{R_1}\right)}{C_1} + \dfrac{1}{C_1 C_2 R_2 R_3}} \tag{4.90}$$

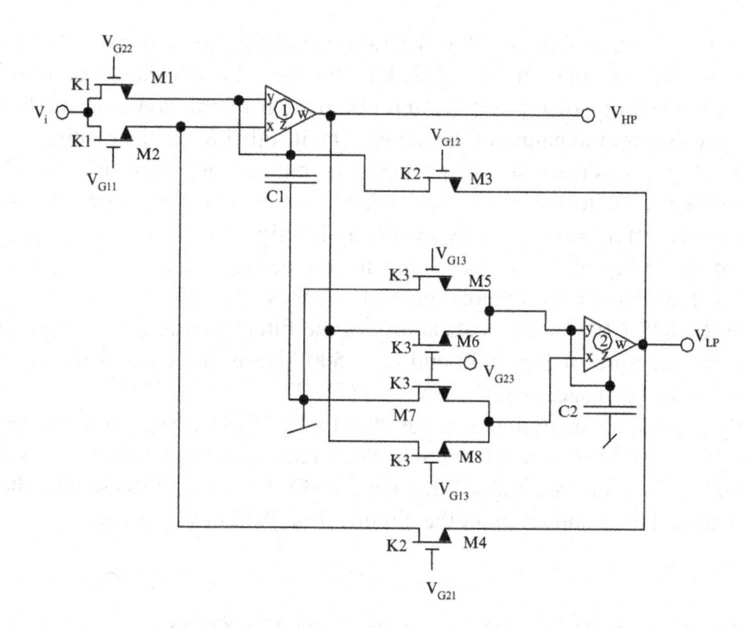

Fig. 4.31 Another MOSFET-C biquad (adapted from [72] © 1999 IEEE)

$$\frac{V_{LP}}{V_i} = \frac{\dfrac{1}{C_1C_2R_1R_3}}{s^2 + s\dfrac{\left(\frac{1}{R_2} - \frac{1}{R_1}\right)}{C_1} + \dfrac{1}{C_1C_2R_2R_3}} \tag{4.91}$$

where

$$\omega_o = \sqrt{\frac{1}{C_1C_2R_2R_3}},\quad Q_0 = \sqrt{\frac{C_1}{C_2}\frac{{R_1}^2R_2}{R_3(R_1 - R_2)^2}} \tag{4.92}$$

$R_i = \frac{1}{K_i(V_{G1i} - V_{G2i})} = \frac{1}{K_iV_{G12i}}$, for $i = 1, 2$ and 3 is the equivalent resistance of i_{th} MOS transistor and

$$K_i = \mu_n C_{ox}\left(\frac{W}{L}\right)i \quad \text{for } i = 1, 2 \text{ and } 3 \tag{4.93}$$

For $R_2 = R_3 = R$, $C_1 = C_2 = C$, the angular frequency and quality factor are simplified to

$$\omega_o = \frac{1}{RC} \quad \text{and} \quad Q_0 = \frac{R_1}{R_1 - R} \tag{4.94}$$

The fully-differential filter of Fig. 4.30 although employs as many as six CFOAs, four capacitors and only three MOSFETs but has the advantage of providing orthogonal tunability of ω_0 and Q_0 and elimination of the common mode noise due to fully differential nature of the circuit. On the other hand, the configuration of Fig. 4.31 employs as many as eight MOSFETs, but has only two CFOAs and only two grounded capacitors. It is a single-ended structure and hence, does not provide the advantage offered by the fully differential design but there is a considerable degree of flexibility in tuning the circuit for the desired values of the parameters ω_0 and Q_0 through various external control voltages.

Using CMOS CFOAs, the realisability of the filters having $f_0 = 1$ MHz in the case of the structure of Fig. 4.30 and $f_0 = 500$ kHz in the case of the circuit of Fig. 4.31, have been successfully achieved [71, 72].

Lastly, it must be mentioned that the MOSFET-C CFOA-based filters have been essentially evolved for fully integratable electronically tunable filter designs and in this context, it is, therefore implied that integratable CMOS CFOAs such as those in [73–75] should be assumed in all the circuits described in this section.

4.7 Design of Higher Order Filters Using CFOAs

Although first order filter sections and universal second order biquadratic filter functions constitute basic building blocks which by themselves may be adequate for some filtering applications not having very stringent specifications, on the other hand, there are many applications in which the rate of rejection/selection in stop band/pass band offered by a second order filter (i.e., 40 dB/decade) may not be adequate enough. In such cases, higher order active filters are needed. Higher order active filters can usually be made from a cascade of first order and/or a number of second order biquads. Alternatively, biquads can also be employed to synthesize higher order filters through the so called coupled-biquad topology [76]. Besides this, there are several other methods of designing higher order filters which are based upon a direct synthesis of the given nth order transfer function. In this section, we discuss designing CFOA-based higher order filters.

While there has been a lot of activity on universal VM and CM biquad filter realization using CFOAs, comparatively only a few researchers have explored the methods of designing higher order filters using CFOAs. In [77], Acar and Ozoguz have presented a signal flow graph (SFG) based approach for synthesizing an arbitrary nth-order transfer function. Rathore and Khot in [78] have given a systematic method of deriving CFOA-based all-grounded-capacitor filter from current mode RLC prototype ladders. Said et al. in [79] proposed a new technique for current mode realization of doubly terminated LC ladder filters in which a higher order filter is designed by using element transfer method. Besides second order biquads, third order Butterworth filters have continued to attract the attention of the researchers from time to time. Nandi et al. [78] presented a CFOA-based configuration of a third order Butterworth low-pass filter using internal device transadmittance and parasitic components along with passive RC elements. Koukiou and Psychalinos [80] have presented modular filter structures using CFOAs. More recently, multiple loop feedback filters using CFOAs have been reported by Katopodis and Psychalinos [81].

In the following, we describe some of the prominent methods for designing VM/CM higher order filters using CFOAs from amongst those outlined above.

4.7.1 *Signal Flow Graph Based Synthesis of nth Order Transfer Function Using CFOAs*

This method involves a signal-flow graph approach, with multiple-loop-feedback topology, in which any nth-order voltage transfer function using active-RC circuit employing n grounded capacitors, (n + 2) CFOAs and at most (3n + 2) resistors (which may be either grounded or virtually grounded) are used.

4.7.2 *Doubly Terminated Wave Active Filters Employing CFOA-Based on LC Ladder Prototypes*

This technique [79] relies on current mode realization of doubly terminated LC ladder filters using CFOA-based linear transformation of port variables. A design example of 3rd order low pass filter has been considered and compared with the systematic approach for deriving CFOA-based all-grounded-capacitor filters from current mode RLC prototype ladder introduced by Rathore and Khot [78]. Figure 4.32 shows the realization of two-port networks for the series and shunt elements, while Figs. 4.33, 4.34, 4.35, 4.36 show the realization of all possible combinations for the one port network in the input and output of the doubly terminated ladder filter using CFOA as an active element.

Figure 4.37 shows the active realisation of a 3rd order low pass filter employing CFOA-based circuit equivalents for each passive element.

PSPICE simulation results show [79] that this realization has low power dissipation, low total harmonic distortion and smaller active component-count as compared to one introduced in [78] (shown here in Fig. 4.38) The structure of Fig. 4.37 has successfully achieved f_0 of the order of 1 MHz.

4.7.3 *Higher Order Modular Filter Structures Using CFOAs*

CFOAs can be employed to realize higher order wave active filters using lossy integration-subtraction and simple summation and subtraction building blocks.

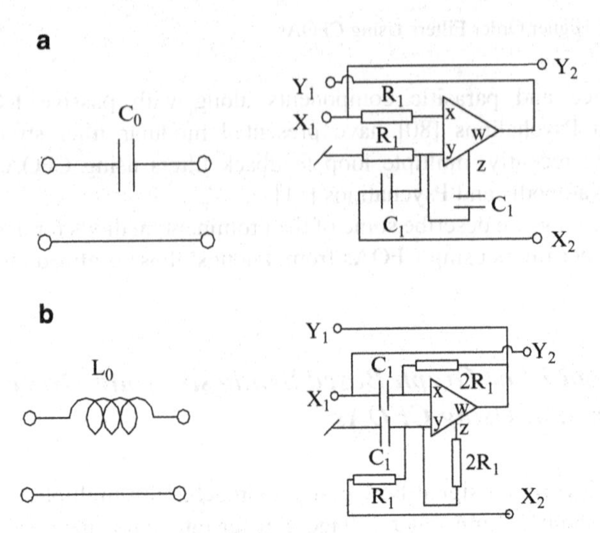

Fig. 4.32 (**a**) Floating capacitor realization. (**b**) Floating inductor realization(adapted from [79] © 2011 Elsevier)

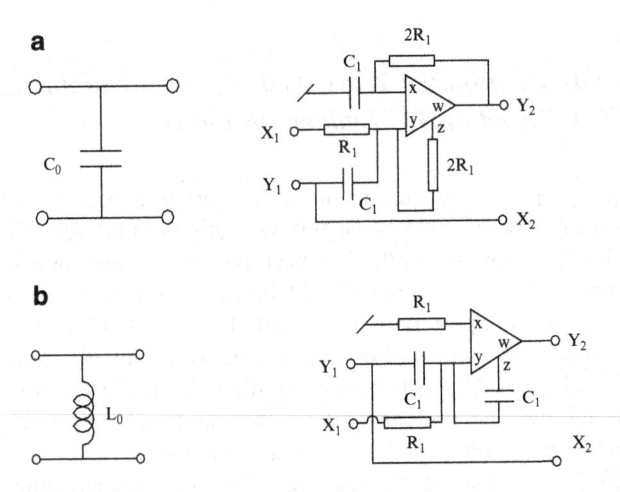

Fig. 4.33 (**a**) Grounded capacitor realization. (**b**) Grounded inductor realization (adapted from [79] © 2011 Elsevier)

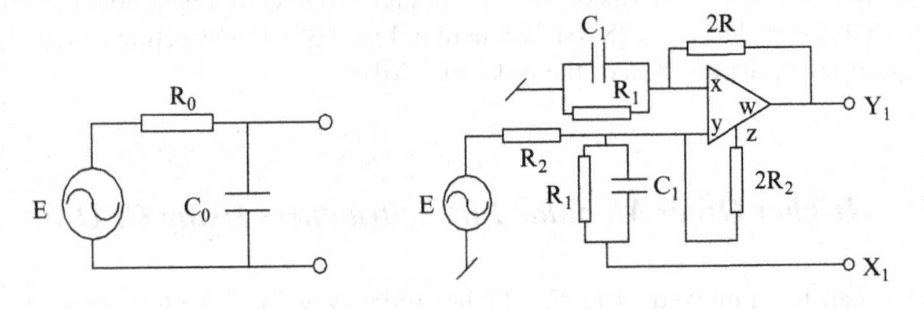

Fig. 4.34 Active realisation for input terminal with shunt capacitor and source resistance (adapted from [79] © 2011 Elsevier)

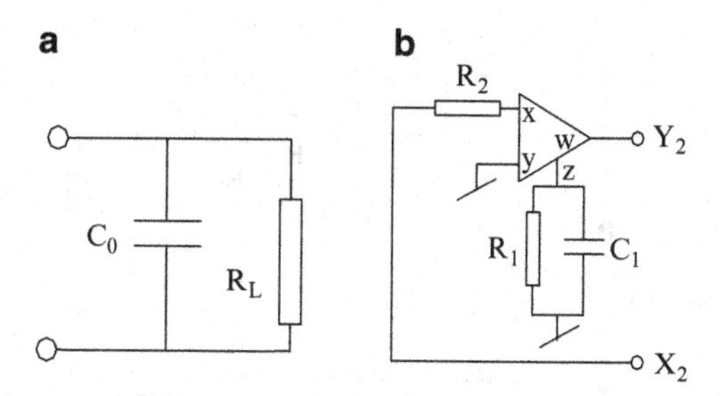

Fig. 4.35 Active realisation for output terminal with shunt capacitor and load resistance (adapted from [79] © 2011 Elsevier)

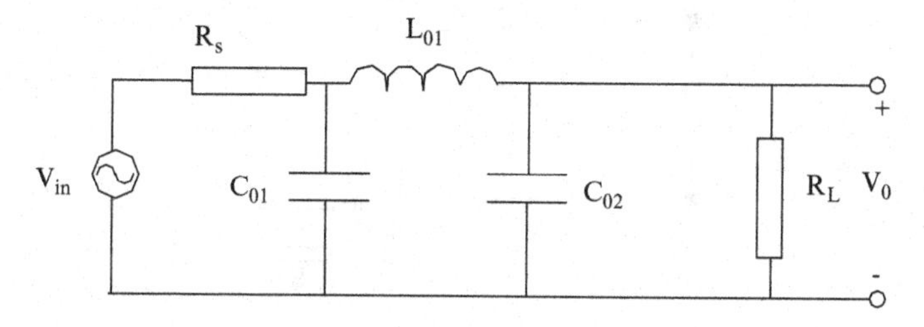

Fig. 4.36 Third order Chebyshev lowpass ladder

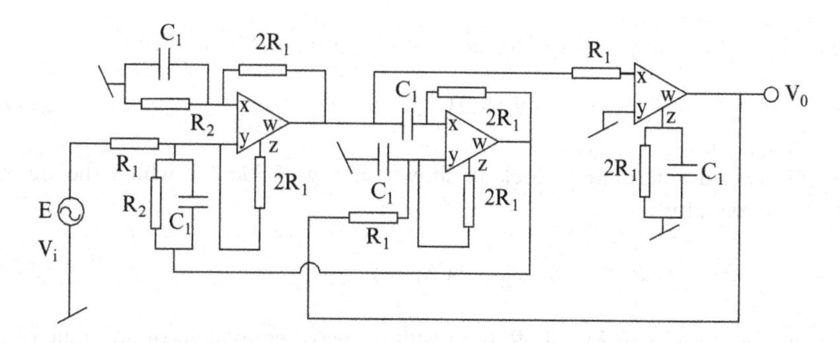

Fig. 4.37 Active realization of a third order low pass filter using CFOA-based circuits (adapted from [79] © 2011 Elsevier)

Figure 4.39a shows an implementation of lossy integration-subtraction circuit and Fig. 4.39b gives the realization of subtraction of two input voltages.

Assuming ideal CFOAs, the output voltage V_o for the circuit shown in Fig. 4.39a is given by

$$V_0 = \frac{1}{1+\tau s}(V_{in1} - V_{in2}) \quad \text{where} \quad \tau = R_a C_a \text{ is the time constant} \tag{4.95}$$

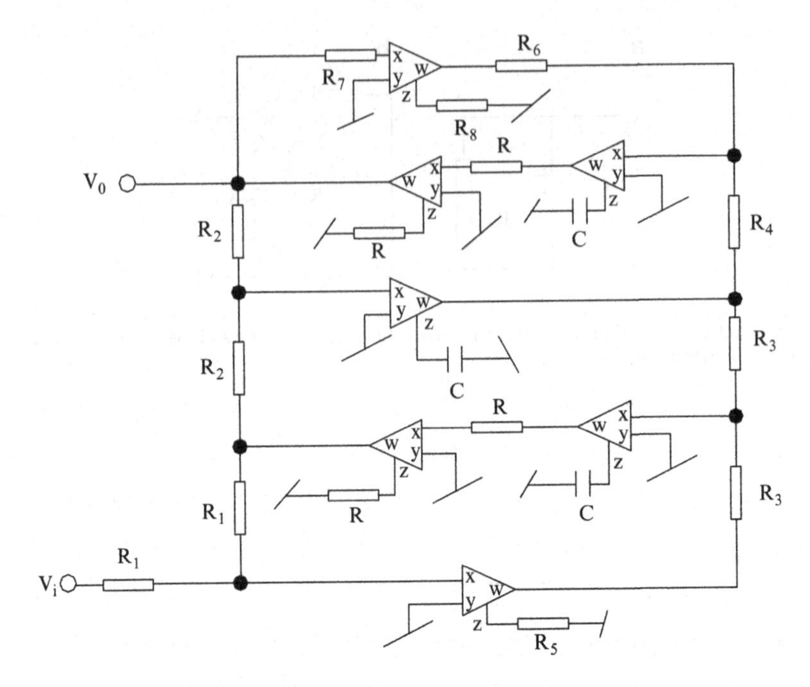

Fig. 4.38 Active realization of third order low pass filter using CFOA based circuits (adapted from [79] © 2008 John Wiley & Sons, Ltd)

Similarly, the output voltage for the circuit of Fig. 4.39b is given by

$$V_0 = (V_{in1} - V_{in2}) \tag{4.96}$$

CFOA-based summation block is shown in Fig. 4.39c for which the output voltage is given by

$$V_0 = (V_{in1} + V_{in2}) \tag{4.97}$$

Using the circuits of Fig. 4.39, the resulting wave equivalent of an inductor in series-branch is as shown in Fig. 4.40.

The complete set of wave equivalents derived in [80] has been shown in Tables 4.2 and 4.3 respectively.

For the construction of the complete wave filter (1) equal port resistances are assumed and (2) cross-cascade connection of the incident and reflected waves has been considered(because the incident wave at each port is equal to the reflected wave of the preceding port). For a 3rd order low pass LC ladder filter as shown in Fig. 4.41, the block diagram representation shown in Fig. 4.42 is derivable.

The validity of this method for a 3rd order low pass filter (cut-off frequency 100 kHz) has been confirmed using CFOA-based wave active equivalents. The commercially available AD844 devices were used as CFOAs.

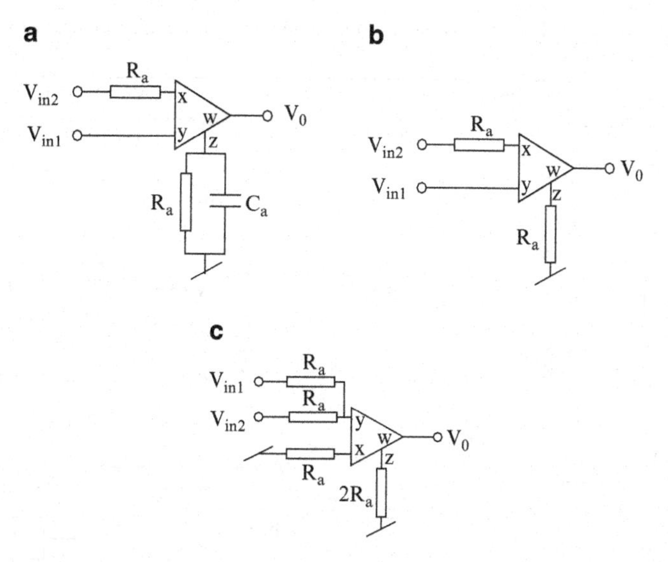

Fig. 4.39 (**a**) CFOA-based lossy integration-subtraction, (**b**) CFOA-based subtraction circuit (adapted from [80] © 2010 Radioengineering), (**c**) CFOA-based subtraction block (adapted from [80] © 2010 Radioengineering)

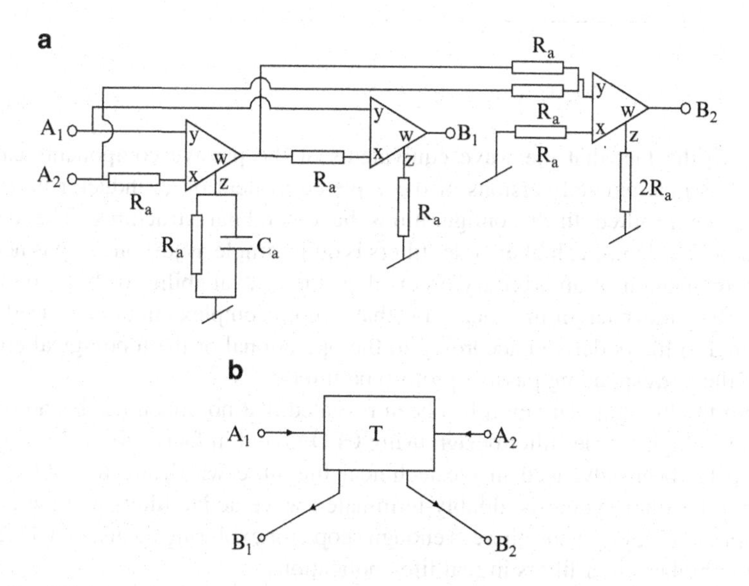

Fig. 4.40 CFOA-based wave equivalent of an inductor in series-branch (**a**) circuit level representation, (**b**) Symbolic notation (adapted from [80] © 2010 Radioengineering)

Table 4.2 Wave equivalents of two-port sub networks in series-branch (adapted from [80] © 2010 Radioengineering)

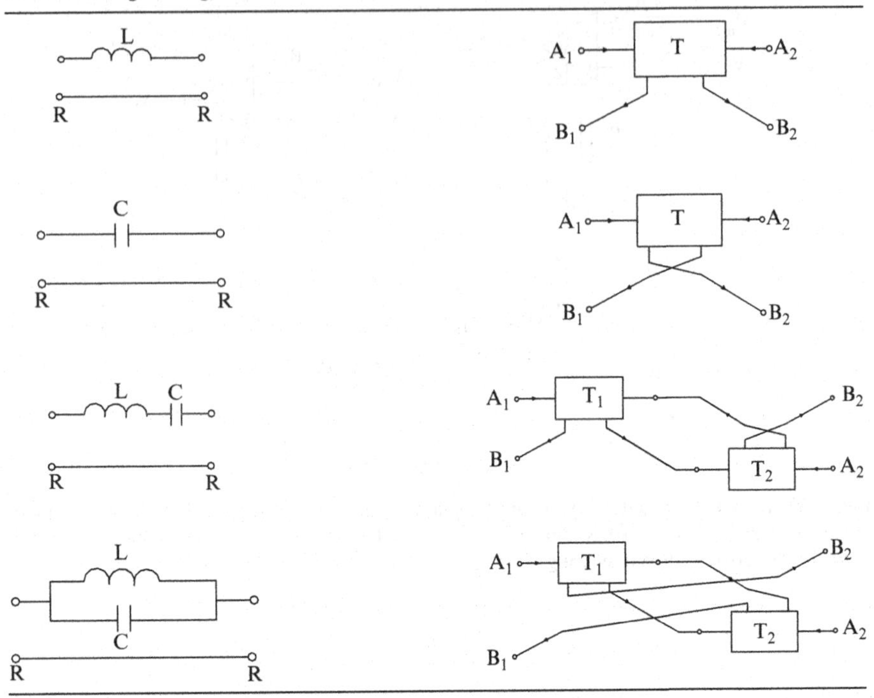

Due to the fact that the wave equivalents of the passive components can be realized as manipulated versions of the wave equivalent of an inductor in series-branch, the resulted filter configurations have modular structures. The design procedure for obtaining higher-order filters is quite simple as just one step is needed for the realization of an arbitrary order filter due to availability of basic building blocks. The drawback of this method is that a more complex circuitry is needed as compared to filters derived according to the operational or the topological emulation of the corresponding passive proto-type filters.

From the literature survey it has been revealed that not much has been done in the area of higher order filter design using CFOAs and in fact, only three methods have so far been advanced in the technical literature which are the SFG based higher order filter synthesis, doubly-terminated wave active filters and the higher order modular filters. Thus, there is enough scope for exploring the use of CFOAs in realizing higher order filters in real life applications.

Table 4.3 Wave equivalents of two-port sub networks in shunt-branch (adapted from [80] © 2010 Radioengineering)

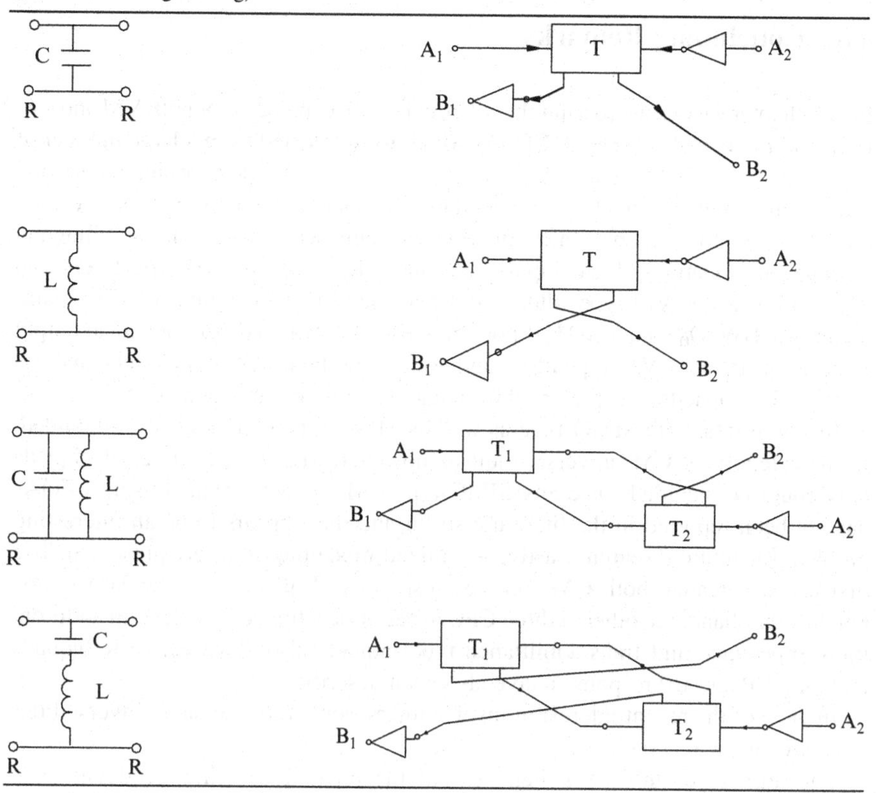

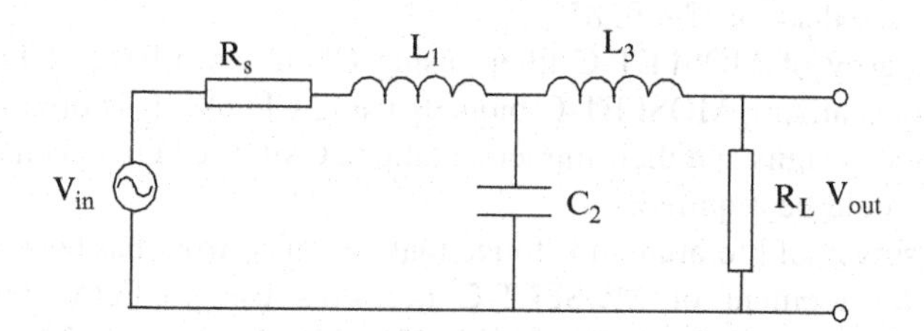

Fig. 4.41 Third order LC ladder prototype filter (adapted from [80] © 2010 Radioengineering)

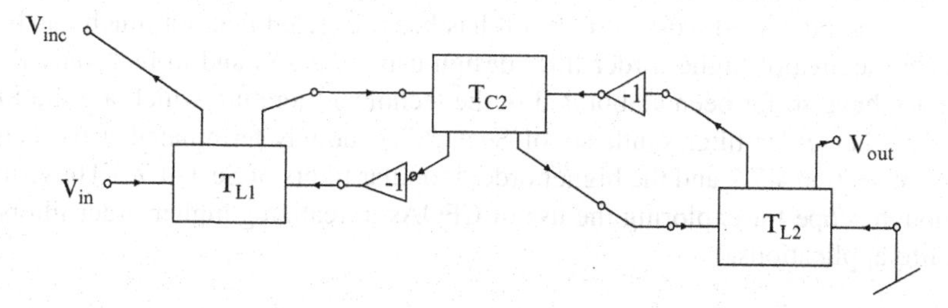

Fig. 4.42 Block diagram of the corresponding wave active filter (adapted from [80] © 2010 Radioengineering)

4.8 Concluding Remarks

In this chapter we discussed a number of circuits and techniques for both VM and CM universal filters realizable with CFOAs which were selected from a large number of such circuits [2–49, 85] available in technical literature. While choosing the second order biquad filter circuits to be included in this chapter, the main focus has been on only those configurations which provide as many as possible of the following advantageous features: Ideal input impedance ($R_{in} = 0$, for CM-type filters and $R_{in} =$ infinite, for VM-type filters), independent/sequential tuning of filter parameters ω_0, BW, Q_0 (and also H_0 if possible), use of only two capacitors, ideal output impedance ($R_0 = 0$, VM-type filters and, $R_0 =$ infinite, for CM-type filters) and use of grounded capacitors as preferred for integrated circuit implementation.

In VM filters, both SIMO-type as well as MISO-type structures were included. In the category of CM universal biquads however, MISO-type universal biquads only could be included since any SIMO-type CM universal biquad using CFOAs has not been reported in the literature so far and this appears to be an interesting problem for future research. Lastly, two mixed-mode biquads were presented. The first one can realize both CM and VM responses of all the five standard filters. On the other hand, the other, a three CFOA-based structure, can realize, in addition, trans-impedance and trans-admittance type biquads also. Two active-R biquads realizing LP and BP responses have also been described.

In view of the recent interest in inverse filters, some representative inverse filter structures were also included.

The authors of this book believe that the area of devising good universal CM/VM biquads using CFOAs is not exhausted yet and there is ample scope of devising newer configurations having better features than those available in the circuits discovered so far [2–49, 85].

In the category of MOSFET-C filters using CFOAs, we have elaborated two techniques for realizing MOSFET-C biquads using CFOAs. It is obvious that for fully integrated designs and their implementations CMOS-CFOAs such as those of [74, 75, 82] would be required.

From the survey of literature, it is found that not much work has been carried out by on the development of MOSFET-C networks using CFOAs beyond that contained in references [63, 64, 73, 75, 82, 83] which has been highlighted in this chapter. One can expect that lot of new configurations and ideas concerning new CFOA-based MOSFET-C circuits might be waiting to be explored.

From the published works [76–84] it has been revealed that not much has been done in the area of higher order filter design using CFOAs and in fact, only three methods have so far been elaborated in the technical literature which are the SFG based higher order filter synthesis of Sect. 4.7.1, doubly-terminated wave active filters of section 4.7.2 and the higher order modular filters of Sect. 4.7.3 Thus, there is enough scope for exploring the use of CFOAs in realizing higher order filters in real life applications.

References

1. Ibrahim MA, Minaei S, Kuntman H (2005) A 22.5 MHz current-mode KHN-biquad using differential voltage current conveyor and grounded passive elements. Int J Electron Commun (AEU) 59:311–318
2. Soliman AM (1998) Generation of CCII and CFOA filters from passive RLC filters. Int J Electron 85:293–312
3. Senani R (1998) Realization of a class of analog signal processing/signal generation circuits: novel configurations using current feedback op-amps. Frequenz 52:196–206
4. Bhaskar DR (2003) Realisation of second-order sinusoidal oscillator/filters with non-interacting Controls using CFAs. Frequenz 57:1–3
5. Chang CM, Hwang CS, Tu SH (1994) Voltage-mode notch, low pass and band pass filter using current-feedback amplifiers. Electron Lett 30:2022–2023
6. Singh AK, Senani R (2005) CFOA-based state-variable biquad and its high-frequency compensation. IEICE Electron Express 2:232–238
7. Soliman AM (1996) Applications of current feedback operational amplifiers. Analog Integr Circ Sign Process 11:265–302
8. Abuelma'atti MT, Al-Zaher HA (1998) New universal filter with one input and five outputs using current-feedback amplifiers. Analog Integr Circ Sign Process 16:239–244
9. Abuelma'atti MT, Al-Zaher HA (1998) New universal filter with one input and five outputs using current-feedback amplifiers. Proc Natl Sci Counc ROC (A) 22:504–508
10. Horng JW, Chang CK, Chu JM (2002) Voltage-mode universal biquadratic filter using single current-feedback amplifier. IEICE Trans Fundament E85-A:1970–1973
11. Abuelma'atti MT, Al-Shahrani SM (1996) New universal filter using two current-feedback amplifiers. Int J Electron 80:753–756
12. Abuelma'atti MT, Al-Zaher HA (1997) New universal filter using two current-feedback amplifiers. Active Passive Electron Comp 20:111–117
13. Liu SI, Wu DS (1995) New current-feedback amplifier-based universal biquadratic filter. IEEE Trans Instrum Meas 44:915–917
14. Wu DS, Lee HT, Hwang YS, Wu YP (1994) CFA-based universal filter deduced from a mason graph. Int J Electron 77:1059–1065
15. Liu SI (1995) High input impedance filters with low components spread using current-feedback amplifiers. Electron Lett 31:1042–1043
16. Topaloglu S, Sagbas M, Anday F (2012) Three-input single-output second-order filters using current-feedback amplifiers. Int J Electron Commun (AEU) 66:683–686
17. Nikoloudis S, Psychalinos C (2010) Multiple input single output universal biquad filer with current feedback operational amplifiers. Circ Syst Sig Process 29:1167–1180
18. Horng JW, Chou PY, Wu JU (2010) Voltage/Current-mode multifunction filters using current-feedback amplifiers and grounded capacitors. Active Passive Electron Comp 5:785631. doi:10.1155/2010/785631
19. Sharma RK, Senani R (2003) Multifunction CM/VM biquads realized with a single CFOA and grounded capacitors. Int J Electron Commun (AEU) 57:301–308
20. Sharma RK, Senani R (2004) On the realization of universal current mode biquads using a single CFOA. Analog Integr Circ Sign Process 41:65–78
21. Sharma RK, Senani R (2004) Universal current mode biquad using a single CFOA. Int J Electron 91:175–183
22. Chang CM, Soliman AM, Swamy MNS (2007) Analytical synthesis of low-sensitivity high-order voltage-mode DDCC and FDCCII-grounded R and C All-pass filter structures. IEEE Trans Circ Syst-I 54:1430–1443
23. Singh VK, Singh AK, Bhaskar DR, Senani R (2005) Novel mixed-mode universal biquad configuration. IEICE Electron Express 2:548–553
24. Toumazou C, Payne A, Pookaiyaudom S (1995) The active-R filter technique applied to current-feedback op-amps. IEEE Int Symp Circ Syst 2:1203–1206

25. Singh AK, Senani R (2001) Active-R design using CFOA-poles: new resonators, filters, and oscillators. IEEE Trans Circ Syst-II 48:504–511
26. Singh AK, Senani R, Tripathi MP (1999) Low-component-count high frequency resonators and their applications using op-amp compensation-poles. Frequenz 53:161–169
27. Horng JW, Hou CL, Huang WS, Yang DY (2011) Voltage/current-mode multifunction filters using one current feedback amplifier and grounded capacitors. Circ Syst 2:60–64
28. Liu SI (1995) Universal filter using two current-feedback amplifiers. Electron Lett 31:629–630
29. Horng JW, Lee MH (1997) High input impedance voltage-mode low pass, band pass and high pass filter using current-feedback amplifiers. Electron Lett 33:947–948
30. Senani R, Gupta SS (1997) Universal voltage-mode/current-mode biquad filter realised with current feedback op-amps. Frequenz 51:203–208
31. Soliman AM (1998) A new filter configuration using current feedback op-amp. Microelectron J 29:409–419
32. Horng JW (2000) New configuration for realizing universal voltage-mode filter using two current feedback amplifiers. IEEE Trans Instrum Meas 49:1043–1045
33. Abuelma'atti MT, Al-Zaher HA (2000) New grounded-capacitor grounded-resistor controlled universal filter using current-feedback amplifiers. Proc Natl Sci Counc ROC (A) 24:205–209
34. Horng JW (2001) Voltage-mode multifunction filter using one current feedback amplifier and one voltage follower. Int J Electron 88:153–157
35. Shah NA, Malik MA (2003) Multifunction filter using current feedback amplifiers. Frequenz 59:264–268
36. Gift SJG, Maundy B (2004) High-performance active band pass filter using current-feedback amplifiers. Int J Electron 91:563–570
37. Shah NA, Iqbal SZ, Rather MF (2005) Versatile voltage-mode CFA-based universal filter. Int J Electron Commun (AEU) 59:192–194
38. Mita R, Palumbo G, Pennisi S (2005) Non-idealities of Tow-Thomas biquads using VOA- and CFOA-based miller integrators. IEEE Trans Circ Syst-II 52:22–27
39. Shah NA, Rather MF, Iqbal SZ (2005) A novel voltage-mode universal filter using a single CFA. J Active Passive Electron Devices 1:183–188
40. Djebbi M, Assi A, Sawan M (2005) Design of monolithic tunable CMOS band-pass filter using current feedback operational amplifiers. Analog Integr Circ Sign Process 45:143–154
41. Singh VK, Singh AK, Bhaskar DR, Senani R (2006) New universal biquads employing CFOAs. IEEE Trans Circ Syst-II 53:1299–1303
42. Sagbas M, Koksal M (2007) Voltage-mode three-input single-output multifunction filters employing minimum number of components. Frequenz 61:87–93
43. Bhaskar DR, Prasad D (2007) New current mode biquad filter using CFOAs. J Active Passive Electron Devices 2:292–298
44. Manhas PS, Pal K, Sharma S, Mangotra LK, Jamwal KKS (2007) Realization of high-Q band pass filter using low voltage current feedback amplifiers. J Active Passive Electron Devices 4:13–20
45. Ferri G, Guerrini N, Piccirilli MC (2003) CFA based fully integrable KHN Biquad. Int Symp Sig Circ Syst 2:569–572
46. Palumbo G, Pennisi S (1999) Filter circuits synthesis with CFOAs-based differentiators. 16th IEEE Instrument Measurement Technology Conference (IMTC). pp 546–550
47. Yuce E (2010) Fully integrable mixed-mode universal biquad with specific application of the CFOA. Int J Electron Commun (AEU) 64:304–309
48. Hou CL, Huang CC, Lan YS, Shaw JJ, Chang CM (1999) Current-mode and voltage-mode universal biquads using a single current-feedback amplifier. Int J Electron 86:929–932
49. Dostal T (1995) Correspondence: Insensitive voltage-mode and current-mode filters from commercially available transimpedance op-amps. IEE Proc Circ Devices Syst 142:140–143
50. Gupta SS, Bhaskar DR, Senani R, Singh AK (2009) Inverse active filters employing CFOAs. Electr Eng 91:23–26

51. Gupta SS, Bhaskar DR, Senani R (2011) New analogue inverse filters realized with current feedback op-amps. Int J Electron 98:1103–1113
52. Wang HU, Chang SH, Yang TY, Tsai PY (2011) A novel multifunction CFOA-based inverse filter. Circ Syst 2:14–17
53. Kerwin WJ, Huelsman LP, Newcomb RW (1967) State-variable synthesis for insensitive integrated circuit transfer functions. IEEE J Solid State Circ SC-2:87–92
54. Bohn DA (1986) Constant-Q graphic equalizer. J Audio Eng Soc 34:16
55. Baker BC (1999) Anti-aliasing, Analog filter for data acquisition systems. Application notes no. AN699 of Microchip Tech Inc 1-10
56. Banu M, Tsividis Y (1982) Floating voltage-controlled resistors in CMOS technology. Electron Lett 18:678–679
57. Banu M, Tsividis Y (1983) Fully integrated active RC filters in MOS technology. IEEE J Solid State Circ SC-18:644–651
58. Tsividis Y, Banu M, Khoury J (1986) Continuous-time MOSFET-C filters in VLSI. IEEE Trans Circ Syst 33:125–140
59. Ismail M, Smith SV, Beale RG (1988) A new MOSFET-C universal filter structure for VLSI. IEEE J Solid State Circ 23:183–194
60. Sakurai S, Ismail M, Michel JY, Sanchez-Sinencio E, Brannen R (1992) A MOSFET-C variable equalizer circuit with simple on-chip automatic tuning. IEEE J Solid State Circ 27:927–934
61. Liu SI, Tsao HW, Lin TK (1990) MOSFET capacitor filters using unity gain CMOS current conveyors. Electron Lett 26:1430–1431
62. Liu SI, Tsao HW, Wu J (1991) CCII-based continuous-time filters with reduced gain-bandwidth sensitivity. IEE Proc Circ Devices Syst 138:210–216
63. Meng XR, Yu ZH (1996) CFA based fully integrated Tow-Thomas biquad. Electron Lett 32:722–723
64. Gunes EO, Anday F (1997) CFA based fully integrated nth-order lowpass filter. Electron Lett 33:571–573
65. Salama KN, Elwan HO, Soliman AM (2001) Parasitic-capacitance-insensitive voltage-mode MOSFET-C filters using differential current voltage conveyor. Circ Syst Sig Process 20:11–26
66. Schmid HP, Moschytz GS (2000) Active- MOSFET-C single-amplifier biquadratic filters for video frequencies. IEE Proc Circ Devices Syst 147:35–41
67. Chiu W, Tsay JH, Liu SI, Tsao H, Chen JJ (1995) Single-capacitor MOSFET-C integrator using OTRA. Electron Lett 31:1796–1797
68. Chen JJ, Tsao HW, Liu SI, Chiu W (1995) Parasitic-capacitance-insensitive current-mode filters using operational transresistance amplifiers. IEE Proc Circuits Devices Syst 142:186–192
69. Chen JJ, Tsao HW, Liu SI (2001) Voltage-mode MOSFET-C filters using operational transresistance amplifier (OTRAs) with reduced parasitic capacitance effect. IEE Proc Circ Devices Syst 148:242–249
70. Hwang YS, Wu DS, Chen JJ, Shih CC, Chou WS (2007) Realization of higher-order OTRA-MOSFET-C active filters. Circ Syst Sig Process 26:281–291
71. Mahmoud SA, Soliman AM (1998) Novel MOS-C balanced-input balanced-output filter using the current feedback operational amplifier. Int J Electron 84:479–485
72. Mahmoud SA, Soliman AM (1999) New MOS-C biquad filter using the current feedback operational amplifier. IEEE Trans Circ Syst-I 46:1510–1512
73. Mahmoud SA, Soliman AM (2000) Novel MOS-C oscillators using the current feedback op-amp. Int J Electron 87:269–280
74. Manetakis K, Toumazou C (1996) Current-feedback op-amp suitable for CMOS VLSI technology. Electron Lett 32:1090–1092
75. Soliman AM, Madian AH (2009) MOS-C Tow-Thomas filter using voltage op amp, current feedback op amp and operational transresistance amplifier. J Circ Syst Comput 18:151–179

76. Schaumann R, Ghausi MS, Laker KR (1990) Design of analog filters: active RC and switched capacitor. Prentice Hall, Englewood Cliffs, NJ
77. Acar C, Ozoguz S (2000) Nth-order voltage transfer function synthesis using a commercially available active component, CFA: signal-flow graph approach. Frequenz 54:134–137
78. Rathore TS, Khot UP (2008) CFA-based grounded-capacitor operational simulation of ladder filters. Int J Circ Theor Appl 36:697–716
79. Said LA, Madian AH, Ismail MH, Soliman AM (2011) Active realization of doubly terminated LC ladder filters using current feedback operational amplifier (CFOA) via linear transformation. Int J Electron Commun (AEU) 65:753–762
80. Koukiou G, Psychalinos C (2010) Modular filter structures using current feedback operational amplifiers. Radioengineering 19:662–666
81. Katopodis V, Psychalinos C (2011) Multiple-loop feedback filters using feedback amplifiers. Int J Electron 98:833–846
82. Mahmoud SA, Awad IA (2005) Fully differential CMOS current feedback operational amplifier. Analog Integr Circ Sign Process 43:61–69
83. Soliman AM, Madian AH (2009) MOS-C KHN filter using voltage op amp, CFOA, OTRA and DCVC. J Circ Syst Comput 18:733–769
84. Nandi R, Sanyal SK, Bandyopadhyay TK (2008) Third order lowpass Butterworth filter function realization using CFA. Int J Electron 95:313–318
85. Fabre A (1995) Comment and reply: Insensitive voltage-mode and current-mode filters from transimpedance op amps. IEE Proc Circ Dev Syst 142:140–143

Chapter 5
Synthesis of Sinusoidal Oscillators Using CFOAs

5.1 Introduction

Sinusoidal oscillators find numerous applications in various electronic, instrumentation, measurement, control and communication systems as test oscillators or signal generators. Since the classical Wien Bridge oscillator (WBO), RC phase shift oscillator, Twin-T oscillator etc. do not have the provision for varying the frequency of oscillation through a single variable passive element (resistance or capacitance; preferably the former), considerable research effort has been made in the eighties and nineties on evolving single element controlled sinusoidal oscillators using IC op-amps. Thus, the area of RC-active oscillators using the conventional voltage-mode op-amps (VOA) had been a very prominent area of analog research before the advent of the Current Conveyors, CFOAs and other modern active circuit building blocks. A large number of VOA-based sinusoidal oscillators were published in the technical literature during 1976–2001 for instance, see [1–9] and the references cited therein.

During the past two decades, there have been numerous investigations, intuitive as well systematic, on the generation of a variety of sinusoidal oscillators employing CFOAs. The aim of this chapter is to give an exposition of some of the prominent CFOA-based sinusoidal oscillators.

5.2 The Evolution of Single Element Controlled Oscillators: A Historical Perspective

Since the traditional WBO requires either ganged variable capacitors or ganged variable resistors for realizing variable frequency sinusoidal oscillations, the problem of realizing a single element controlled oscillator had been a very popular problem among researchers in the 1980s. In 1976, Hribsek and Newcomb [1] presented, for the first time, two single-resistance controlled oscillators each

R. Senani et al., *Current Feedback Operational Amplifiers and Their Applications*, Analog Circuits and Signal Processing, DOI 10.1007/978-1-4614-5188-4_5,

using two op-amps and two grounded capacitors as preferred for IC implementation; see [11, 63] and references cited therein.

The first real single-element-controlled oscillator using a single op-amp without any constraints was introduced by Soliman and Awad [2] in 1978 but in this circuit, the oscillation frequency could be controlled only through a variable capacitor—not a very convenient option as variable capacitors have a very limited tuning range. A single-op-amp based single-resistor-controlled-oscillator (SRCO), capable of being operated as (1) a variable frequency oscillator, (2) a voltage-controlled-oscillator (VCO) and (3) a very low frequency (VLF) oscillator, without any constraints, was proposed by Senani in 1979 in [3].[1] This circuit employs only a single op-amp, six resistors and two capacitors and provides independent single resistance controls of both condition of oscillation (CO) and frequency of oscillation (FO) through two separate grounded resistors. The circuit could also be employed to realize a VCO by replacing the frequency-controlling grounded resistor by a FET used as a voltage-controlled-resistor and is convenient for incorporating the additional amplitude stabilizing/control circuitry easily due to the grounded nature of the resistor governing the CO of the oscillator [4]. Bhattacharyya and Darkani in [5] derived the complete family of sixteen such single-op-amp-RC canonical SRCOs. Methods of generating equivalents of such op-amp based oscillators have also been evolved; for instance, see [12] and [6] and references cited therein.

Interestingly, the problem of devising newer SRCOs employing one or more active building blocks (ABB) has continued to attract the attention and imagination of researchers even now and a large number of SRCOs have been evolved using a variety of other ABBs too during the last two decades. The other ABBs considered have been second generation Current Conveyors (CCII) (and their many variants), operational transconductance amplifiers (OTA), Four terminal floating nullors (FTFN), Current differencing buffered amplifiers (CDBA), Current differencing transconductance amplifiers (CDTA), Operational Trans-resistance amplifiers (OTRA) etc. In fact, the search for newer topologies of SRCOs is aimed at ultimately achieving more and more or all of the following desirable features: employment of grounded capacitors as preferred for IC implementation, use of a minimum possible number of active and/or passive components, suitability for VCO realization, achieving quadrature signal generation, providing explicit voltage mode as well as current mode outputs, achieving a high frequency-stability, exhibiting higher operational frequency range and minimization of the effects of parasitic impedances or non-ideal parameters etc.

In this chapter, we present a variety of SRCOs employing CFOAs with a focus on the works exhibiting a synthesis approach and limiting to only some representative circuits in various categories from the vast amount of literature accumulated during the past two decades in this area (for instance, see [10, 13–67] and the references cited therein).

[1] It has come to the attention of the first author only at the time of finalizing this chapter (13–17 September 2012) that a quite similar single op-amp SRCO employing only five resistors and two capacitors, was proposed by Soliman and Awad in 1978 in [87].

5.3 Advantages of Realizing Wien Bridge Oscillator Using CFOA vis-à-vis VOA

The interest in using CFOAs for realizing sinusoidal oscillators grew after it was demonstrated by Martinez et al. [13, 14] that the use of CFOA rather than VOA in the classical Wien Bridge oscillator offers improved performance, as compared to its VOA-based counterpart, in terms of frequency accuracy, dynamic range, distortion level and frequency span. In the following, we show, as demonstrated in [14] that in the CFOA-version, the condition of oscillation (CO) and frequency of oscillation (FO) become decoupled.

Consider now the Wien bridge oscillator (WBO) using a conventional VOA (see Fig. 5.1a), an ideal analysis gives the closed loop characteristic equation (CE) as

$$s^2 + s\frac{(3-k)}{RC} + \left(\frac{1}{RC}\right)^2 = 0 \tag{5.1}$$

from where the (CO) and (FO) are given by

$$\mathrm{CO} : \mathrm{k} \geq 3 \tag{5.2}$$

$$\mathrm{FO} : \omega_0 = \frac{1}{RC} \tag{5.3}$$

When VOA is assumed to have a one-pole open loop gain function characterized by $A_v(s) \cong \frac{\omega_t}{s}$ for $\omega \gg \omega_p$ where ω_p is the pole frequency and ω_t is the gain-bandwidth product of the op-amp, through a re-analysis of the circuit [14], the following non-ideal FO ($\hat{\omega}_0$) and CO are obtained:

$$(\hat{\omega}_0)^2 = {\omega_0}^2\left(\frac{1}{1+3\tau k\omega_0}\right) \quad \text{and} \quad k \geq 3\left(\frac{1}{1-\tau\omega_0(1-\frac{\omega^2}{{\omega_0}^2})}\right) \tag{5.4}$$

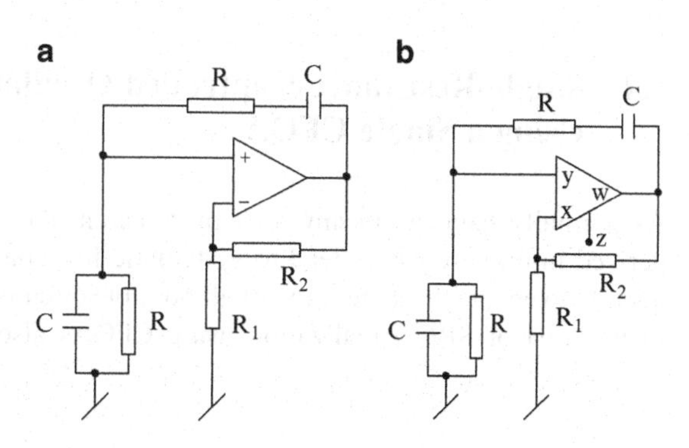

Fig. 5.1 Wien bridge oscillators: (**a**) realized with a VOA, (**b**) realized with a CFOA

where

$$\tau = \frac{1}{\omega_t} \quad \text{and} \quad k = \left(1 + \frac{R_2}{R_1}\right) \tag{5.5}$$

From the above equations, it is seen that because the closed loop amplifier gain k appears in the expressions of FO and CO both, therefore, any change in the signal amplitude calibration (distortion) by changing k, disturbs the oscillation frequency also and vice versa.

In the case of the CFOA-based WBO circuit of Fig. 5.1b on the other hand, the non-inverting amplifier gain is given by

$$k(s) = \frac{k}{1 + s\tau} \tag{5.6}$$

where τ = RpCp with Rp//(1/sCp) being the parasitic output impedance looking into terminal-Z of the CFOA. The non-ideal FO and CO are given by

$$(\hat{\omega})^2 = \frac{{\omega_0}^2}{1 + 3\tau\omega_0} \tag{5.7}$$

and

$$\kappa \geq 3 + \tau\omega_0\left(1 - \frac{\hat{\omega}^2}{\omega_0^2}\right) \tag{5.8}$$

From the above equations, difference can be seen in the behavior of the CFOA-version of the Wien bridge oscillator as compared to its VOA counterpart. It may be noted that CO and FO in the CFOA-version are de-coupled in the sense that k does not appear in (5.7) hence, any change in adjusting the CO by changing k, does not have any effect on FO.

5.4 Single-Resistance-Controlled Oscillators (SRCO) Using a Single CFOA

As normally happens in any area of research, the initial results are quite often derived intuitively which lead to systematic formulation of methodologies subsequently to enable the generation of all possible circuits belonging to a specific class. In the area of SRCO realization using CFOAs also, a number of circuits were

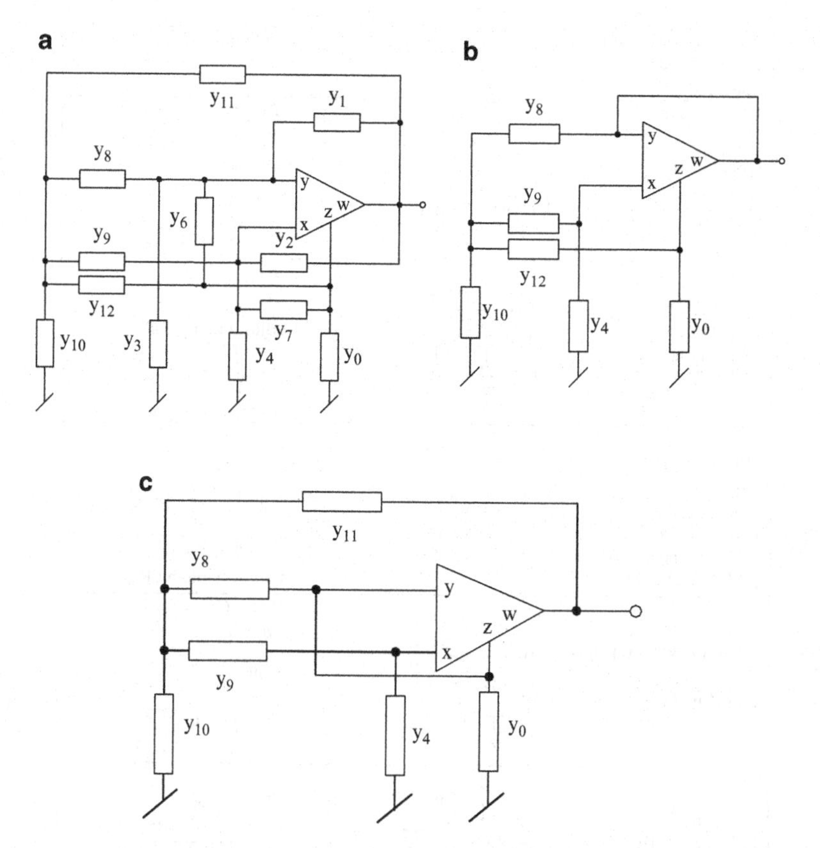

Fig. 5.2 Generalized single CFOA configurations for systematic generation of canonic SRCOs [86]. (**a**) Generalized six-node structure. (**b**) First converted five-node structure obtained from the circuit of (**a**). (**c**) Second converted five node structure obtained from the circuit of (**a**)

initially reported in a piece-meal manner by various researchers. It was only after the research carried over about a decade that systematic approaches started being formulated by a number of research groups for deriving systematically all possible CFOA-based SRCOs belonging to specific classes.

One such systematic approach was formulated in [15, 19] which was based upon the constitution of the most general single-CFOA-based six-node twelve-admittance structure shown in Fig. 5.2a and two converted five-node structures obtained therefrom, shown in Fig. 5.2b, c, whose general characteristic equations were found wherefrom all canonic single-CFOA-based SRCOs i.e. circuits using no more than three resistors and two capacitors were enumerated. A class of Single-CFOA-based SRCOs resulting from this approach has been shown in Fig. 5.3.

In the set of eight SRCOs displayed in Fig. 5.3, it may be noted that oscillators 6 and 8 do not permit independent adjustability of CO although FO can be varied independently; oscillators 1, 2, 3, 5, 7 provide the control of CO through a single grounded resistor which is attractive from the point of view of ease of incorporating

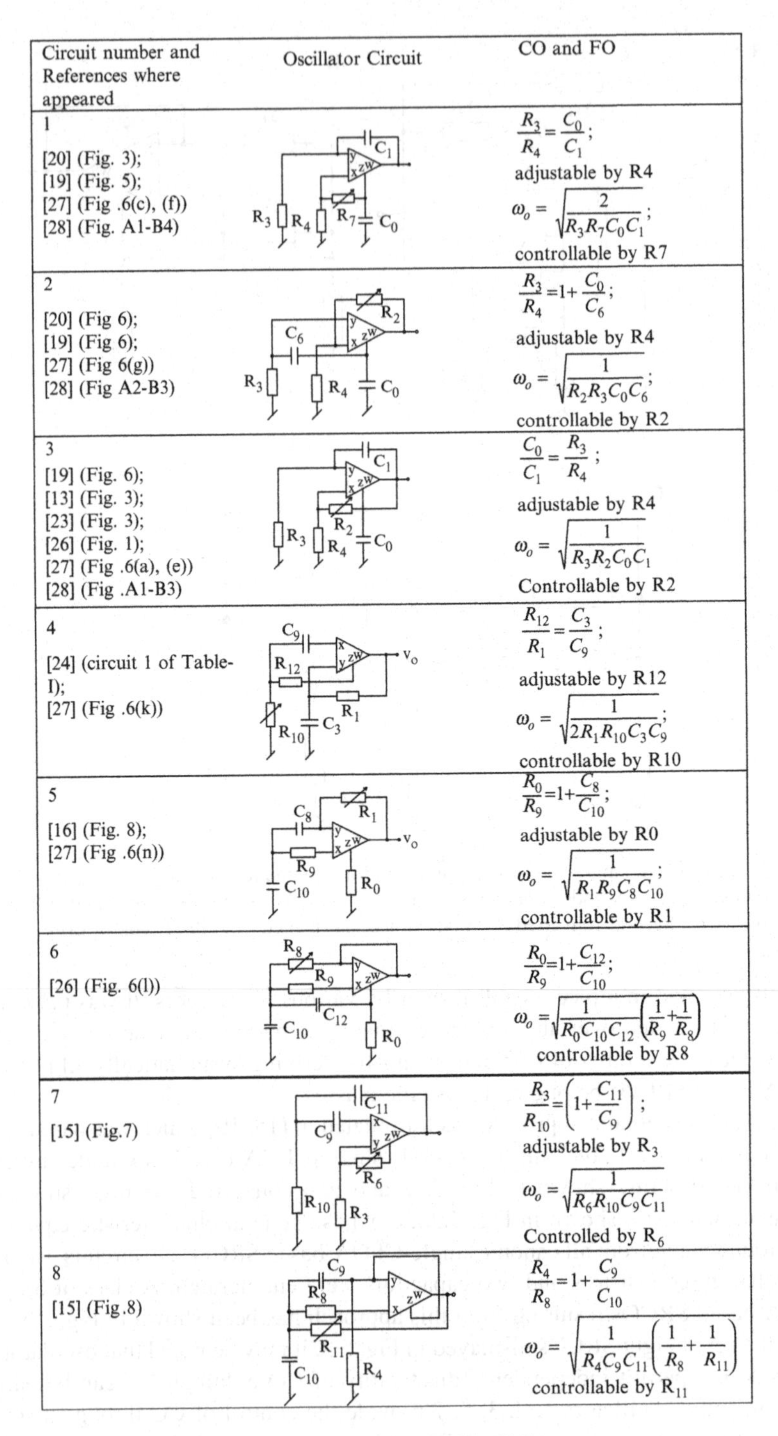

Circuit number and References where appeared	Oscillator Circuit	CO and FO
1 [20] (Fig. 3); [19] (Fig. 5); [27] (Fig .6(c), (f)) [28] (Fig. A1-B4)		$\frac{R_3}{R_4}=\frac{C_0}{C_1}$; adjustable by R4 $\omega_o=\sqrt{\frac{2}{R_3R_7C_0C_1}}$; controllable by R7
2 [20] (Fig 6); [19] (Fig 6); [27] (Fig 6(g)) [28] (Fig A2-B3)		$\frac{R_3}{R_4}=1+\frac{C_0}{C_6}$; adjustable by R4 $\omega_o=\sqrt{\frac{1}{R_2R_3C_0C_6}}$; controllable by R2
3 [19] (Fig. 6); [13] (Fig. 3); [23] (Fig. 3); [26] (Fig. 1); [27] (Fig .6(a), (e)) [28] (Fig .A1-B3)		$\frac{C_0}{C_1}=\frac{R_3}{R_4}$; adjustable by R4 $\omega_o=\sqrt{\frac{1}{R_3R_2C_0C_1}}$ Controllable by R2
4 [24] (circuit 1 of Table-I); [27] (Fig .6(k))		$\frac{R_{12}}{R_1}=\frac{C_3}{C_9}$; adjustable by R12 $\omega_o=\sqrt{\frac{1}{2R_1R_{10}C_3C_9}}$; controllable by R10
5 [16] (Fig. 8); [27] (Fig .6(n))		$\frac{R_0}{R_9}=1+\frac{C_8}{C_{10}}$; adjustable by R0 $\omega_o=\sqrt{\frac{1}{R_1R_9C_8C_{10}}}$; controllable by R1
6 [26] (Fig. 6(l))		$\frac{R_0}{R_9}=1+\frac{C_{12}}{C_{10}}$; $\omega_o=\sqrt{\frac{1}{R_0C_{10}C_{12}}\left(\frac{1}{R_9}+\frac{1}{R_8}\right)}$ controllable by R8
7 [15] (Fig.7)		$\frac{R_3}{R_{10}}=\left(1+\frac{C_{11}}{C_9}\right)$; adjustable by R_3 $\omega_o=\sqrt{\frac{1}{R_6R_{10}C_9C_{11}}}$ Controlled by R_6
8 [15] (Fig.8)		$\frac{R_4}{R_8}=1+\frac{C_9}{C_{10}}$ $\omega_o=\sqrt{\frac{1}{R_4C_9C_{11}}\left(\frac{1}{R_8}+\frac{1}{R_{11}}\right)}$ controllable by R_{11}

Fig. 5.3 The class of Single-CFOA-based canonic SRCOs [86]

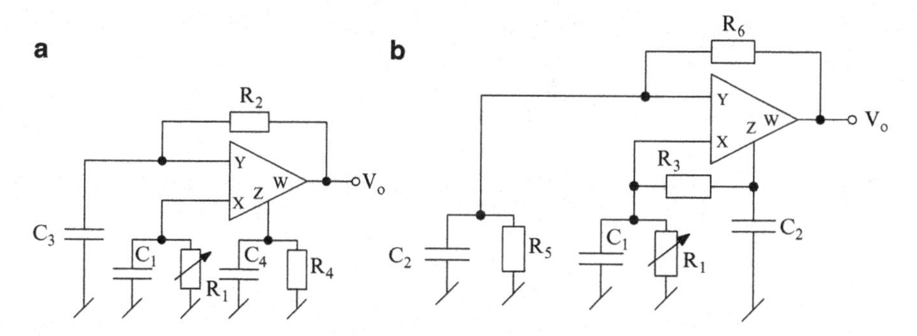

Fig. 5.4 Single-CFOA-three-GC SRCOs proposed by Toker et al. (adapted from [27] © 2002 Elsevier)

amplitude stabilization/control circuitry. On the other hand, oscillator 4 provides FO control through a grounded resistor thereby making it a suitable choice for easy conversion into a VCO by replacing this frequency-controlling grounded resistor by a FET or linearized VCR [68].

It may be mentioned that in Fig. 5.3 we have included only canonic SRCOs i.e. those circuits which use no more than three resistors and two capacitors. For other interesting SRCOs using more than three resistors and more than two capacitors the reader is referred to the work reported in [27, 28] from where it is found that permitting more than two capacitors makes it possible to realize single-CFOA SRCOs using all grounded capacitors—a feature which is not possible with canonic-single-CFOA oscillators. Two such three-GC SRCOs devised by Toker et al. [27] are shown in Fig. 5.4.

The CO and FO for the circuits of Fig. 5.4 are as follows:

For the circuit of Fig. 5.4a

$$\text{CO}: \quad C_2\left(\frac{1}{R_5}+\frac{1}{R_6}\right)-C_1\left(\frac{1}{R_6}\right)+C_3\left(\frac{2}{R_3}\right)\leq 0 \tag{5.9}$$

$$\text{FO}: \quad \text{f}_0=\frac{1}{2\pi}\sqrt{\frac{\frac{2}{R_3R_5}-\frac{1}{R_1R_6}}{C_2C_3}} \tag{5.10}$$

and for the circuit of Fig. 5.4b

$$\text{CO}: \quad \left\{\frac{1}{R_2}(C_4-C_1)+C_3\left(\frac{1}{R_4}\right)\right\}\leq 0 \tag{5.11}$$

$$\text{FO}: \quad \text{f}_0=\frac{1}{2\pi}\sqrt{\frac{\frac{1}{R_2}\left(\frac{1}{R_4}-\frac{1}{R_1}\right)}{C_3C_4}} \tag{5.12}$$

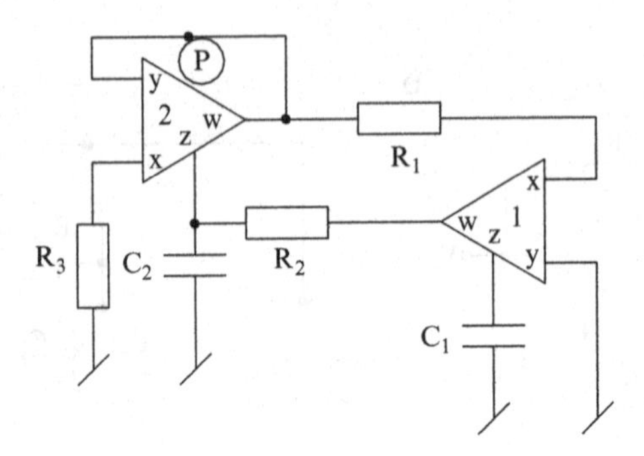

Fig. 5.5 A two-CFOA-GC SRCO proposed by Senani and Singh (adapted from [70])

Thus, in both cases FO can be controlled by a single variable resistor R_1 although an independent control of CO is not available.

It may be seen that out of the two circuits, that of Fig. 5.4a has the advantage that both the Z-pin parasitic impedances $R_p//(1/sC_p)$ can be absorbed in the external components R_4 and C_4. To the best knowledge of the authors till date not even a single SCRO is known which can have only two grounded capacitors while using only a single CFOA. In the next three sections we show, however, that given two CFOAs, two-GC SRCOs are possible which can also be synthesized quite systematically.

5.4.1 A Novel SRCO Employing Grounded Capacitors

A novel SRCO employing two CFOAs and both grounded capacitors (GC) as preferred for integrated circuit implementation was introduced by Senani and Singh in 1996 [25]. This circuit is shown in Fig. 5.5 and is characterized by the following CO and FO:

$$R_3 = R_2 \tag{5.13}$$

and

$$f_0 = \frac{1}{2\pi}\sqrt{\frac{1}{R_1 R_2 C_1 C_2}} \tag{5.14}$$

Thus, the CO can be satisfied by adjusting R_3 whereas the FO can be *independently* varied by R_1.

It must be pointed out that although frequency controlling resistor R_1 is not physically connected to ground; nevertheless, since it is connected to virtual ground, there is no difficulty in replacing R_1 by FET-based voltage-controlled resistor (VCR) conventional or otherwise such as Senani's floating VCR (FVCR) in [69] to obtain voltage controlled oscillations.

In designing this oscillator for providing a specified frequency range, it must be kept in mind that the various external RC components are to be selected to have appropriate values so that the parasitic impedances appearing at the x-input and compensation pins z of the CFOAs have the least effect on the performance. Alternatively, these internal compensating capacitances can be absorbed into the main capacitances as they appear in shunt with them, while r_x of both CFOAs can be easily accommodated in the resistors R_1 and R_3.

The frequency stability is considered to be an important figure of merit for evaluating/comparing the performance of sinusoidal oscillators. It is defined as $S_F = -d\phi/du$ evaluated at $u = 1$ where $u = \omega/\omega_o$. Thus, S_F can be determined by finding the open loop transfer function of the circuit of Fig. 5.5 which is given by

$$T(s) = \frac{s\frac{1}{R_3C_2}}{s^2 + s\frac{1}{R_2C_2} + \frac{1}{R_1R_2C_1C_2}} \tag{5.15}$$

Taking the values of various passive components as: $R_2 = R_3 = R$, $C_1 = C_2 = C$, $R_1 = R/n$ we get $\omega_0 = \sqrt{(n)}/RC = \sqrt{(n)}/\tau$, while the open loop transfer function becomes

$$T(s) = \frac{s/\tau}{s^2 + s/\tau + \omega_0^2} = \frac{s\frac{\omega_0}{\sqrt{n}}}{s^2 + \frac{s\omega_0}{\sqrt{n}} + \omega_0^2} \tag{5.16}$$

by putting $s = j\omega$, we get

$$T(j\omega) = \frac{j\frac{\omega}{\omega_0}\frac{1}{\sqrt{n}}}{\left[1 - \left(\frac{\omega}{\omega_0}\right)^2\right] + j\frac{\omega}{\omega_0}\frac{1}{\sqrt{n}}} \tag{5.17}$$

From the above, the S_F has been found to be $S_F = 2\sqrt{n}$ which can be made large by keeping n large. It has been shown in [25] that by breaking the link at P the resulting open loop circuit can be used as a lowpass/bandpass filter. Also, by removing external capacitors C_1 and C_2 and incorporating the Z-pin parasitic capacitances into design, the circuit can be used as an active-R oscillator with ω_0 still controllable through R_1. In active-RC mode, this SRCO works well in generating oscillation frequencies of the order of 500 kHz while in active-R mode, it has been possible to extend the generated frequencies till 9.85 MHz.

5.5 Two-CFOA-Two-GC SRCOs: The Systematic State Variable Synthesis

Subsequent to the publication of the two CFOA-SRCO of [25], it occurred to the first author of this monograph that if one formulates state equations of this circuit then with some algebraic manipulations it should be possible to convert these into node equations which could then be synthesized using CFOAs and RC components. It was soon realized that a chosen [A] matrix of the state variable characterization of an SRCO could, therefore, lead to more than one circuit. Furthermore, since for any specified CO and FO, many different state space representations leading to different [A] matrices (but all leading to the same characteristic equation (CE)) could be evolved, this methodology appeared to have the potential of generating a large number of SRCO (quite likely all possible) circuits. In this section, we give an account of the state variable methodology from [36–38, 70] and outline some selected circuits from the complete family of 14 two-CFOA-two-GC SRCOs generated therefrom.

A canonic second-order (i.e. employing only two capacitors) oscillator can, in general, be characterized by the following autonomous state equation:

$$\begin{bmatrix} \dot{x}_1 \\ \dot{x}_2 \end{bmatrix} = \begin{bmatrix} a_{11} & a_{12} \\ a_{21} & a_{22} \end{bmatrix} \begin{bmatrix} x_1 \\ x_2 \end{bmatrix} \tag{5.18}$$

From the above, the characteristic equation (CE)

$$s^2 - (a_{11} + a_{22})s + (a_{11}a_{22} - a_{12}a_{21}) = 0 \tag{5.19}$$

gives the condition of oscillation and frequency of oscillation as

$$CO: \quad (a_{11} + a_{22}) = 0 \tag{5.20a}$$

$$FO: \quad \omega_0 = \sqrt{(a_{11}a_{22} - a_{12}a_{21})} \tag{5.20b}$$

The methodology of [36–38] involves: (1) a selection of the parameters a_{ij}, $i = 1, 2$; $j = 1, 2$, in accordance with the required features (e.g. non-interacting controls for frequency of oscillation and condition of oscillation through separate resistors), (2) conversion of the resulting state equation into node equations (NE) and finally, (3) constitution of a physical circuit from these node equations.

Different circuits are expected to be generated by making different choices of parameters a_{11}, a_{12}, a_{21} and a_{22}. For non-interactive controls of condition of oscillation and frequency of oscillation, let us assume that condition of oscillation is to be controlled by R_1 (independent of R_2) and frequency of oscillation is to be controlled by R_2 (independent of R_1; with the third resistor R_3 featuring in both condition of oscillation and frequency of oscillation). These conditions can be satisfied in a number of ways leading to different [A] matrices. It has been shown in [37, 38] that a set of 14 different matrices can be conceived.

To illustrate the procedure, consider now the following [A] matrix which satisfies the above requirements:

$$[A] = \begin{bmatrix} 0 & \frac{1}{C_1 R_2} \\ -\frac{1}{C_2 R_3} & \frac{1}{C_2}\left(\frac{1}{R_3} - \frac{1}{R_1}\right) \end{bmatrix} \tag{5.21}$$

From the above matrix, the CO and FO are given by

$$\mathrm{R}_3 = \mathrm{R}_1 \tag{5.22}$$

and

$$f_o = \frac{1}{2\pi\sqrt{C_1 C_2 R_2 R_3}} \tag{5.23}$$

From the above matrix, the following node equations can be written

$$C_1 \frac{dx_1}{dt} = \frac{x_2}{R_2} \tag{5.24}$$

$$C_2 \frac{dx_2}{dt} = \frac{(x_2 - x_1)}{R_3} - \frac{x_2}{R_1} \tag{5.25}$$

The synthesis of the final circuit using node equations (5.24) and (5.25) is shown in Fig. 5.6a which is self-explanatory.

Following the above explained procedure, a large number of circuits are derived in [70] out of which a set of 14 SRCOs are demonstrated in [37, 38]. Some exemplary circuits possessing interesting properties are shown here in Fig. 5.6 (FO is same for all oscillators as given by (5.23)).

The circuits shown in Fig. 5.6 have a number of interesting properties which are as follows.

Single resistance control (SRC) of frequency of oscillation through a grounded resistor makes it easier to incorporate FET-based voltage controlled resistors (VCR) thereby leading to VCO realizations whereas SRC control of condition of oscillation through a grounded resistor is desirable from the viewpoint of easy incorporation of amplitude stabilization/ control circuitry. The circuit of Fig. 5.6a is seen to provide controls of condition of oscillation and frequency of oscillation both through separate grounded-resistors R_1 and R_2, respectively and is, therefore, superior to the remaining SRCOs of Fig. 5.6 SRCOs from this view point.

In case of the circuits of Fig. 5.6a–c, f the z-pin parasitic capacitances can be easily merged with the main external capacitances and hence, these parasitics do not affect the circuit behavior adversely. In the circuits of Fig. 5.6d, e, g, the

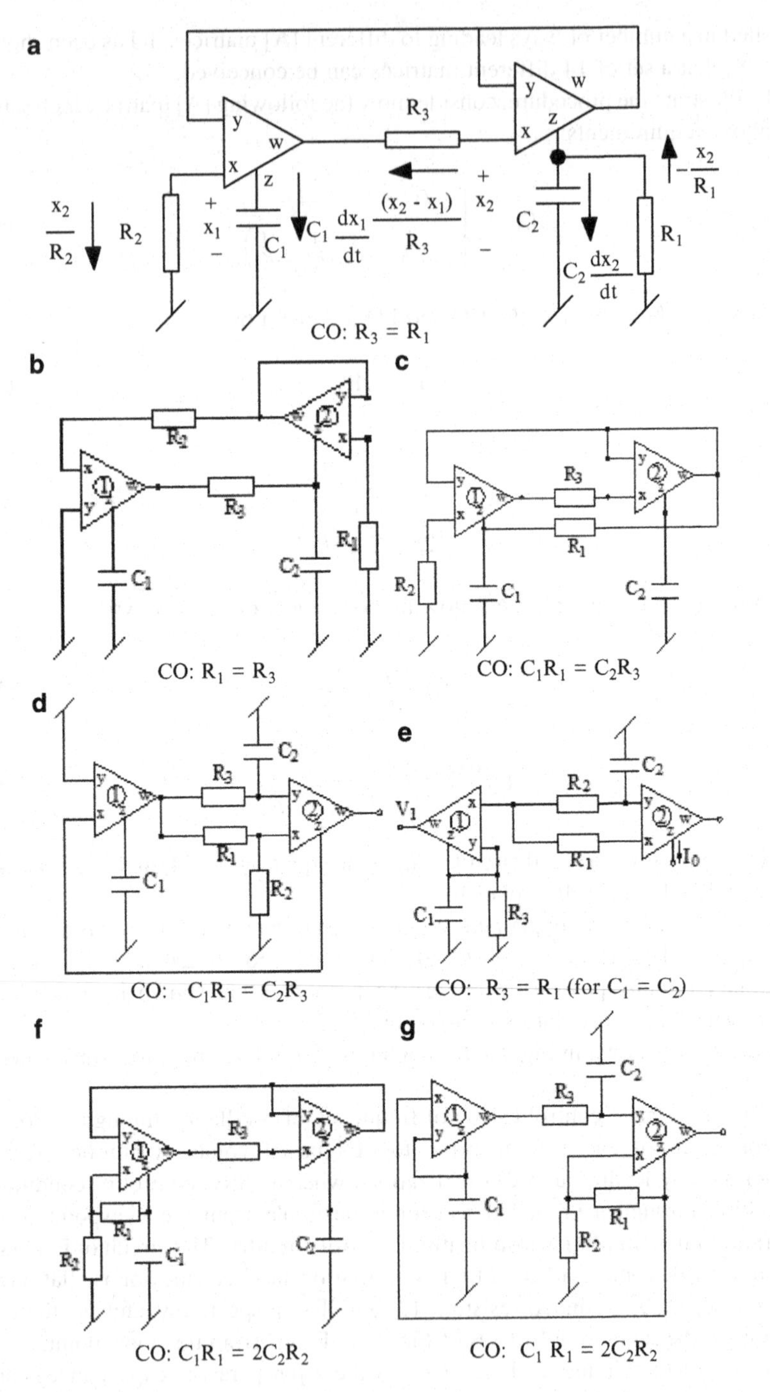

Fig. 5.6 Some exemplary circuits synthesized through the state variable methodology [70]. (**a**) CO: $R_3 = R_1$. (**b**) CO: $R_1 = R_3$. (**c**) CO: $C_1 R_1 = C_2 R_3$. (**d**) CO: $C_1 R_1 = C_2 R_3$. (**e**) CO: $R_3 = R_1$ (for $C_1 = C_2$). (**f**) CO: $C_1 R_1 = 2C_2 R_2$. (**g**) CO: $C_1 R_1 = 2C_2 R_2$

capacitor C_1 is connected to z-terminal of CFOA1 and no capacitor is connected at the z-terminal of CFOA2. However, the parasitic capacitance at the z-terminal of the CFOA2 is made ineffective by z-terminal being connected to virtual ground (as in Fig. 5.6d) or is ineffective as the current through z-terminal of CFOA2 is not coming into picture as in case of Fig. 5.6e. However, in the circuit shown in Fig. 5.6g, the z-pin parasitic capacitance at the z-terminal of the CFOA2 cannot be accounted for.

It is found that with $C_1 = C_2 = C$ and $R_1 = R_3 = R$, $R_2/R = n$ for the circuits of Fig. 5.6a–d, f–g the frequency stability factor S_F can be made sufficiently large as 'n' can always be kept greater than unity and therefore, these circuits enjoy excellent frequency stability properties.

The circuit of Fig. 5.6e is notable due to the availability of an explicit current output.

The family of 14 two-CFOA-GC SRCOs presented in [37, 38, 70] has been found to work quite well for generating sinusoidal signals up to several hundred kHz.

5.6 Other Two-CFOA Sinusoidal Oscillator Topologies

The SRCOs were shown to have been derived in the previous section through a systematic synthesis procedure with the objective of possessing the following features: (a) use of two GCs, (b) use of two CFOAs and (c) independent control of CO and FO through two separate resistors. All the SRCOs were based upon the tuning laws of the type

$$R_1 = R_3 \tag{5.26}$$

$$f_0 = \frac{1}{2\pi\sqrt{C_1 C_2 R_2 R_3}} \tag{5.27}$$

which give the resulting circuits, the control of CO trough R_1 and that of FO by R_2.

Although, major attention has been received in the literature on the above kind of SRCOs, oscillators governed by other type of tuning laws, which thereby provide CO control through a single variable capacitor or FO control through a single variable capacitor or provide an expression for FO containing a difference term are also useful due to the following:

1. Oscillators providing single element control (SEC) of FO through a single variable capacitor can be used as a transducer oscillator in conjunction with capacitive transducers.
2. Oscillators providing CO control through a capacitor can be used in some capacitance measurement schemes, for instance see [71–73].
3. Oscillators having a difference term in the expression of FO may be usefully employed as very low frequency oscillators [41].

In view of the above, therefore, the catalogue of 14 two CFOA oscillators and their variants [37, 38, 70] some of which were described in the previous section does not really exhaust all possible GC-SECOs realizable from two CFOAs.

It, therefore, turns out that given only two CFOAs only two GCs, along with two to three resistors, a number of *other* sinusoidal oscillator circuits should be possible which may have tuning law different than those considered so far and yet satisfy the single-element-controllability conditions. In Fig. 5.7 , we have presented a number of such two-CFOAs-GC SECOs. These circuits too are derived by the state variable methodology by framing new tuning laws, determining the required [A] matrices, converting the [A] matrices into node equations and finally synthesizing the resulting node equations by physical circuits using CFOAs and RC elements. The details of the derivation are given in [48]. The following features of the circuits of the Fig. 5.7 may now be noted.

- Circuits 1–2 have tuning laws that do not conform to (5.26) and (5.27) and yet these circuits do possess features (a) and (b).
- Circuit 3 is the only oscillator circuit realizable with a bare minimum of only four passive components. It may be pointed out that this circuit can be treated to be the CFOA-version of a similar circuit using CCIIs described earlier in references [74] and [75] but by contrast, this CFOA-version has the advantage of providing buffered outlets from the output of either CFOA.
- Circuit 4 although employs three grounded capacitors but still qualifies for feature (c).

It is worth mentioning that like most CFOA-based circuits, the influence of the parasitic impedances of the CFOAs can be reduced by selecting the external resistors to be much larger than the input resistance r_x of the X-terminal and smaller than the parasitic output resistance R_p looking into the compensation terminal-Z of the CFOA and the external capacitances to be larger than the parasitic output capacitance C_p of the CFOAs.

Analysis of the frequency stability of the circuits reveals that the frequency stability factors are quite large for the circuits shown in Fig. 5.7 similar to other circuits contained in [16, 25, 44].

It is worth noticing that oscillator 1 contains a difference term in the expression for FO of type $\omega_0 = \frac{\sqrt{1-n}}{RC}$ where n is the frequency controlling resistors ratio.

This oscillator qualifies to be used for generating very low frequency oscillations (i.e. 1 Hz or lower) by choosing n such that (1 − n) can be made as small as possible so that lower values of FO are achievable. On the other hand, oscillators 3 and 4 appear to be suitable for capacitance measurement methods such as those of [71–73]. In such a case, the unknown capacitance can be connected in place of C_1 and then the known variable capacitance C_2 is to be varied until the circuit just starts (or stops) oscillating as described in [71–73].

One more two-CFOAs-two-GC-based SRCO but with an additional frequency scaling factor in the expression for f_0 was introduced by Liu and Tsay in [22] which is shown in Fig. 5.8.

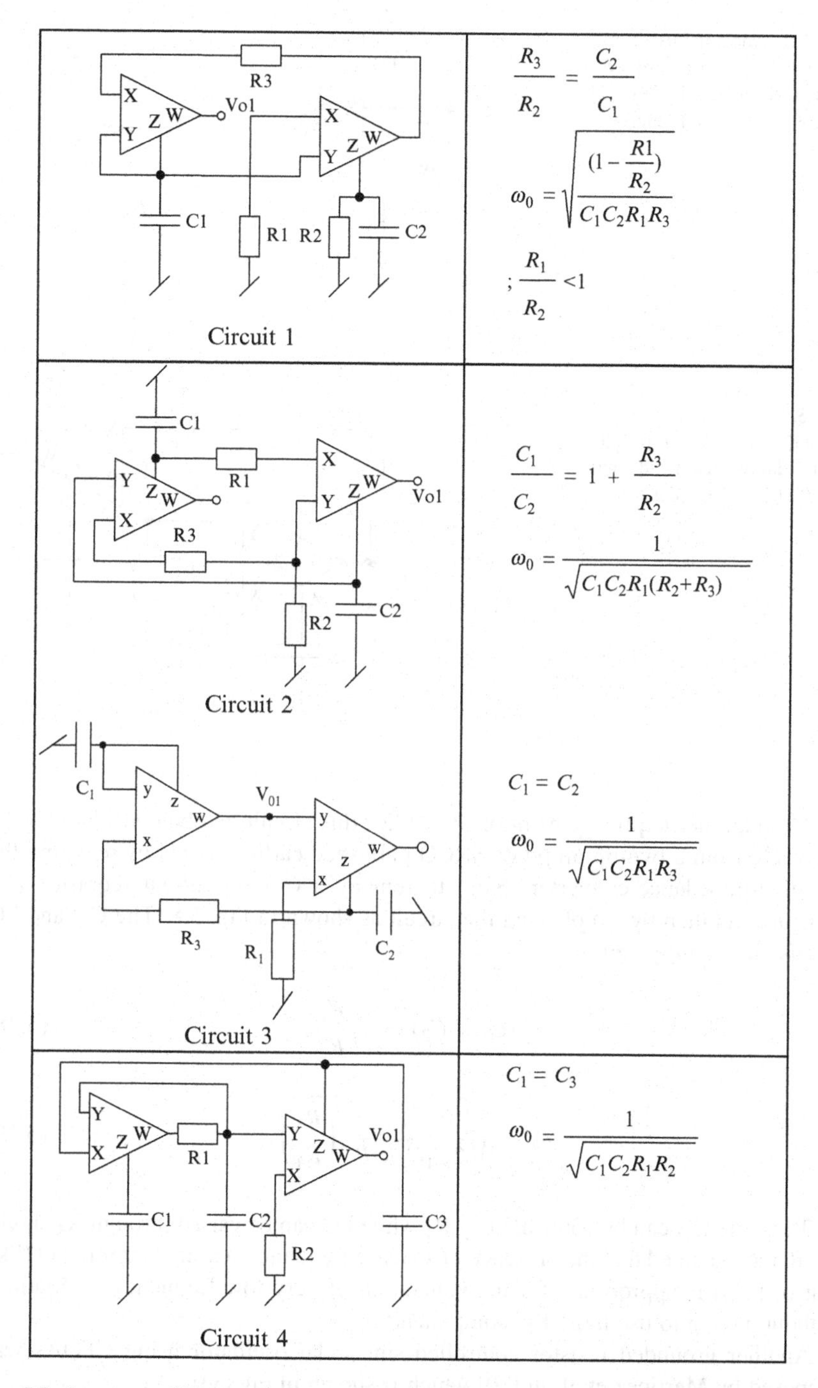

Fig. 5.7 Some two-CFOA oscillators with different tuning laws (adapted from [48] © 2006 IEEE)

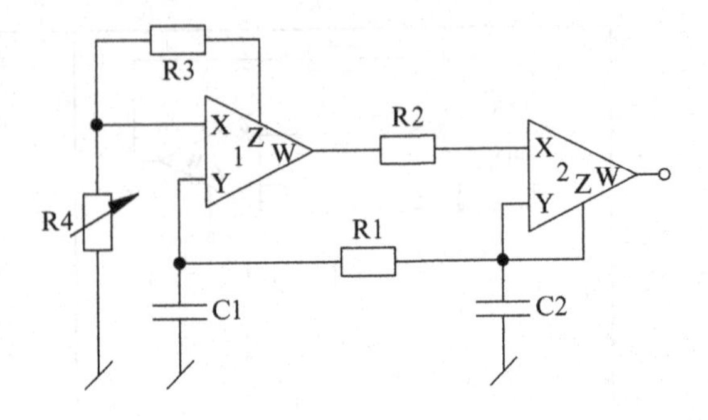

Fig. 5.8 Another SRCO grounded-resistor controlled using GCs (adapted from [22] © 1996 Taylor & Francis)

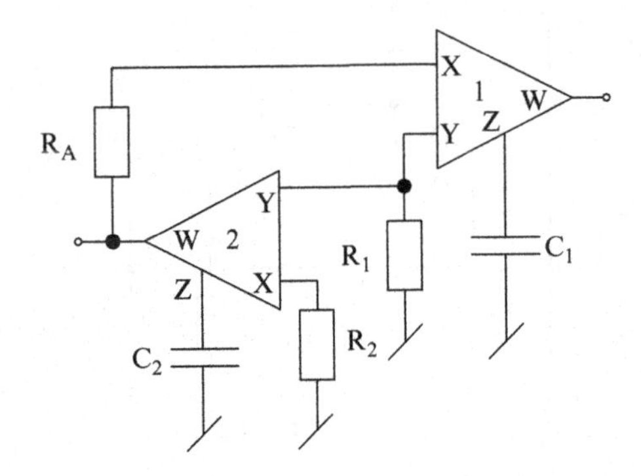

Fig. 5.9 Yet another grounded resistor controlled sinusoidal oscillator (adapted from [10] © 1997 IEE)

Although not explicitly mentioned in [22], this circuit can be considered to be derivable from a two-op-amp-GC SRCO published earlier in [11] by realizing the negative-impedance-converter (NIC) therein by a CFOA without requiring any resistors and thereby simplifying the circuit as shown in Fig. 5.8. The CO and FO of this circuit are given by

$$(C_1 + C_2) = C_1 \frac{R_1}{R_2} \tag{5.28}$$

$$\omega_0 = \sqrt{\frac{1}{2C_1C_2R_1R_2}\left(\frac{R_3}{R_4}\right)} \tag{5.29}$$

Thus, the CO can be controlled by R_1 while FO can be varied through R_4 and/or R_3. It may be noted that the presence of an extra frequency scaling factor in FO (like that of [11]) is appropriate for the generation of very low frequency oscillations without having to use large RC components.

Another grounded resistor controlled sinusoidal oscillator using CFOAs was proposed by Martinez et al. in [10] which is shown in Fig. 5.9.

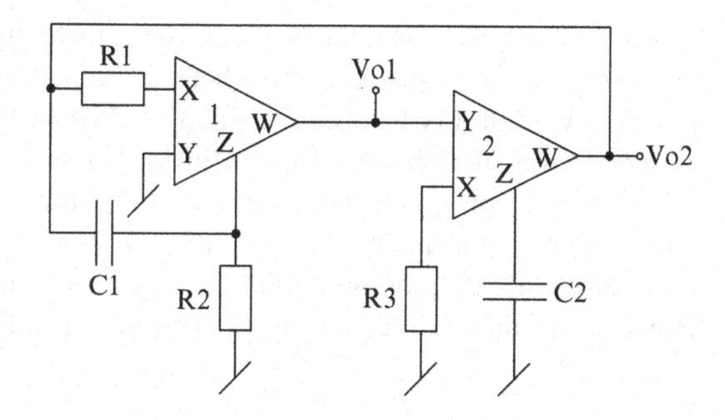

Fig. 5.10 SRCO proposed by Tangsrirat and Surakampontorn (adapted from [51] © 2009 Elsevier)

The CO and FO for this circuit are given by

$$R_A = R_1 \quad \text{and} \quad \omega_0 = \sqrt{\frac{1}{C_1 C_2 R_2 R_A}} \tag{5.30}$$

In this oscillator also, both CO and FO can be controlled independently through grounded resistors R_1 and R_2 respectively.

Martinez and Sanz in [29] gave a method for generation of variable frequency sinusoidal oscillators based on two integrator loops and presented a sinusoidal oscillator using two CFOAs. However, this circuit is quite similar to the SRCO circuit of Senani and Singh [25].

Tangsrirat and Surakampontorn in [51] proposed a single resistance controlled quadrature oscillator using two CFOAs which is shown in Fig. 5.10.

The CO and FO for this circuit are given by

$$R_2 C_1 = R_3 C_2 \tag{5.31}$$

and

$$\omega_0 = \sqrt{\frac{1}{R_1 R_3 C_1 C_2}} \tag{5.32}$$

In this case, the CO can be adjusted by R_2 without affecting the oscillation frequency, while ω_0 can be adjusted by R_1.

Another quadrature oscillator using two CFOAs has been proposed by Hou and Wang in [31], which has been obtained by a circuit transformation proposed by them through which OTA-C circuits can be transformed into CFOA-RC circuits.

A systematic method of realization of low frequency oscillators was proposed by Elwakil [41], which requires a passive resistor and one active (negative) resistor which modifies the expression for oscillation-frequency such that it contains a difference term. By keeping this difference term as small as possible, low frequency oscillations can be achieved.

Three types of sinusoidal oscillators each using three CCII+, two to four resistors and two grounded capacitors have been presented by Martinez et al. in [62], whose practical workability has been verified by AD844 type CFOAs.

Apart from the above, Soliman in [42] has presented a number of sinusoidal oscillators out of which only one is two-resistor-two capacitor-Two CFOA-based circuit which is a quadrature oscillator. The other two sinusoidal oscillators use two CFOA-4R-2C configurations having identical expressions for CO and FO. However, in these circuits, only CO can be controlled by a single resistance.

5.7 Design of Active-R SRCOs

By taking into account the parasitic capacitance of the Z-terminal of the CFOA into design, a variety of active-R oscillators have been reported in technical literature. Some prominent circuits in this category are highlighted in this section.

5.7.1 Active-R Sinusoidal Oscillators Using CFOA-Pole

Such circuits were first presented by Liu et al. in [34], an exemplary circuit out of which is shown in Fig. 5.11.

If $R_1, R_3 \gg R_x$ and $R_2, R_4 \ll R_p$, the condition of oscillation of this oscillator is given by

$$\frac{R_6}{R_3(R_5 + R_6)} = \frac{1}{R_4} + \frac{1}{R_2} \tag{5.33}$$

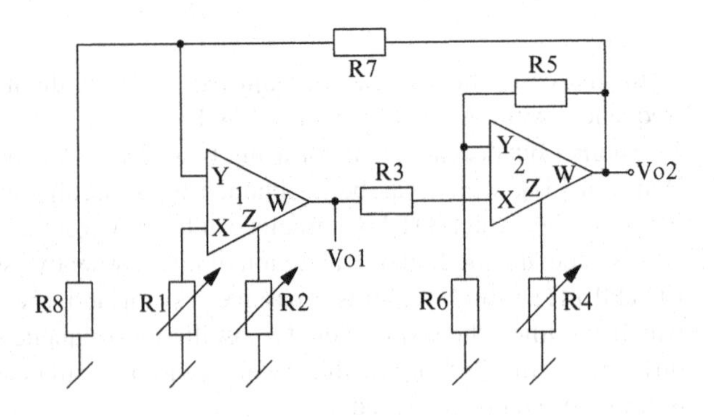

Fig. 5.11 Active-R sinusoidal oscillator (adapted from [34] © 1994 Taylor & Francis)

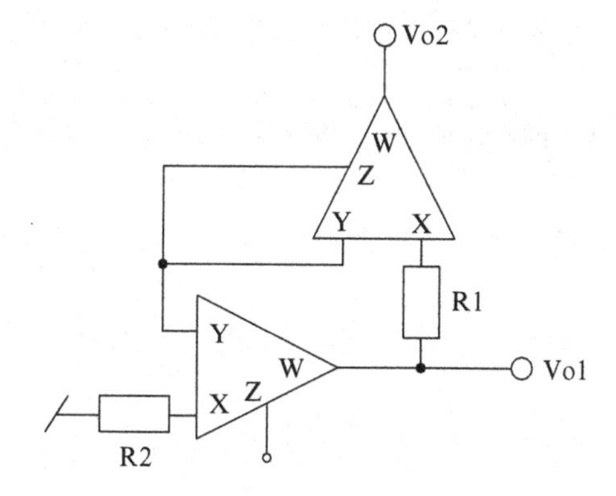

Fig. 5.12 Active-R SRCO proposed by Singh and Senani (adapted from [43] © 2001 IEEE))

whereas the frequency of oscillation is given by

$$\omega_0 = \frac{1}{C_p}\sqrt{\frac{1}{R_2R_4} + \frac{R_8}{R_1R_3(R_7+R_8)} - \frac{R_6}{R_2R_3(R_5+R_6)}} \tag{5.34}$$

It is, thus, seen that the oscillation frequency can be independently controlled by the resistor R_1 which does not appear in the condition of oscillation.

5.7.2 Low-Component-Count CFOA-Pole Based Active-R SRCOs

Two such circuits, each employing two CFOAs and only two resistors, were proposed by Singh and Senani in [43], one of which is shown in Fig. 5.12.

The CO and FO for this circuit are given by

$$R'_2 = \frac{R_p}{\left(\frac{2R_p}{R_y}+1\right)+\left(1+\frac{2C_y}{C_p}\right)} \tag{5.35}$$

and

$$f_0 = \frac{1}{2\pi C_pR_p}\sqrt{\frac{\frac{2R_p}{R_y}+\frac{R_p}{R'_2}\left(\frac{R_p}{R'_1}-1\right)+1}{\left(1+\frac{2C_y}{C_p}\right)}} \tag{5.36}$$

where $R'_1 = (R_1 + R_{x1})$, $R'_2 = (R_2 + R_{x2})$, R_p, C_p are the Z-port parasitics, R_y, C_y are the y-port parasitics and R_x is the x-port parasitic input resistance.

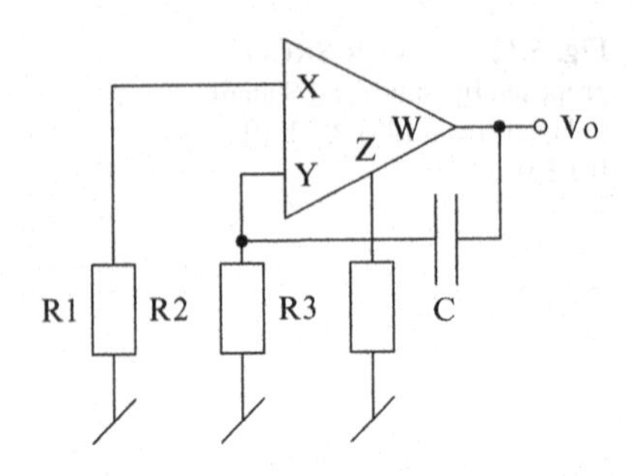

Fig. 5.13 CFOA-pole based sinusoidal oscillator introduced by Liu et al. (adapted from [34] © 1994 Taylor & Francis)

5.7.3 *Other Two-CFOA Based Active-R SRCOs*

A number of Two-CFOA active-R SRCO circuits can be easily obtained from those oscillators presented in Sects. 5.6 and 5.7 where each CFOA has a capacitor connected from its Z-pin to ground. Thus, from such circuits, active-R VCOs can be obtained by deleting the external capacitors connected at the Z-terminals of the CFOAs and employing in their places the Z-pin parasitic capacitances in the design.

5.7.4 *CFOA-Pole-Based RC Oscillator*

A low component oscillator using one external capacitor and the pole of the CFOA was presented by Liu et al. in [34] which is shown in Fig. 5.13.

If $R_1 \gg R_x$ and $R_3 \ll R_p$, the CO and FO for this circuit are given by

$$\frac{C}{R_1} = \frac{C}{R_3} + \frac{C_p}{R_2} \tag{5.37}$$

$$\omega_0 = \sqrt{\frac{1}{C_p C R_2 R_3}} \tag{5.38}$$

Although this oscillator has the advantage of using only four passive elements, but oscillation frequency cannot be independently controlled.

On the other hand, four sinusoidal oscillators each consisting of two capacitors, a single CFOA with its pole accounted in the design, were proposed by Abuelma'atti et al. in [21]. A single resistance controlled oscillator out of this set is shown in Fig. 5.14 for which the CO and FO are given by

$$\frac{1}{C_p}(C_4R_3 - C_2R_1) = R_1\left(1 + \frac{C_4R_3}{C_pR_p}\right) \tag{5.39}$$

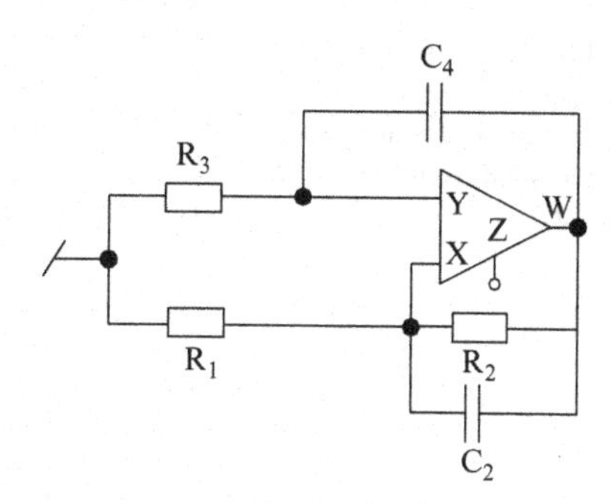

Fig. 5.14 A SRCO using CFOA-pole proposed by Abuelma'atti-Farooqi-Alshahrani (adapted from [21] © 1996 IEEE)

and

$$\omega_0 = \sqrt{\frac{1}{C_p C_4 R_p R_3}\left(1 + \frac{R_p}{R_2}\right)} \tag{5.40}$$

It is thus seen that in this circuit CO can be controlled by R_1 whereas FO is independently controlled by R_2. It has been demonstrated in [21] that using $R_1 = R_3 = 100\ \Omega$, $C_2 = 4.7$ pF, $C_4 = 22$ pF and using AD844 type CFOA biased with ±15 V DC power supplies, variable frequency oscillations up to 27.5 MHz, with peak to peak voltage of 2–9 V, were successfully obtained using this circuit.

Yet another variable frequency oscillator consisting of single CFOA-pole and single capacitor was proposed by Martinez et al. in [13], which enjoys independent tunability of oscillation frequency and condition of oscillation. In the same year, Abuelma'atti and Al-Shahrani in [23] and Abuelma'atti and Farooqi in [26] also proposed a number of SRCOs using the CFOA-pole.

It is worth pointing out that in [49], Nandi has reported an interesting active-R oscillator using a building block termed as CFA (realized from two CFOAs of the normal kind) along with only three external resistors which is capable of generating sinusoidal oscillations over a tuning range of 2.8–40 MHz using AD844 type CFOAs.

5.7.5 *A Simple Multiphase Active-R Oscillator Using CFOA Poles*

A multiphase oscillator circuit was proposed by Wu et al. in [32] which is shown in Fig. 5.15. This circuit has the following merits: (1) uses only parasitic poles of CFOAs thereby making it suitable for high frequency oscillations and monolithic IC fabrication due to complete elimination of external capacitors, (2) exhibits large output voltage swing and (3) has moderately low Total Harmonic Distortion (THD).

Assuming all the CFOAs to be identical, the loop gain for an n-phase oscillator can be expressed as

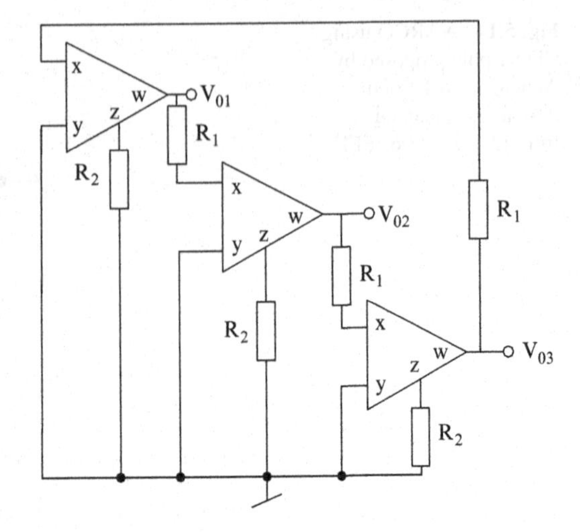

Fig. 5.15 Multiphase oscillator (for n = 3) proposed by Wu et al. (adapted from [32] © 1995 IEE)

$$L(s) = \left(\frac{-G_0}{1 + \frac{s}{\omega_b}}\right)^n \quad \text{where } \omega_b = 1/\mathrm{R_b C_p}, \tag{5.41}$$

$$\mathrm{R_b = R_2//R_p \ \ and \ \ G_0 = R_b/R_a, \quad R_a = (R_1 + R_x)}$$

The frequency and condition of oscillation are given by

$$\omega_0 = \omega_b \tan\frac{\pi}{n} \ \text{ and } \ R_b \geq R_a \sec\frac{\pi}{n} \tag{5.42}$$

As a special case, the condition of oscillation, for frequency of oscillation of a three phase sinusoidal oscillator (n = 3), can be expressed as

$$\omega_0 = \omega_b\sqrt{3} \text{ and } R_b \geq 2R_a \text{ or } R_1 \leq \frac{R_2 R_p}{2(R_2 + R_p)} - R_x \tag{5.43}$$

Thus, the circuit of Fig. 5.15 produces the maximum and the minimum oscillation frequencies when $R_1 = 0$ and $R_1 = (R_p/2) - R_x$ respectively.

5.8 SRCOs Providing *Explicit* Current Output

In view of the proliferation of current-mode filters and other signal processing circuits, the design of oscillators providing an explicit-current-output (ECO) from a high output impedance node has also become important. Sinusoidal oscillators with ECO would be useful as signal generators to test various current-mode circuits. Although, there have been a number of investigations [12, 46, 76–79] on realizing oscillators with ECO using other building blocks, such as first generation current conveyor [46, 76], differential difference current conveyors [77], differential difference complementary current feedback amplifier [78], four terminal floating nullor [12], unity-gain voltage and current followers [79], however, none of these building blocks are available commercially yet.

On the other hand, because a CFOA of AD844 type does have a current output terminal and is commercially available, ECO oscillators made from CFOAs are of practical importance. In this section, we show how the state-variable approach of synthesis [36] of oscillators can be extended to synthesize systematically current-mode sinusoidal oscillators with *explicit current output* using CFOAs as active building blocks. Of course, current-mode oscillators based on CCII+ can also be implemented by AD844 however; oscillators using exclusively CCII+ which have the capability providing explicit current output are known to employ three CCII+ whereas none of the circuits in earlier works [12, 76–79] have been realized with CFOAs.

The state variable methodology described earlier may be easily tailored to suit the evolution of the SRCOs with explicit current output. As already described in Sect. 5.5, the various conditions for non-interacting controls of CO and FO have some requirements which are repeated here for convenience:

(a) The expression of (a_{11} + a_{22}) should either not have terms containing R_2 or they should be cancelled out. Thus, in (a_{11} + a_{22}), there should be two terms left with opposite signs involving R_1 and R_3.
(b) Similarly, to have FO independent of R_1, the expression ($a_{11}a_{22}$ − $a_{12}a_{21}$) should either not have the terms containing R_1 or they should be cancelled out. Thus, FO should be a function of resistors R_2 and R_3 only (along with C_1 and C_2).

Let us now construct the required [A] matrix by choosing $a_{11} = \frac{1}{C_1R_1}$, $a_{22} = -\frac{1}{C_2R_3}$ which satisfy the requirement (a). Now, choosing $a_{12} = -\frac{1}{C_1}\left(\frac{1}{R_1} + \frac{1}{R_2}\right)$, $a_{21} = \frac{1}{C_2R_3}$, we can satisfy the requirement (b). The required [A] matrix, therefore, takes the following form:

$$[A] = \begin{bmatrix} \frac{1}{C_1R_1} & -\frac{1}{C_1}\left(\frac{1}{R_1} + \frac{1}{R_2}\right) \\ \frac{1}{C_2R_3} & -\frac{1}{C_2R_3} \end{bmatrix} \tag{5.44}$$

which results in the following CO and FO:

$$CO: \quad R_1 = \frac{C_2}{C_1}R_3 \tag{5.45}$$

$$\omega_0 = \frac{1}{\sqrt{C_1C_2R_2R_3}} \tag{5.46}$$

By substitution of (5.44) into (5.18), the following node equations are obtained

$$C_1 \frac{dx_1}{dt} = \frac{x_1 - x_2}{R_1} - \frac{x_2}{R_2} \tag{5.47}$$

$$C_2 \frac{dx_2}{dt} = \frac{x_1 - x_2}{R_3} \tag{5.48}$$

For meeting the specific objective of having an *explicit current-mode output*, we consider implementing (5.47) and (5.48) using CFOAs keeping in mind that the z-terminal of at least one of the CFOAs has to be left unutilized to make the current output available from a high output impedance node. The circuit, thus formulated from (5.47), (5.48) takes the form of circuit (a) of Fig. 5.16 where the mechanism of constructing the circuit can be understood by following the various current segments of the (5.47) and (5.48) as marked in the circuit of Fig. 5.16a. Two other circuits, similarly derived, are shown in Fig. 5.16b, c.

It may be seen that as intended, the explicit current output is available from the z-terminal of the first CFOA in all the three circuits.

Like most of the CFOA-based oscillators, the parasitics of the CFOA make the non-ideal expressions of the oscillation frequencies of the circuit of Fig. 5.16 to be different than their ideal counterparts. Thus, parasitics would limit the operation of the oscillators at higher frequencies. However, it has been shown in [64] that with judicious choice of component values, oscillations around 1 MHz range are attainable with these circuits. As an example, Fig. 5.17 shows a typical waveform (1.06 MHz, 2.3 V (p–p)) obtained from the oscillator of Fig. 5.16c using AD 844 CFOAs biased with ±15 V DC supplies.

Some other interesting explicit current output oscillators (ECO) using CFOA are discussed next.

1. Two single CFOA-based ECO oscillators were presented by Senani and Sharma in [47]. One of the circuits from [47] is shown in Fig. 5.18.
   The CO and FO for this circuit are given by

$$R_3 = 6(R_1 + R_2); \text{ provided } C_1 = C_2 = C_3 = C \tag{5.49}$$

$$f_0 = \frac{1}{2\pi C\sqrt{3R_1R_2}} \tag{5.50}$$

   Although the circuit has the advantage of using a single CFOA, a drawback of this circuit is that it has three capacitors. On the other hand, f_0 can be varied through a potentiometer by changing 'n' (which is the ratio R_1/R_2) while their sum ($R_1 + R_2$), and hence, the CO remains invariant. However, CO can be adjusted independently through the resistor R_3.
2. Another two-CFOA-based SRCO family with explicit current output has been proposed recently by Lahiri et al. [58], out of which an exemplary SRCO is

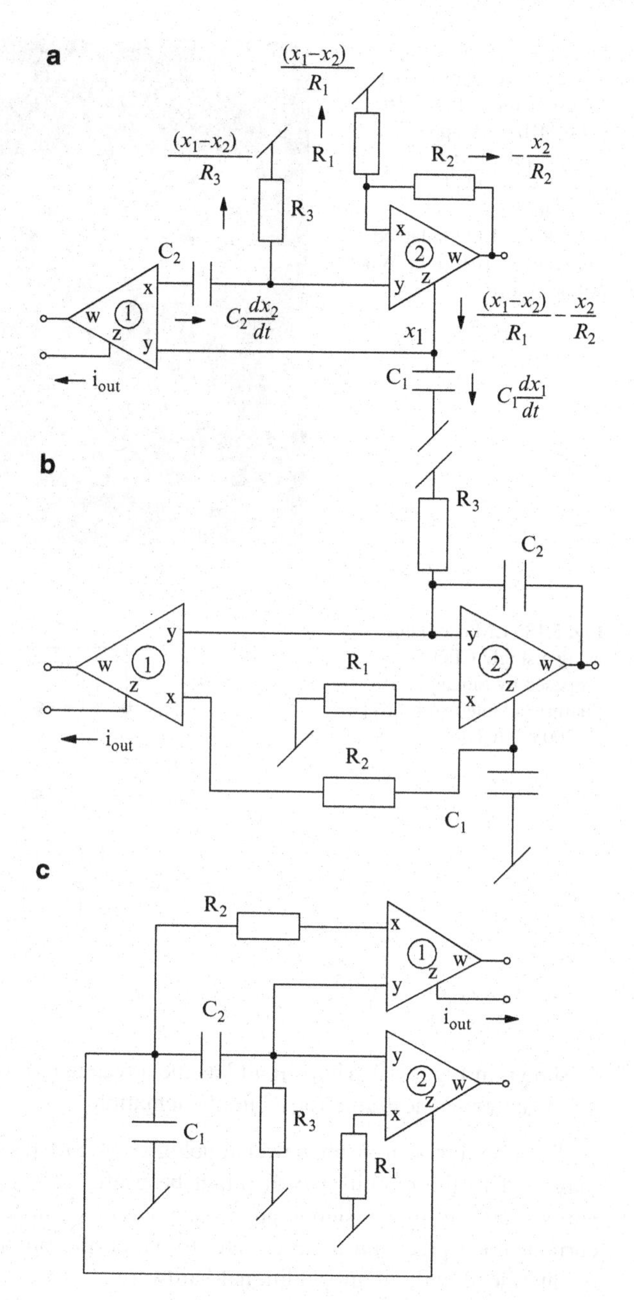

Fig. 5.16 Some exemplary SRCOs providing explicit current output (adapted from [64] © 2010 Jojn Wiley & Sons Ltd.)

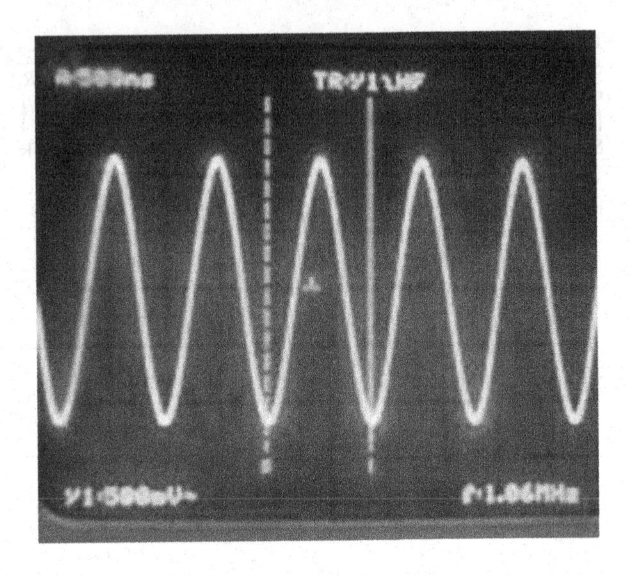

Fig. 5.17 A typical waveform generated from the oscillator of Fig. 5.16c (1.06 MHz, 2.3 Vpp). Component values: $C_1 = C_2 = 100$ pF, $R_1 = 404\ \Omega$, $R_2 = R_3 = 1$ kΩ (adapted from [64] © 2010 John Wiley & Sons Ltd.)

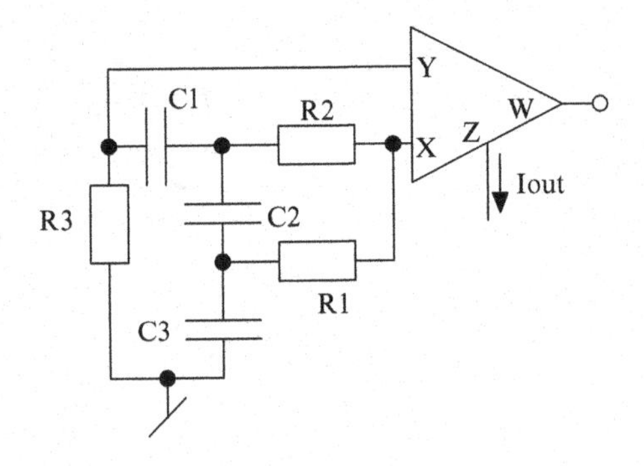

Fig. 5.18 CM oscillator using a single CFOA proposed by Senani and Sharma (adapted from [47] © 2005 IEICE)

shown in Fig. 5.19. This circuit has the advantage of employing both grounded capacitors as desirable for IC implementation.

The ongoing search for newer topologies of SRCOs with ECO using CFOAs may lead to the circuits which might be useful as test oscillators for verifying various current-mode signal processing circuits[2] such as current-mode filters, current-mode precision rectifiers etc. to which the proposed kind of circuits can be interfaced without any additional hardware.

[2] In spite of the criticism of [88], the current-mode techniques have given way to a number of important analog signal processing/signal generation circuits over the past three decades.

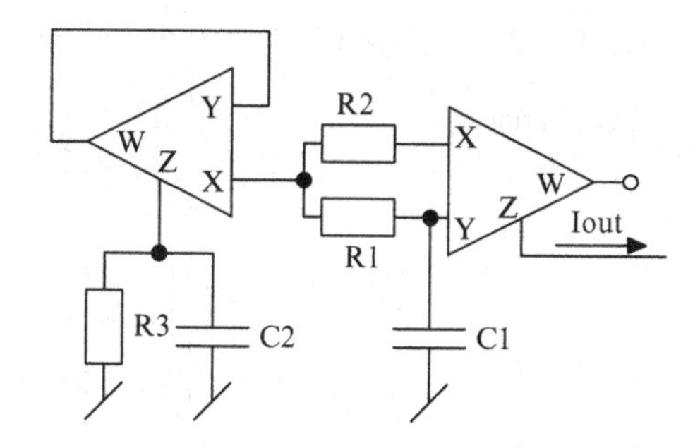

Fig. 5.19 Explicit-current output second order sinusoidal oscillator (adapted from [58] © 2011 Elsevier)

5.9 Fully-Uncoupled SRCOs Using CFOAs

CFOA-based canonic SRCOs which employ a minimum of five passive components, namely, three resistors and two grounded capacitors (as desirable from the view point of IC implementation) and possess tuning laws such that both CO and FO can be controlled /adjusted by two independent resistors, require at least two CFOAs. A major drawback of such topologies is that as soon as various non-idealities /parasitics of the CFOAs are accounted for, the theoretically derived independence of CO and FO vanishes due to the frequency-controlling resistor also getting involved in the non-ideal expression for the CO. Those oscillators are called fully-decoupled in which CO and FO are decided by two completely different sets of components, such that none of the components involved in CO are also involved in FO and vice versa. Such SRCOs are characterized by tuning laws of the type

$$\text{CO}: \quad (R_1 - R_2) \leq 0 \tag{5.51}$$

and

$$\text{FO} \quad f_0 = \frac{1}{2\pi}\sqrt{\frac{1}{C_1C_2R_3R_4}} \tag{5.52}$$

which shows that such circuits would, need at least four resistors along with two capacitors. Such ‘fully-uncoupled’ SRCOs, however, are not feasible with only two active elements and call for the employment of at least three active elements as in [42, 44].

There appear to be only two circuits known in earlier literature employing CFOAs which belong to the category of fully uncoupled oscillators, namely, the circuit presented by Soliman in [42] and the one proposed by Bhaskar in [44]. The Solimans’ circuit from [42] is shown in Fig. 5.20, whereas the circuit presented by Bhaskar [44] is shown in Fig. 5.21.

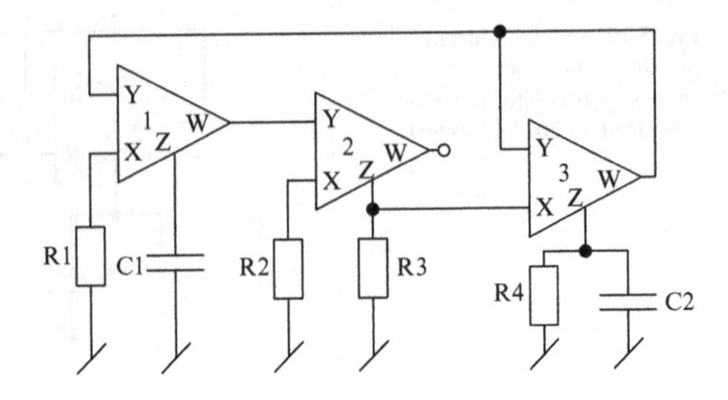

Fig. 5.20 Fully uncoupled oscillator (adapted from [42] © 2000 Springer)

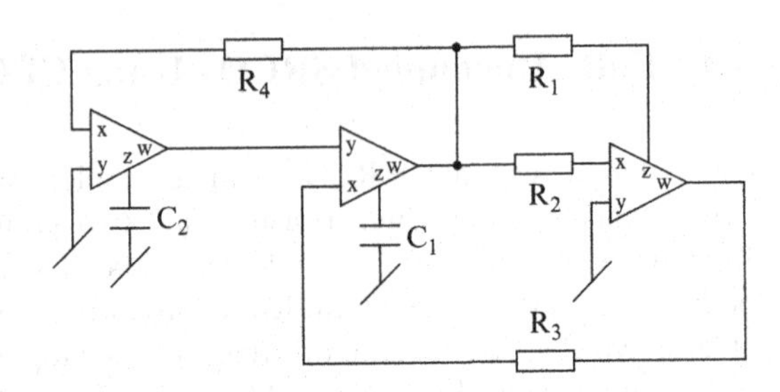

Fig. 5.21 Fully uncoupled oscillator (adapted from [44] © 2003 Frequenz)

It is interesting to note that both the circuits employ exactly the same number of active and passive components. The ideal CO and FO for the circuit of Fig. 5.20 are given by

$$R_3 = R_4$$

and

$$fo = \frac{1}{2\pi\sqrt{C_1C_2R_1R_2}} \tag{5.53}$$

whereas the CO and FO for the circuit of Fig. 5.21 are given by

$$R_1 = R_2$$

$$fo = \frac{1}{2\pi\sqrt{C_1C_2R_3R_4}} \tag{5.54}$$

It has been shown in [59] that the above described fully-uncoupled oscillators from [42, 44] also fail to retain the independent controllability of FO under the influence of non-ideal parasitic impedances of CFOAs as all the four resistors employed in the oscillators appear in the non-ideal expressions of both CO and FO, thereby completely disturbing the intended property.

In this section, we show two circuits from [59] which retain the independent controllability of FO even under the influence of CFOA parasitic impedances. These circuits are shown in Fig. 5.22.

Assuming that the CFOAs are characterized by: $i_y = 0,\ v_x = v_y,\ i_z = i_x$ and $v_w = v_z$, both the circuits are governed by a common characteristic equation (CE) given by:

$$s^2 + \frac{s}{C_1}\left(\frac{1}{R_0} - \frac{1}{R_1}\right) + \frac{1}{C_1 C_2 R_2 R_3} = 0 \tag{5.55}$$

From this characteristic equation, the CO and FO are found to be

$$\text{CO}: \quad (R_1 - R_0) \le 0$$
$$\text{FO}: \quad \omega_0 = \frac{1}{\sqrt{C_1 C_2 R_2 R_3}} \tag{5.56}$$

For an evaluation of the non-ideal performance of these circuits, we consider the finite input resistance R_{xi} at the x-port, i = 1–3, parasitic components R_{yi} in parallel with $1/sC_{yi}$ at the y-port and parasitic components R_{zi} in parallel with $1/sC_{zi}$ at the z-port of all the CFOAs i = 1–3. Analysis reveals that in both the cases, the non-ideal CE of both the circuits continues to remain second order. The non-ideal CO and FO for both the circuits have been found to be as under

For the circuits of Fig. 5.22a, b

CO:

$$(C_2 + C_{z2})\left\{\frac{1}{R_0} - \frac{1}{R_1 + R_{x1}} + \frac{1}{R_{y1}} + \frac{1}{R_{z1}} + \frac{1}{R_{z3}}\right\} + \frac{C_1 + C_{z1} + C_{y1} + C_{z3}}{R_{z2}} \le 0 \tag{5.57}$$

FO:

$$f'_0 = \frac{1}{2\pi}\sqrt{\frac{1}{C_1 C_2 R_2 R_3}}\left(\frac{1}{\left(1 + \frac{C_{y1}+C_{z1}+C_{z3}}{C_1}\right)\left(1 + \frac{C_{z2}}{C_2}\right)}\right)^{1/2} \left[\frac{1}{\left(1 + \frac{R_{x2}}{R_2}\right)\left(1 + \frac{R_{x3}}{R_3}\right)} + \frac{R_2 R_3}{R_{z2}}\left\{\frac{1}{R_0} + \frac{1}{R_{y1}} + \frac{1}{R_{z1}} + \frac{1}{R_{z3}} - \frac{1}{R_1 + R_{x1}}\right\}\right]^{1/2} \tag{5.58}$$

From (5.57)–(5.58), it is observed that in both the circuits, the frequency controlling resistors R_2 and R_3 do not come into the non-ideal expressions for CO; therefore, the independent controllability of FO remains intact even under the influence of the non-ideal parameters/parasitic of the CFOAs employed.

It is worth mentioning that if the circuits are to be converted into voltage controlled oscillators by replacing the frequency-controlling resistors R_2 and/or R_3 by FET-based or CMOS voltage-controlled-resistors (VCR), this does not pose

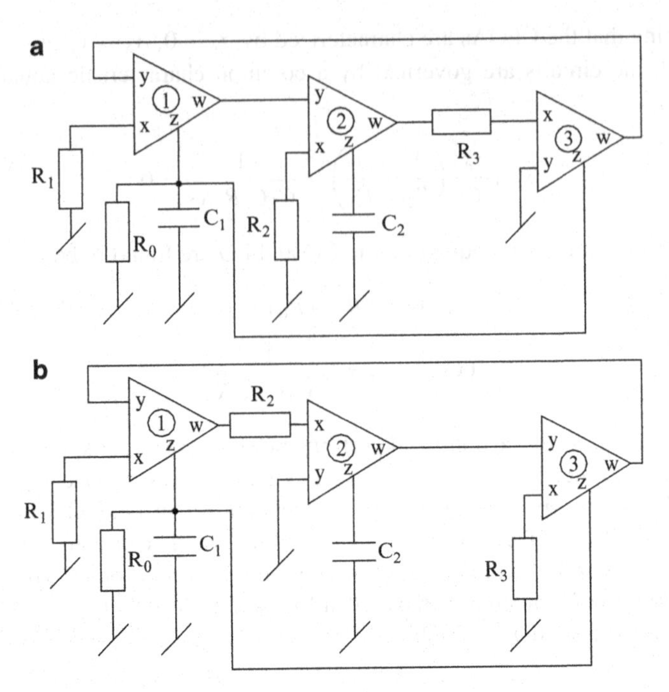

Fig. 5.22 Fully-uncoupled SRCOs proposed by Bhaskar et al. (adapted from [59] © 2012 Springer)

any difficulty since it is well known that grounded/floating VCRs using any of the above mentioned devices could be realized with exactly the same amount of hardware, for instance, see [69, 80–82].

Using the definition of frequency stability factor (S_F) $S_F = \left.\frac{d\phi(u)}{du}\right|_{u=1}$ where $u = \frac{\omega}{\omega_o}$ is the normalized frequency and $\phi(u)$ denotes the phase function of the open loop transfer function, with $C_1 = C_2 = C$, $R_0 = R_1 = R_2 = R$ and $R_3 = R/n$, S_F for the above oscillators is found to be $S_F = 2\sqrt{n}$. On the other hand, if both the resistors R_2 and R_3 are varied simultaneously i.e., $R_2 = R_3 = R/n$, then S_F becomes 2n. This figure appears to be the highest (like that of [44]) attained so far as compared to all SRCOs known so far. Thus, both the circuits of Fig. 5.22 offer very high frequency stability factors for larger values of n.

Thus, the circuits of Fig. 5.22 possess interesting and practically important properties not available in any of the earlier known CFOA-based sinusoidal oscillators.

Lastly, it must be mentioned that the generation of any new three-CFOA-two-GC-four-resistor fully-uncoupled oscillators which, apart from retaining independent controllability of FO, can also retain independent controllability of CO even under the influence of the non-ideal parameters/parasitics of the CFOAs, appears to be an interesting but challenging open problem.

5.10 Voltage-Controlled-Oscillators Using CFOAs and FET-Based VCRs

Voltage-controlled oscillators (VCOs) are important building blocks in many instrumentation, electronic and communication systems.

A well-known method of realizing a sinusoidal VCO is to realize a single-resistance-controlled-oscillator (SRCO) and then replace the frequency controlling resistor by a FET-based voltage-controlled-resistor (VCR). This section discusses a number of CFOA-based VCO configurations which offer different advantageous features.

A number of CFOA-based SRCOs have been described in the earlier sections of this chapter in which frequency of oscillation (FO) can be independently controlled through a single variable resistor without affecting the condition of oscillation (CO). Thus, in a specific CFOA-based SRCO, this frequency controlling resistor may be either grounded (having one terminal connected to ground) or floating (none of the resistor terminals connected to ground).

From a careful examination of the family of 14 SRCOs presented earlier in [38], it is found that in all, only seven structures are suitable for being converted into VCOs by replacing frequency controlling resistors by appropriate grounded and floating VCRs. Out of this set of seven, only five circuits can be realized using no more than two CFOAs, which are shown here in Fig. 5.23. The CO and FO of these VCOs are given in Table 5.1. It may be noted that in all the cases, the CO can be adjusted by R_1 without affecting FO whereas FO is controllable independently by the resistance Rm and hence, by V_C.

Taking into consideration the finite X-terminal input resistance Rx and parasitic impedance at the Z-terminal (consisting of a resistance Rp in parallel with a capacitance Cp) it is found that the influence of CFOA parasitics on the performance of these oscillators can be reduced by choosing external resistances to be much greater than Rx and much smaller than Rp and selecting external capacitors to be much larger than Cp.

From the frequency stability analysis it has been found [50] that all the VCOs of Fig. 5.23 enjoy good frequency stability properties.

Some experimental results for the oscillators of Fig. 5.23a, e from [50] are shown here in Figs. 5.24, 5.25 and 5.26.

5.11 State-Variable Synthesis of Linear VCOs Using CFOAs

In this section, we describe a systematic approach of synthesizing sinusoidal *linear* VCOs (i.e. an oscillator providing *linear* tuning law of the type $f_0 \propto V_C$ between the oscillation frequency f_0 and an external control voltage V_C).

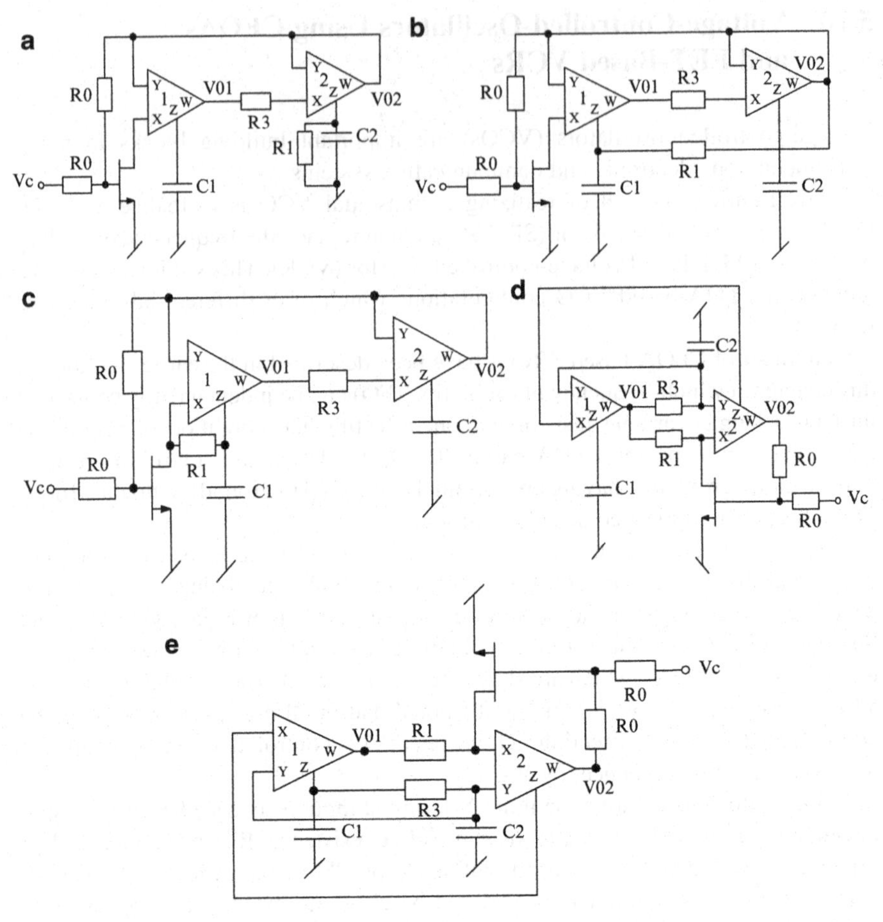

Fig. 5.23 Voltage controlled oscillators proposed by Gupta et al. (adapted from [50] © 2009 Elsevier)

Table 5.1 CO and FO for the oscillators of Fig. 5.23

VCO number	Condition of oscillation	Frequency of oscillation
(a)	$R_1 = R_3$	$f_0 = \frac{1}{2\pi\sqrt{C_1C_2R_mR_3}}$
(b), (d)	$R_1 = \frac{C_2}{C_1}R_3$	$f_0 = \frac{1}{2\pi\sqrt{C_1C_2R_mR_3}}$
(c)	$R_1 = 2\frac{C_2}{C_1}R_3$	$f_0 = \frac{1}{2\pi\sqrt{C_1C_2R_mR_3}}$
(e)	$R_1 = R_3\left(\frac{C_1+C_2}{C_2}\right)$	$f_0 = \frac{1}{2\pi\sqrt{C_1C_2R_mR_3}}$

where $R_m = r_{DS} = \frac{2V_p^2}{I_{DSS}(V_C-2V_p)}$

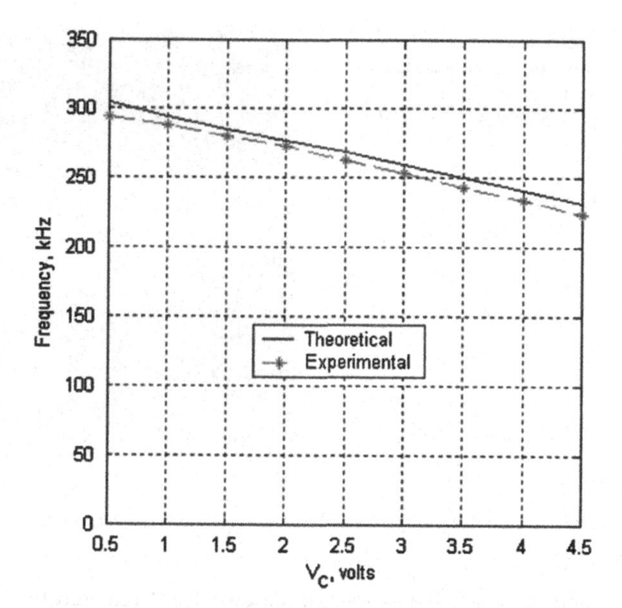

Fig. 5.24 Variation of oscillation frequency with control voltage V_C for the VCO (a) of Fig. 5.23 (adapted from [50] © 2009 Elsevier).

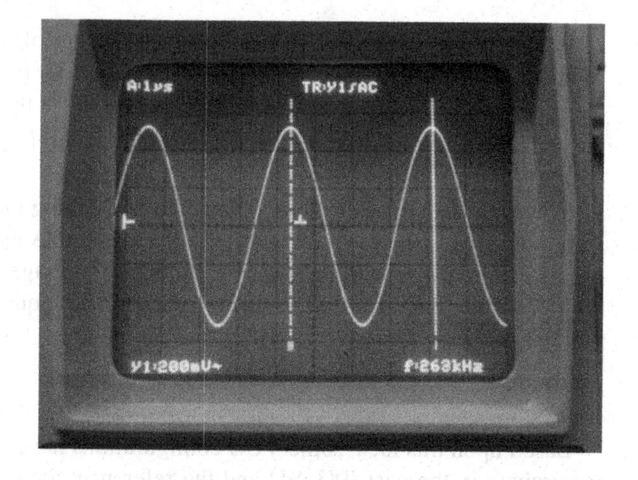

Fig. 5.25 A typical waveform generated from VCO (a) of Fig. 5.23, $f_0 = 263$ kHz, $V_0 = 1$ V (p–p); $V_{CC} = \pm 6$ V DC (adapted from [50] © 2009 Elsevier)

It may be noted that the VCOs, presented in Sect. 5.11 are although simple, however, they do not provide a linear tuning law between the control voltage (say, V_C) and the oscillation frequency f_0, since the tuning law for such VCOs is of the form

$$f_0 \propto \sqrt{\frac{1}{r_{ds}}} \tag{5.59}$$

where $r_{ds} = \dfrac{V_p^2}{I_{DSS}(V_c - 2V_p)}$ (in case of VCR realized by JFET) (5.60)

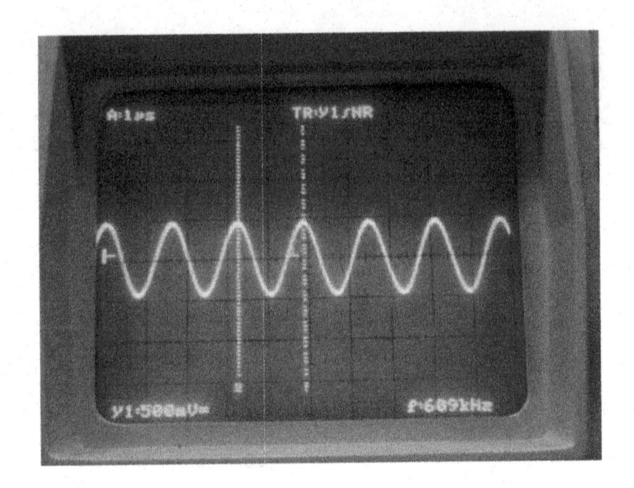

Fig. 5.26 A typical waveform generated from VCO (e) of Fig. 5.23, $f_0 = 609$ kHz, $V_0 = 1$ V (p–p); $V_{CC} = \pm 5$ V DC (adapted from [50] © 2009 Elsevier)

$$\text{and } r_{ds} = \frac{1}{2k(V_{gs} - V_{th})} \quad \text{(in case of VCR realized by MOSFET)} \tag{5.61}$$

In (5.60) and (5.61), V_p is the pinch off voltage of the JFET and I_{DSS} is the saturated drain current (at $V_{gs} = 0$) of the FET, and V_C is the control voltage, V_{th} is the threshold voltage and $K = \mu_S C_{OX}\left(\frac{W}{L}\right)$ where μ_s is the surface mobility, C_{ox} is the capacitance of the gate electrode per unit area and $\left(\frac{W}{L}\right)$ is the aspect ratio of the MOSFET.

However, if an oscillator is evolved with two analog multipliers (AM) appropriately embedded into a circuit configuration, to enable independent control of the oscillation frequency through an external control voltage V_C applied as a common multiplicative input to both the multipliers, this technique may give rise to a linear tuning law of the form

$$f_0 \propto V_C \tag{5.62}$$

Based upon this idea, some VCO configurations have been proposed by various researchers in the past ([83–85] and the reference cited therein) employing traditional voltage-mode op-amps (VOA) and AMs. These circuits, however, require larger number of resistors (5–12) and their usability is limited to low frequency ranges due to the finite GBP and limited slew rate of VOAs.

In [65] Gupta et al. derived a class of VCOs using state variable technique through which different circuits could be generated by making different choices of the parameters of the [A] matrix (i.e. a_{11}, a_{12}, a_{21} and a_{22}) of the state variable characterization.

Based upon the already described technique let us construct the [A] matrix of the oscillator to be synthesized in the following form:

$$[A]_1 = \begin{bmatrix} 0 & -\frac{1}{C_1 R_2} \\ \frac{1}{C_2 R_3} & \frac{1}{C_2}\left(\frac{1}{R_1} - \frac{1}{R_3}\right) \end{bmatrix} \tag{5.63}$$

The CE formulated from the above matrix gives the following CO and FO:

$$CO: \quad R_1 = R_3 \tag{5.64}$$

$$\text{FO}: \quad \omega_0 = \frac{1}{\sqrt{C_1 C_2 R_2 R_3}} \tag{5.65}$$

The oscillators to be derived have to incorporate at least two analog multipliers (characterized by $V_o = K\left(\frac{V_1 V_2}{V_{ref}}\right)$, where V_1 and V_2 are two inputs, V_{ref} is the reference voltage set internally, usually at 10 V in case of AD534 and K can be set up +1 or −1 by grounding appropriate input terminals). In order to provide linear control of oscillation frequency through an external control voltage V_C (to be applied as a common multiplicative input to both the analog multipliers) the selection of the matrix parameters outlined above needs to be modified to include the term β $\left(\beta = \frac{V_C}{V_{ref}}\right)$. However, it needs to be done in such a way that the final expression of the CO does not contain the term β and the expression of FO is modified to $f_0 = \frac{\beta}{2\pi\sqrt{C_1 C_2 R_2 R_3}}$ so that we can get $f_0 \propto \beta$ and hence, $f_0 \propto V_C$.

The parameters of the matrix [A] given in (5.63) can now be *modified* in one of the following ways:

(i) By including β^2 as a factor of a_{12} or a_{21}
(ii) β as a factor a_{12} as well as that of a_{21}
(iii) β as a factor in all the parameters of matrix [A]

Using the *modification* (i) we get the following node equations

$$C_1 \frac{dx_1}{dt} = -\frac{\beta^2 x_2}{R_2} \tag{5.66}$$

$$C_2 \frac{dx_2}{dt} = \frac{x_1 - x_2}{R_3} + \frac{x_2}{R_1} \tag{5.67}$$

If we employ two AMs and two CFOAs and try to implement above NEs we can synthesize VCO-1 as shown in Fig. 5.27. Various current components of (5.66) and (5.67) have been marked in VCO-1 to make the synthesis clear.

If we apply *modification* (iii), we get the following node equations:

$$C_1 \frac{dx_1}{dt} = -\frac{\beta x_2}{R_2} \tag{5.68}$$

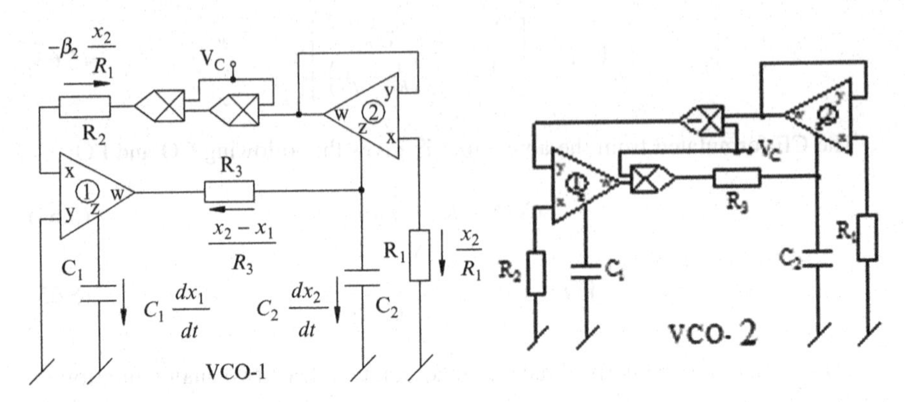

Fig. 5.27 VCOs derived from matrix [A] CO: $R_1 \geq R_3$, FO: $f_0 = \frac{\beta}{2\pi\sqrt{C_1C_2R_2R_3}}$ (adapted from [65] © 2011 World Scientific Publishing Company)

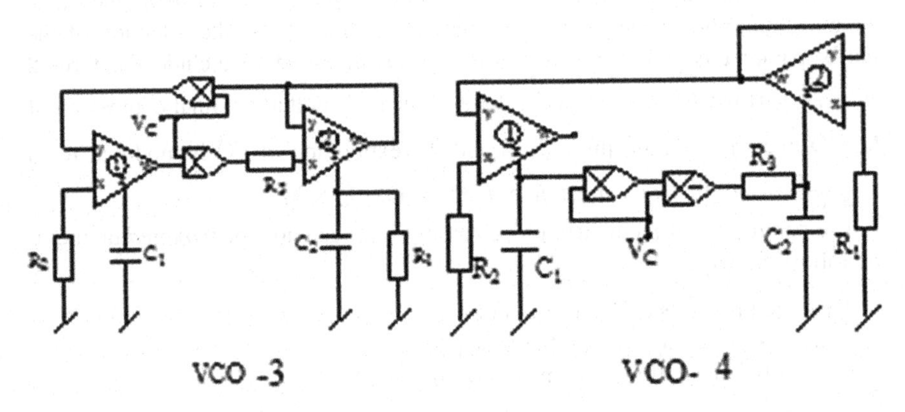

Fig. 5.28 Additional VCOs CO: $R_1 \geq R_3$, FO: $f_0 = \frac{\beta}{2\pi\sqrt{C_1C_2R_2R_3}}$ (adapted from [65] © 2011 World Scientific Publishing Company)

$$C_2 \frac{dx_2}{dt} = \frac{\beta\, x_1 - x_2}{R_3} + \frac{x_2}{R_1} \tag{5.69}$$

The implementation of (5.68) and (5.69) gives us VCO-2 which is shown in Fig. 5.27. It may be noted that if polarity of β is inverted in both the AMs, the synthesized circuit still remains VCO with the same CO and FO. Based on the state variable methodology explained above, along with any or all the modifications (i)–(iii) suggested above, a number of VCOs have been generated from the suitable matrices in [65]. Two other circuits from the set of 12 VCOs generated in [65] are shown here as VCO-3 and VCO-4 in Fig. 5.28.

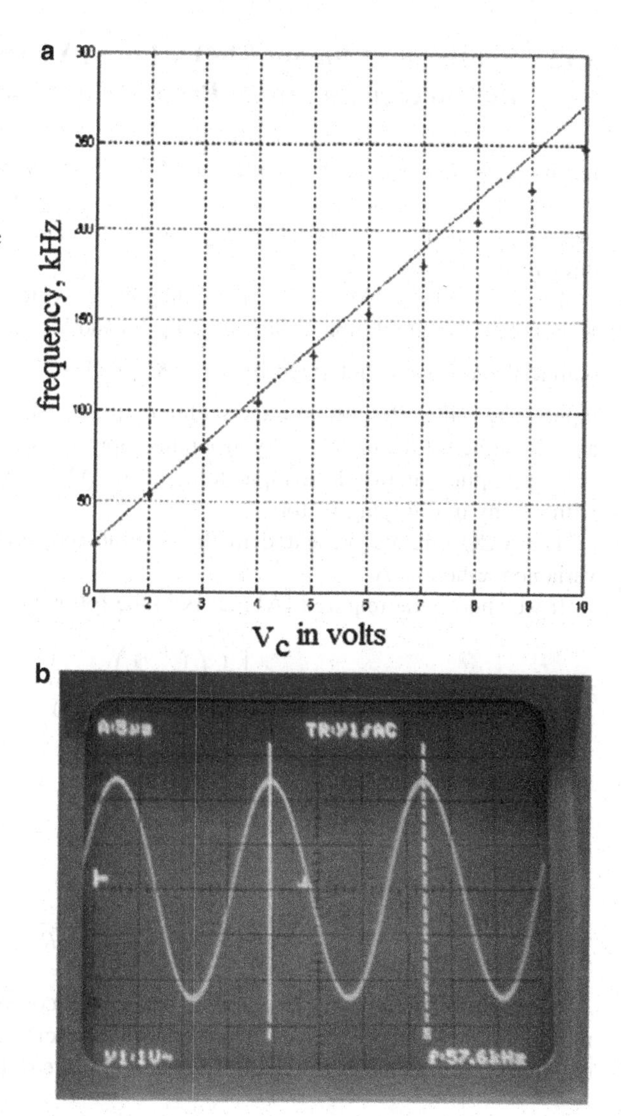

Fig. 5.29 Experimental results for VCO-1 of Fig. 5.27. (**a**) Variation of oscillation frequency with control voltage V_C. (**b**) A typical waveform generated for $V_C = 2$ V (adapted from [65] © 2011 World Scientific Publishing Company)

The expressions for the frequency stability factors for all the VCOs have been evaluated in [65] and it has been found that in all the VCOs, S^F can be made large.

Variation of oscillation frequency with control voltage V_C (for $C_1 = C_2 = 50$ pF, $R_2 = R_3 = 10\,\text{k}\Omega$, with DC biasing $\pm V_{CC} = \pm 6$ V for CFOAs and $\pm V_{CC} = \pm 15$ V AMs) and a typical waveform generated by VCO-1 of Fig. 5.27 is shown in Fig. 5.29.

5.12 Synthesis of Single-CFOA-Based VCOs Incorporating the Voltage Summing Property of Analog Multipliers

In this section, we present a family of CFOA-based VCOs which employ a bare minimum number of active and passive components namely only one CFOA, only two multipliers (essential for obtaining linear control of oscillation frequency), two/three resistors and two capacitors.

The AD534 type AM is a 4-port building block symbolically shown in Fig. 5.30 with three differential inputs (shown as V_1, V_2 and V_{Z1} in the present case) and one output V_0 and is characterized by $V_o = K\left(\frac{V_1 V_2}{V_{ref}}\right) + V_Z$ where V_1 and V_2 are two inputs, V_{ref} is the reference voltage set internally, usually at 10 V in case of AD534 and K can be set up +1 or −1 by grounding appropriate input terminals, V_Z is the voltage applied at the third input terminal of AM which appears at the output without any multiplying factor.

The VCO circuits presented in this section too are derived using the state-variable methodology.

If we choose the required [A] matrix in the following form:

$$[A] = \begin{bmatrix} \frac{1}{C_1}\left(\frac{1}{R_3} - \frac{1}{R_1}\right) & \frac{1}{C_1 R_2} \\ -\frac{1}{C_2 R_3} & 0 \end{bmatrix} \tag{5.70}$$

The CE formulated from the above matrix gives the following CO and FO:

$$CO: \quad R_1 = R_3 \tag{5.71}$$

$$FO: \quad \omega_0 = \frac{1}{\sqrt{C_1 C_2 R_2 R_3}} \tag{5.72}$$

Since the oscillators to be derived have to incorporate at least two analog multipliers in order to provide linear control of oscillation frequency through an external control voltage V_C which is applied in place of the second input V_2 of the

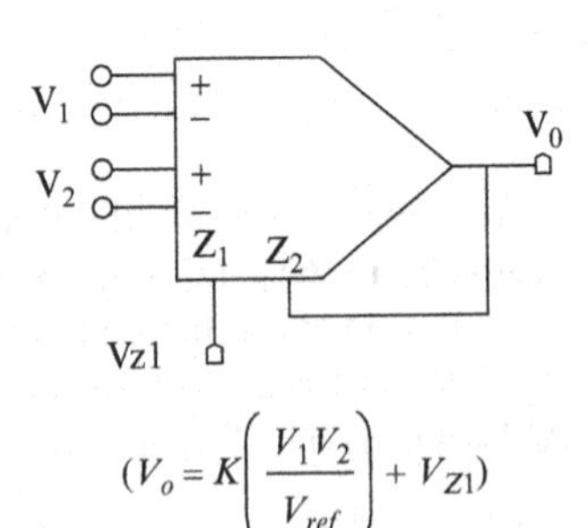

Fig. 5.30 Symbolic notation of an Analog Multiplier of AD 534 type

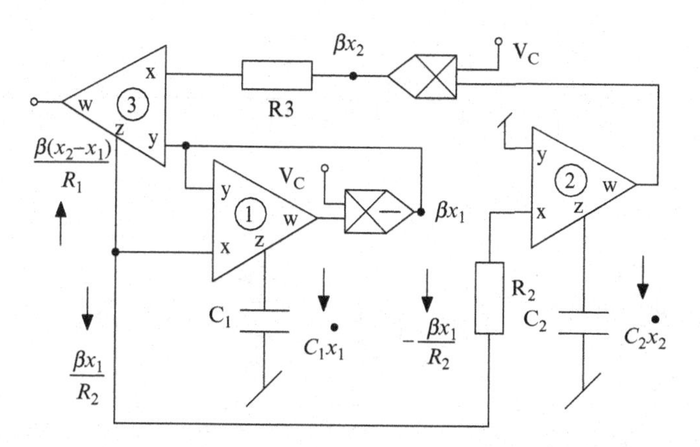

Fig. 5.31 VCO derived from matrix [A] (CO: $R_1 = R_2$, FO: $f_0 = \frac{\beta}{2\pi}\sqrt{\frac{1}{C_1C_2R_1R_2}}$) (adapted from [66])

multipliers (to be applied as a common multiplicative input to both the analog multipliers), the selection of the matrix parameters outlined above needs to be modified to include the term β $\left(\beta = \frac{V_C}{V_{ref}}\right)$. It should be done in such a way that the final expression of the CO does not contain the term β but the expression of FO is modified to $f_0 = \frac{\beta}{2\pi\sqrt{C_1C_2R_2R_3}}$ so that we can get $f_0 \propto \beta$ and hence, $f_0 \propto V_C$.

Consider now the following [A] matrix

$$[A]_2 = \begin{bmatrix} \frac{\beta}{C_1}\left(\frac{1}{R_2} - \frac{1}{R_1}\right) & \frac{\beta}{C_1R_1} \\ -\frac{\beta}{C_2R_2} & 0 \end{bmatrix} \tag{5.73}$$

From matrix [A], we get the following node equations

$$C_1\dot{x}_1 = \frac{\beta(x_2 - x_1)}{R_1} + \frac{\beta x_1}{R_2} \tag{5.74}$$

$$C_2\dot{x}_1 = -\frac{\beta x_1}{R_2} \tag{5.75}$$

Implementation of the above NEs employing two multipliers results in the circuit shown in Fig. 5.31. Various current components of (5.74) and (5.75) have been marked in the circuit to make the synthesis clear.

It may be noted that here we require three CFOAs along with two AMs to implement NEs of (5.74) and (5.75). We now show that by an alternative representation of (5.74) and (5.75) and an appropriate incorporation of the z-terminal of AM it becomes possible to implement the modified NEs with only a single CFOA. Let us add (5.74) and (5.75) to create a new equation ((5.76) in the following) while we keep (5.75) as it is (shown as (5.77) in the following):

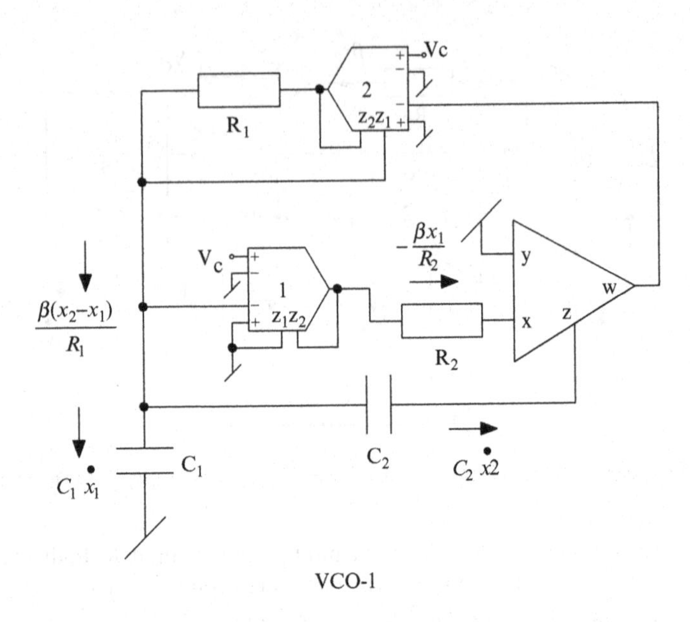

Fig. 5.32 VCO derived from matrix [A] (adapted from [66])

$$C_1\dot{x}_1 + C_2\dot{x}_2 = \frac{\beta(x_2 - x_1)}{R_1} \tag{5.76}$$

$$C_2\dot{x}_2 = -\frac{\beta x_1}{R_2} \tag{5.77}$$

It is interesting to note that when the voltage summing property of the AMs is effectively utilized the implementation of (5.76) and (5.77) then leads to a different circuit (shown as VCO-1 in Fig. 5.32) which requires only a single CFOA in contrast to the circuit of Fig. 5.31 needing three CFOAs.

It has been shown in [66] that in addition to (5.73), the following matrices are suitable for the synthesis of such kind of VCOs.

$$[A]_2 = \begin{bmatrix} \frac{\beta}{C_1}\left(\frac{1}{R_2} - \frac{2}{R_1}\right) & \frac{\beta}{C_1 R_1} \\ -\frac{\beta}{C_2 R_2} & 0 \end{bmatrix} \tag{5.78}$$

$$[A]_3 = \begin{bmatrix} \frac{\beta}{C_1 R_1} & -\frac{\beta}{C_1 R_1} \\ \frac{2\beta}{C_2 R_2} & -\frac{\beta}{C_2 R_2} \end{bmatrix} \tag{5.79}$$

$$[A]_4 = \begin{bmatrix} \frac{\beta^2}{C_1 R_1} & -\frac{\beta^2}{C_1 R_1} \\ \frac{1+\beta^2}{C_2 R_2} & -\frac{\beta^2}{C_2 R_2} \end{bmatrix} \tag{5.80}$$

$$[A]_5 = \begin{bmatrix} 0 & \dfrac{\beta}{C_1R_2} \\ -\dfrac{\beta}{C_2R_3} & \dfrac{1}{C_2}\left(\dfrac{1}{R_3} - \dfrac{1}{R_1}\right) \end{bmatrix} \tag{5.81}$$

The circuits resulting from the synthesis based upon the above matrices are shown in Fig. 5.33.

The following may now be noted:

- VCOs 3 and 4 offer the use of both grounded capacitors as desirable for IC implementation and out of these VCO 4 also has one of the CO controlling resistor R_2 grounded.
- The VCO 5 possesses simultaneously almost all the desirable features namely, completely non-interacting control of CO through R_1 (the CO controlling resistor being grounded), the employment of both grounded capacitors and an additional degree of freedom via R_2 to scale up or down the frequency f_0 which is otherwise linearly controllable by β.

The prominent non-idealities of the CFOAs include—a finite non-zero input resistance R_x at port-X (typically around 50 Ω), y-port parasitic consisting of a parasitic resistance R_y (typically 2 MΩ) in parallel with a parasitic capacitance C_y (typically 2 pF) and Z-port parasitic impedance consisting of a parasitic resistance R_p (typically 3 MΩ) in parallel with a parasitic capacitance C_p (typically, between 4 and 5 pF). In case of an analog multiplier, the finite non-zero output resistance r_{out}, as per datasheet of AD534, is merely 1 Ω and hence, can be ignored in all the cases. On the other hand, the input impedance of the AM, being 10 MΩ, is sufficiently high and hence, its effect can be ignored. The errors caused by the influence of CFOA parasitics can be kept small by choosing all external resistors to be much larger than R_x but much smaller than R_p and choosing both external capacitors to be much larger than C_p.

A non-ideal analysis carried out in [66] shows that the independent control of CO and FO remains intact for VCO-5 even after consideration of the parasitics. Hence, VCO-5 is the best circuit from this viewpoint.

All the VCOs have been experimentally studied in [66] using AD844 type CFOAs and AD534 type AMs biased with ±12 V DC power supplies. The component values chosen were as under: For VCOs $R_1 = R_2 = 2$ kΩ, those for VCO-2 were chosen as $R_1 = 2$ kΩ, $R_2 = 1$ kΩ and for VCO-5 $R_1 = R_2 = R_3 = 1$ kΩ. Capacitor values for all the VCOs were taken as C1 $= C_2 =$1 nF. As per [66] it has been possible to generate oscillation frequencies from tens of kHz to several hundreds of kHz with tolerable errors in the frequency.

In the absence of an automatic amplitude control, it is normally expected that amplitude of oscillation would also vary when the frequency is varied through the external control voltage V_C. This has indeed been the case for VCOs 1, 2. However, in case of VCOs 3, 4 and 5, the peak-to-peak output voltage has been found to be constant, 17 $V_{p\text{-}p}$ in case of VCO 3 and 4 and 10 $V_{p\text{-}p}$ in case of VCO 5 when V_C

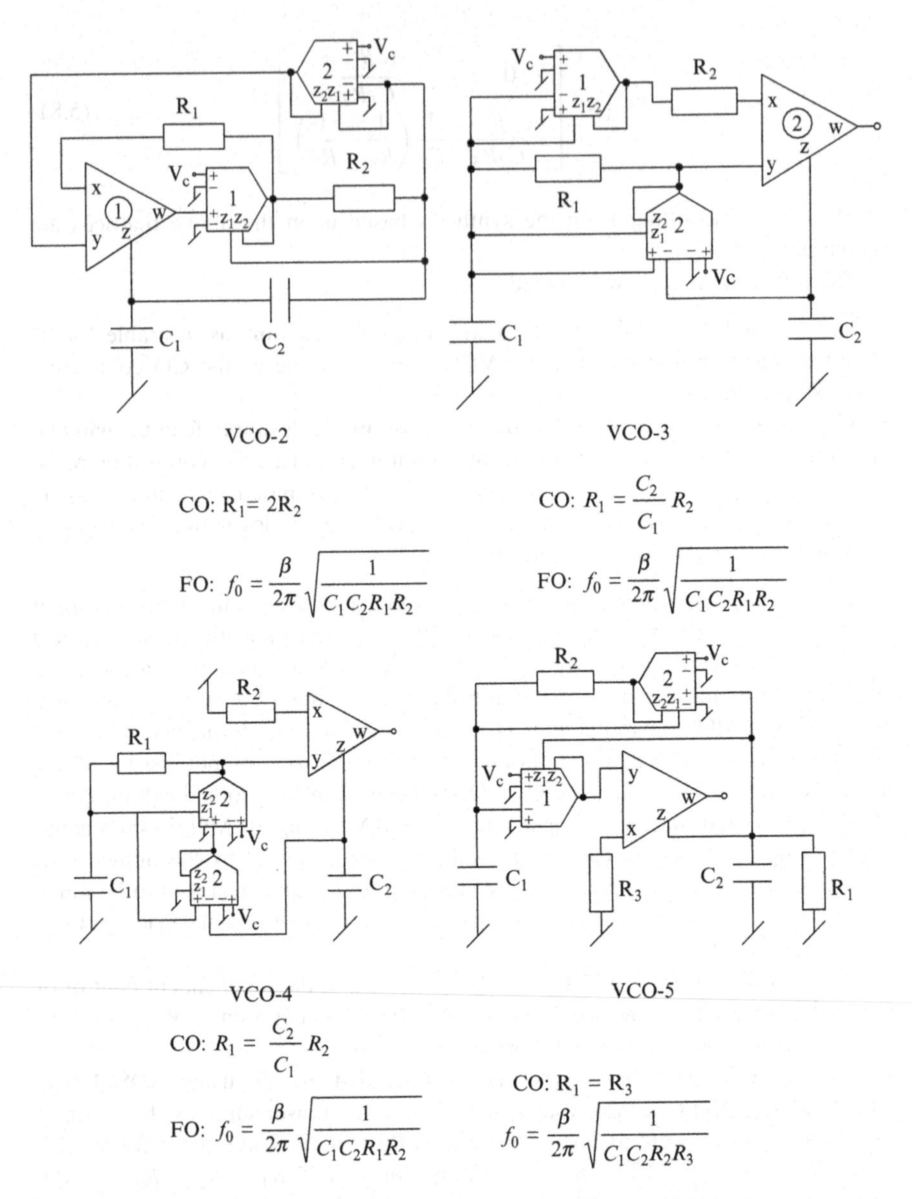

VCO-2
CO: R1= 2R2
FO: $f_0 = \frac{\beta}{2\pi}\sqrt{\frac{1}{C_1C_2R_1R_2}}$
VCO-3
CO: $R_1 = \frac{C_2}{C_1}R_2$
FO: $f_0 = \frac{\beta}{2\pi}\sqrt{\frac{1}{C_1C_2R_1R_2}}$
VCO-4
CO: $R_1 = \frac{C_2}{C_1}R_2$
FO: $f_0 = \frac{\beta}{2\pi}\sqrt{\frac{1}{C_1C_2R_1R_2}}$
VCO-5
CO: R1 = R3
$f_0 = \frac{\beta}{2\pi}\sqrt{\frac{1}{C_1C_2R_2R_3}}$

Fig. 5.33 GC-VCOs derived from matrices $[A]_2 - [A]_5$ (adapted from [66])

was varied from 1–10 V. Thus, VCOs 3, 4 and 5 have been found to be superior than the other VCOs in this respect.

Some sample results of the various VCOs are shown in Fig. 5.34a, b which show the variation of oscillation frequency with control voltage V_C for VCOs 1 and 5 respectively.

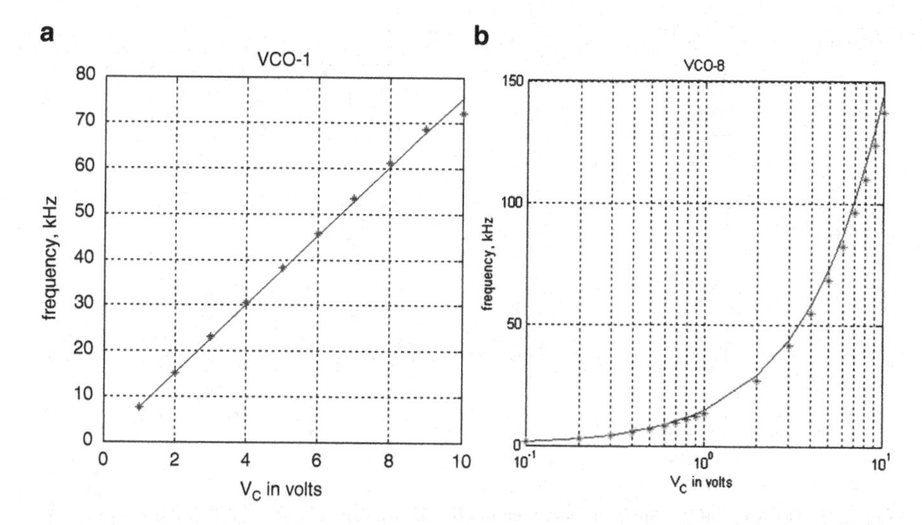

Fig. 5.34 Experimental results of the VCOs. (**a**) Variation of frequency with V_C for VCO-1, (**b**) Variation of frequency with V_C for VCO-5 (adapted from [66])

In view of a number of new CMOS CFOA and CMOS multiplier architectures being evolved in the recent literature, it may be expected that these ideas could possibly be carried over to the design of completely CMOS-based linear VCOs in future.

5.13 MOSFET-C Sinusoidal Oscillator

Since filters and oscillators are closely related, it is obvious that the techniques which are employed to synthesize a biquad filter can as well be employed to synthesize a sinusoidal oscillator. In principle, an active RC band pass filter can be easily converted into a MOS-C oscillator by replacing resistors by MOS-VCRs. For this purpose, the classical two-integrator-loop is a natural choice. From the previous chapter it is known that both lossy and lossless MOSFET-C integrators can be realized in a number of ways.

Three popular nonlinearity cancellation techniques involving one, two and four MOSFETs are shown in Fig. 5.35.

Assuming that all the MOS transistors are operating in triode region, the current I in case of Fig. 5.35a is given by

$$I = 2K(V_G - V_{TH}) \quad \text{for } (V_G - V_{TH}) \geq |V_1| \quad \text{where } K = \mu_n C_{OX}\left(\frac{W}{L}\right) \tag{5.82}$$

and symbols have their usual meaning.

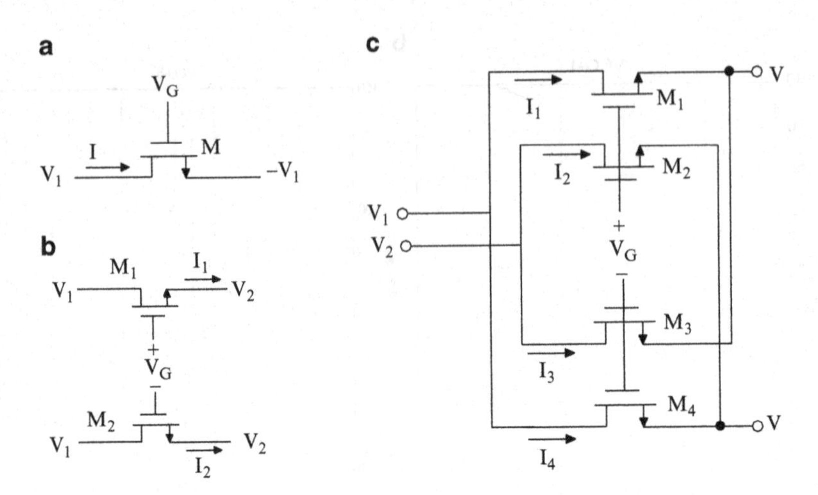

Fig. 5.35 (**a**) An NMOS transistor with even nonlinearities cancellation. (**b**) Two MOS transistors circuit with full nonlinearities cancellation. (**c**) Four MOS transistors circuit with full nonlinearities cancellation (adapted from [67] © 2000 Taylor & Francis)

For the circuit of Fig. 5.35b,

$$I = (I_1 - I_2) = KV_G(V_1 - V_2) \text{ for}(V_G - V_{TH}) \geq \max(V_1, V_2) \tag{5.83}$$

and lastly, for the circuit of Fig. 5.35c,

$$I = (I_5 - I_6) = KV_G(V_1 - V_2) \tag{5.84}$$

Utilizing the two MOSFETs and four MOSFETs implementation, a two CFOA grounded capacitor quadrature oscillator was proposed by Mahmoud and Soliman [67] which is shown in Fig. 5.36.

A straight forward analysis of the oscillator of Fig. 5.36 shows that the condition of oscillation and frequency of oscillation of the circuit are given by

$$g_{m1} = g_{m3} \tag{5.85}$$

$$\omega_0 = \sqrt{\frac{g_{m2}g_{m3}}{C_1C_2}} \tag{5.86}$$

where

$$g_{mi} = K_iV_{Gi} \ (i = 1,\ 2 \text{ and } 3) \tag{5.87}$$

Thus, the condition of oscillation can be controlled by the transconductance g_{m1} and hence, by V_{G1} while frequency of oscillation can independently be tuned by g_{m2} and hence by V_{G2}.

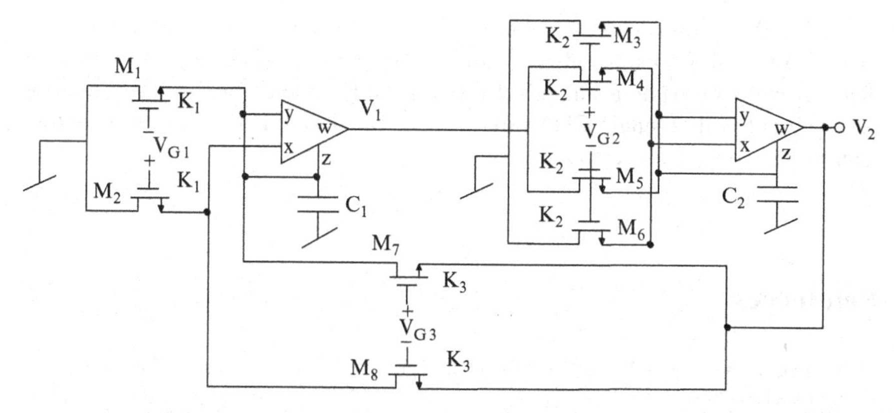

Fig. 5.36 MOS-C CFOA quadrature oscillator proposed by Mahmoud and Soliman (adapted from [67] © 2000 Taylor & Francis)

5.14 Concluding Remarks

In this chapter, we have elaborated a variety of sinusoidal oscillators which included canonic SRCOs using a single CFOA, two-CFOA-based oscillators employing grounded capacitors, SRCOs with explicit current output and fully-uncoupled SRCOs providing independent control of FO and CO through separate resistors exhibiting the notable property that independent control of FO remains intact even under the influence of the various parasitic impedances of a CFOAs. Also presented were VCOs employing nonlinearity-cancelled FETs through a general scheme. Subsequently, two different varieties of linear VCOs based upon the use of analog multipliers in conjunction with CFOAs were described. In the first one, through a systematic state variable formulation, a class of two-CFOA linear VCOs were synthesized which provide linear control FO through an external voltage (i.e. $f_0 \propto V_c$). This was followed by another family of VCOs, again synthesized through a state variable methodology, in which the circuit complexity was reduced by appropriately taking into account the voltage addition feature of the AD534 type analog multipliers. The resulting VCOs not only provide linear tuning law but in contrast to the circuits presented in the earlier section, these circuits can be implemented with only a single CFOA and two analog multipliers. A CFOA-based MOS-C oscillator was also described.

Lastly, we would like to outline two interesting ideas worthy of further investigations and research. In Sect. 5.4 a family of eight single-CFOA oscillators was presented. Unfortunately, till date, there has not been any comparative study of all the eight circuits to determine as to which one of these is the best of the entire class and this problem is still open to investigation.

Yet another problem whose solution has not yet been found is whether or not a Single-CFOA SRCO using only two grounded capacitors exists? In view of the fact that such a circuit using a single VOA does exist [7], the existence of a similar or better circuit with a single CFOA may not be ruled out. This constitutes another interesting problem for research.

References

1. Hribsek M, Newcomb RW (1976) VCO controlled by one variable resistor. IEEE Trans Circ Syst 23:166–169
2. Soliman AM, Awad SS (1978) A novel sine-wave generator using a single operational amplifier. Proc IEEE 66:253–254
3. Senani R (1979) New canonic sinusoidal oscillator with independent frequency control through a single grounded resistor. Proc IEEE 67:691–692
4. Pyara VP, Dutta Roy SC, Jamuar SS (1983) Identification and design of single amplifier single resistance controlled oscillators. IEEE Trans Circ Syst 30:176–181
5. Bhattacharyya BB, Darkani MT (1984) A unified approach to the realization of canonic RC-active, single as well as variable, frequency oscillators using operational amplifiers. J Franklin Inst 317:413–439
6. Singh V (2001) Realisation of operational floating amplifier based 1-op-amp based sinusoidal oscillators. IEEE Trans Circ Syst 48:377–381
7. Singh V (1980) Novel sinusoidal oscillator employing grounded capacitors. Electron Lett 16:757–758
8. Dutta Roy SC, Pyara VP (1979) Single element controlled oscillators: a network synthetic approach. Proc IEEE 67:1565–1566
9. Senani R (1988) Analysis, synthesis and design of new types of RC-active sinusoidal oscillators. Part I and part II. Frequenz 42(223–228):251–256
10. Martinez PA, Sabadell J, Aldea C (1997) Grounded resistor controlled sinusoidal oscillator using CFOAs. Electron Lett 33:346–348
11. Senani R (1980) Novel sinusoidal oscillator employing grounded capacitors. Electron Lett 16:62–63
12. Senani R (1994) On equivalent forms of single op-amp sinusoidal RC oscillators. IEEE Trans Circ Syst-I 41:617–624
13. Martinez PA, Celma S, Sabadell J (1996) Designing sinusoidal oscillators with current-feedback amplifiers. Int J Electron 80:637–646
14. Celma S, Martinez PA, Carlosena A (1994) Current feedback amplifiers based sinusoidal oscillators. IEEE Trans Circ Syst-I 41:906–908
15. Singh VK, Sharma RK, Singh AK, Bhaskar DR, Senani R (2005) Two new canonic single-CFOA oscillators with single resistor controls. IEEE Trans Circ Syst-II 52:860–864
16. Senani R (1998) Realization of a class of analog signal processing/signal generation circuits: Novel configurations using current feedback op-amps. Frequenz 52:196–206
17. Soliman AM (1996) Applications of current feedback operational amplifiers. Analog Integr Circ Sign Process 11:265–302
18. Celma S, Martinez PA, Carlosena A (1994) Approach to the synthesis of canonic RC-active oscillators using CCII. IEE Proc Circ Devices Syst 141:493–497
19. Senani R, Singh VK (1996) Comment: synthesis of canonic single-resistance-controlled-oscillators using a single current-feedback-amplifier. IEE Proc Circ Devices Syst 143:71–72
20. Liu SI, Shih CS, Wu DS (1994) Sinusoidal oscillators with single element control using a current-feedback amplifier. Int J Electron 77:1007–1013

21. Abuelma'atti MT, Farooqi AA, Al-Shahrani SM (1996) Novel RC oscillators using the current-feedback operational amplifier. IEEE Trans Circ Syst-I 43:155–157
22. Liu SI, Tsay JH (1996) Single-resistance-controlled sinusoidal oscillator using current-feedback amplifiers. Int J Electron 80:661–664
23. Abuelma'atti MT, Al-Shahrani SM (1996) A novel low-component count single-element-controlled sinusoidal oscillator using the CFOA pole. Int J Electron 80:747–752
24. Soliman AM (2000) Three oscillator families using the current feedback op-amp. Frequenz 54:126–131
25. Senani R, Singh VK (1996) Novel single-resistance-controlled-oscillator configuration using current feedback amplifiers. IEEE Trans Circ Syst-I 43:698–700
26. Abuelma'atti MT, Farooqi AA (1996) A novel single-element controlled oscillator using the current-feedback-operational amplifier pole. Frequenz 50:183–184
27. Toker A, Cicekoglu O, Kuntman H (2002) On the oscillator implementations using a single current feedback op-amp. Comput Electr Eng 28:375–389
28. Gunes EO, Toker A (2002) On the realization of oscillators using state equations. Int J Electron Commun 56:317–326
29. Martinez PA, Sanz BMM (2005) Generation of two integrator loop variable frequency sinusoidal oscillator. Int J Electron 92:619–629
30. Abuelma'atti MT, Al-Ghazwani A (2000) New quartz crystal oscillators using the current-feedback operational amplifier. Active Passive Electron Comp 23:131–136
31. Hou CL, Wang WY (1997) Circuit transformation method from OTA-C circuits into CFA-based RC circuits. IEE Proc Circ Devices Syst 144:209–212
32. Wu DS, Liu SI, Hwang YS, Wu YP (1995) Multiphase sinusoidal oscillator using the CFOA pole. IEE Proc Circ Devices Syst 142:37–40
33. Abuelma'atti MT, Al-Shahrani AM (1998) Novel CFOA-based sinusoidal oscillators. Int J Electron 85:437–441
34. Liu SI, Chang CC, Wu DS (1994) Active-R sinusoidal oscillator using the CFA pole. Int J Electron 77:1035–1042
35. Abuelma'atti MT, Al-Shahrani SM (1997) New CFOA-based grounded-capacitor single-element-controlled sinusoidal oscillator. Active Passive Electron Comp 20:119–124
36. Senani R, Gupta SS (1997) Synthesis of single-resistance-controlled oscillators using CFOAs: simple state-variable approach. IEE Proc Circ Devices Syst 144:104–106
37. Gupta SS, Senani R (1998) State variable synthesis of single-resistance-controlled grounded capacitor oscillator oscillators using only two CFOAs. IEE Proc Circ Devices Syst 145:135–138
38. Gupta SS, Senani R (1998) State variable synthesis of single-resistance-controlled grounded capacitor oscillator oscillators using only two CFOAs: additional new realizations. IEE Proc Circ Devices Syst 145:415–418
39. Abuelma'atti MT, Al-Zaher HA (1998) New grounded-capacitor sinusoidal oscillators using the current-feedback-amplifier pole. Active Passive Electron Comp 21:23–32
40. Abuelma'atti MT, Al-Shahrani SM (1997) New CFOA-based sinusoidal oscillators. Int J Electron 82:27–32
41. Elwakil AS (1998) Systematic realization of low-frequency oscillators using composite passive–active resistors. IEEE Trans Instrument Measure 47:584–586
42. Soliman AM (2000) Current feedback operational amplifier based oscillators. Analog Integr Circ Sign Process 23:45–55
43. Singh AK, Senani R (2001) Active-R design using CFOA-poles: new resonators, filters and oscillators. IEEE Trans Circ Syst-II 48:504–511
44. Bhaskar DR (2003) Realization of second-order sinusoidal oscillator/filters with non-interacting controls using CFAs. Frequenz 57:12–14
45. Abuelma'atti MT, Al-Shahrani SM (2003) Synthesis of a novel low-component programmable sinusoidal oscillator. Active Passive Electron Comp 26:31–36

46. Bhaskar DR, Prasad D, Imam SA (2004) Grounded-capacitor SRCOs realized through a simple general scheme. Frequenz 58:175–177
47. Senani R, Sharma RK (2005) Explicit-current-output sinusoidal oscillators employing only a single current-feedback op-amp. IEICE Electron Express 2:14–18
48. Bhaskar DR, Senani R (2006) New CFOA-based single-element-controlled sinusoidal oscillators. IEEE Trans Instrument Measure 55:2014–2021
49. Nandi R (2008) Tunable active-R oscillator using a CFA. IEICE Electron Express 5(8):248–253
50. Gupta SS, Bhaskar DR, Senani R (2009) New voltage controlled oscillators using CFOAs. Int J Electron Commun 63:209–217
51. Tangsrirat W, Surakampontorn W (2009) Single-resistance-controlled quadrature oscillator and universal biquad filter using CFOAs. Int J Electron Commun 63:1080–1086
52. Soliman AM (2010) Transformation of oscillators using Op amps, unity gain cells and CFOA. Analog Integr Circ Sign Process 65:105–114
53. Koren V (2002) RF oscillator uses current-feedback op-amp. EDN:83–84
54. Soliman AM (2011) Transformation of a floating capacitor oscillator to a family of grounded capacitor oscillators. Int J Electron 98:289–300
55. Abuelma'atti MT (2010) Identification of a class of two CFOA-based sinusoidal RC oscillators. Analog Integr Circ Sign Process 65:419–428
56. Wangenheim LV (2011) Comment on 'Identification of a class of two CFOA-based sinusoidal RC oscillators'. Analog Integr Circ Sign Process 67:117–119
57. Abuelma'atti MT (2012) Reply to comment on "Identification of class of two CFOA-based sinusoidal RC oscillators". Analog Integr Circ Sign Process 71:155–157
58. Lahiri A, Jaikla W, Siripruchyanun M (2011) Explicit-current-output second-order sinusoidal oscillators using two CFOA's and grounded capacitors. Int J Electron Commun 65:669–672
59. Bhaskar DR, Gupta SS, Senani R, Singh AK (2012) New CFOA-based sinusoidal oscillators retaining independent control of oscillation frequency even under the influence of parasitic impedances. Analog Integr Circ Sign Process 73:427–437
60. Bhaskar DR, Senani R, Singh AK (2010) Linear sinusoidal VCOs: new configurations using current feedback op-amps. Int J Electron 97:263–272
61. Bhaskar DR, Senani R, Singh AK, Gupta SS (2010) Two simple analog multiplier based linear VCOs using a single current feedback Op-amp. Circ Syst 1:1–4
62. Martinez PA, Sabadell J, Aldea C, Celma S (1999) Variable frequency sinusoidal oscillators based on $CCII^{+}$. IEEE Trans Circ Syst-I 46:1386–1390
63. Gupta SS, Senani R (2003) Realizations of current-mode SRCOs using all grounded passive elements. Frequenz 57:26–37
64. Gupta SS, Sharma RK, Bhaskar DR, Senani R (2010) Sinusoidal oscillators with explicit current output employing current-feedback op-amps. Int J Circ Theor Appl 38:131–147
65. Gupta SS, Bhaskar DR, Senani R (2011) Synthesis of linear VCOs: the state-variable approach. J Circ Syst Comput 20:587–606
66. Gupta SS, Bhaskar DR, Senani R (2012) Synthesis of new single CFOA-based VCOs incorporating the voltage summing property of analog multipliers. ISRN Electron: Article ID 463680, 8 p
67. Mahmoud SA, Soliman AM (2000) Novel MOS-C oscillators using the current feedback op-amp. Int J Electron 87:269–280
68. Nay K, Budak A (1983) A voltage-controlled-resistance with wide dynamic range and low distortion. IEEE Trans Circ Syst 30:770–772
69. Senani R (1994) Realization of linear voltage-controlled resistance in floating form. Electron Lett 30:1909–1911
70. Gupta SS (2005) Realization of some class of linear/nonlinear analog electronic circuits using current-mode building blocks. Ph.D. thesis. Faculty of Technology, University of Delhi
71. Natarajan S (1989) Measurement of capacitances and their loss factors. IEEE Trans Instrument Measure 38:1083–1087

72. Ahmad W (1986) A new simple technique for capacitance measurement. IEEE Trans Instrument Measure 35:640–642
73. Awad SS (1988) Capacitance measurement based on an operational amplifier circuit: error determination and reduction. IEEE Trans Instrument Measure 37:379–382
74. Horng JW (2001) A sinusoidal oscillator using current-controlled current conveyors. Int J Electron 88:659–664
75. Fongsamut C, Anuntahirunrat K, Kumwachara K, Surakampontorn W (2006) Current-conveyor-based single-element-controlled and current-controlled sinusoidal oscillators. Int J Electron 93:467–478
76. Senani R, Gupta SS (2000) Novel SRCOs using first generation current conveyor. Int J Electron 87:1187–1192
77. Kilinc S, Jain V, Aggarwal V, Cam U (2006) Catalogue of variable frequency and single-resistance-controlled oscillators employing a single differential-difference-complementary-current-conveyor. Frequenz 60:142–146
78. Gupta SS, Senani R (2005) Grounded-capacitor SRCOs using a single differential difference complementary current feedback amplifier. IEE Proc Circ Devices Syst 152:38–48
79. Gupta SS, Senani R (2006) New single resistance controlled oscillator configurations using unity-gain cells. Analog Integr Circ Sign Process 46:111–119
80. Moon G, Zaghloul ME, Newcomb RW (1990) An enhancement-mode MOS voltage-controlled linear resistor with large dynamic range. IEEE Trans Circ Syst 37:1284–1288
81. Elwan HO, Mahmoud SA, Soliman AM (1996) CMOS voltage-controlled floating resistor. Int J Electron 81:571–576
82. Al-Shahrani SM (2007) CMOS wideband auto-tuning phase shifter cir cuit. Electron Lett 43:804–806
83. Senani R, Bhaskar DR, Tripathi MP (1993) On the realization of linear sinusoidal VCOs. Int J Electron 74:727–733
84. Senani R, Bhaskar DR (1996) New active-R sinusoidal VCOs with linear tuning laws. Int J Electron 80:57–61
85. Bhaskar DR, Tripathi MP (2000) Realization of novel linear Sinusoidal VCOs. Analog Integr Circ Sign Process 24:263–267
86. Singh VK (2004) Realization of a class of analog signal processing/signal generation circuits. Ph.D. thesis. Uttar Pradesh Technical University, Lucknow
87. Soliman AM, Awad SS (1978) A canonical voltage controlled oscillator realized using a single operational amplifier. Frequenz 32:153–154
88. Schmid H (2003) Why 'Current Mode' does not guarantee good performance. Analog Integr Circ Sign Process 35:79–90

Chapter 6
Miscellaneous Linear and Nonlinear Applications of CFOAs

6.1 Introduction

In the preceding chapters we have discussed applications which prove the utility of CFOAs as a versatile building block in realizing a variety of linear circuits. It is not surprising therefore that because of wide spread use of CFOAs they have received attention as attractive building blocks for realizing a variety of non-linear functions as well. In this chapter, we would provide an exposition to the application of CFOAs in realizing miscellaneous linear and non-linear functions and non-sinusoidal waveform generators which include both relaxation and chaotic oscillators.

6.2 Electronically-Variable-Gain Amplifier

AD 844 CFOA is an excellent choice as an output amplifier to be used in conjunction with analog multiplier AD 539 in various connection modes of this multiplier. A voltage variable gain amplifier realized by a combination of an analog multiplier AD 539 and the CFOA AD844 [1] is shown in Fig. 6.1.

The output voltage of the circuit is given by $V_0 = -\frac{V_{in}V_c}{V_{ref}}$ where V_c is the external control voltage and $V_{ref} = 2$ V. The gain V_{out}/V_{in} can be electronically controlled through V_c. In this case V_c is to be taken as a positive voltage which can be varied from 0 to 3.2 V (max.) while the signal voltage V_{in} is to be kept nominally ±2 V full scale but can be extended up to ±4.2 V.

R. Senani et al., *Current Feedback Operational Amplifiers and Their Applications*, Analog Circuits and Signal Processing, DOI 10.1007/978-1-4614-5188-4_6,

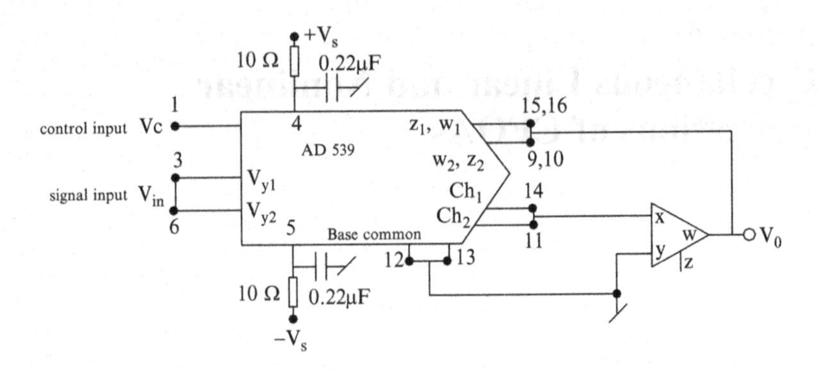

Fig. 6.1 Electronically-variable-gain amplifier using AD539 and AD844

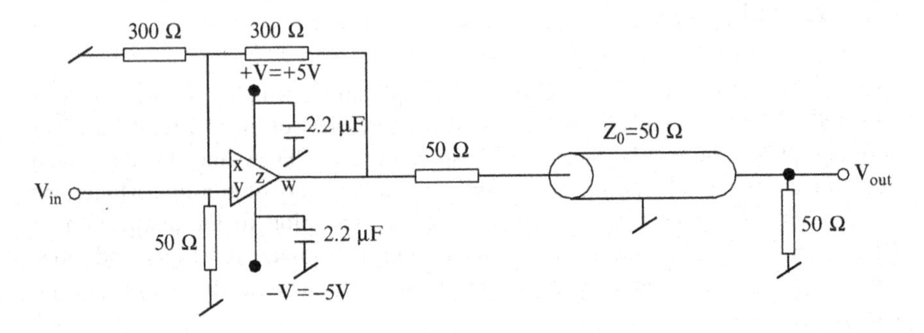

Fig. 6.2 CFOA as a cable driver

6.3 Cable Driver Using CFOA

Most CFOAs find an immediate application to drive low impedance cables. Figure 6.2 shows an illustrative application that provides a gain of +2 by configuring CFOA as a non-inverting amplifier. It is easy to see that the arrangement provides an overall gain of +1 to the signal reaching the load R_L. With a CFOA AD844, the circuit provides −3 dB bandwidth of around 30 MHz.

6.4 Video Distribution Amplifier

Several CFOAs, such as THS 3001, find an excellent application as a video distribution amplifier as shown in Fig. 6.3.

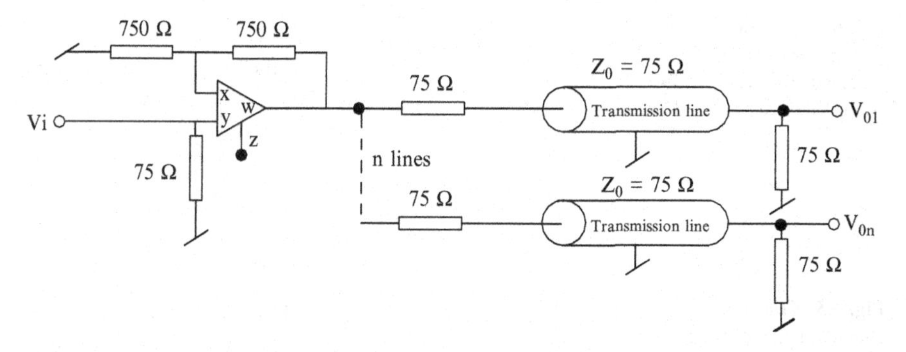

Fig. 6.3 Video distribution amplifier

6.5 Schmitt Triggers and Non-sinusoidal Waveform Generators

At the root of any non-sinusoidal signal generator lies either a comparator or a comparator with hysteresis (using positive feedback) often referred as Schmitt trigger. The first Schmitt trigger using a CCII+ was presented by Di Cataldo et al. [2].

Consider now the Schmitt trigger of Fig. 6.4 which is, in fact, a CFOA version of the CCII-based Schmitt trigger of Cataldo et al. [2]. In a CFOA, the output voltage is ultimately limited to V_{sat+} and V_{sat-} with the current flowing into the Z-terminal being

$$I_{sat+} = -\frac{V_{sat+}}{R_1 + R_2} \tag{6.1}$$

$$I_{sat-} = \frac{V_{sat-}}{R_1 + R_2} \tag{6.2}$$

If the two threshold voltages are V_{TL} and V_{TH}, they can be determined as follows.

If we assume that V_O is in the state V_{sat+} then to change this stable state, the current i_x must satisfy the condition $i_x \geq i_z$ which means

$$\frac{V_{in} - V_y}{R_s} \geq -\frac{V_{sat+}}{R_1 + R_2} \tag{6.3}$$

The higher threshold level V_{TH} is, therefore, given by

$$V_{TH} = \frac{R_1 - R_s}{R_1 + R_2} V_{sat+} \tag{6.4}$$

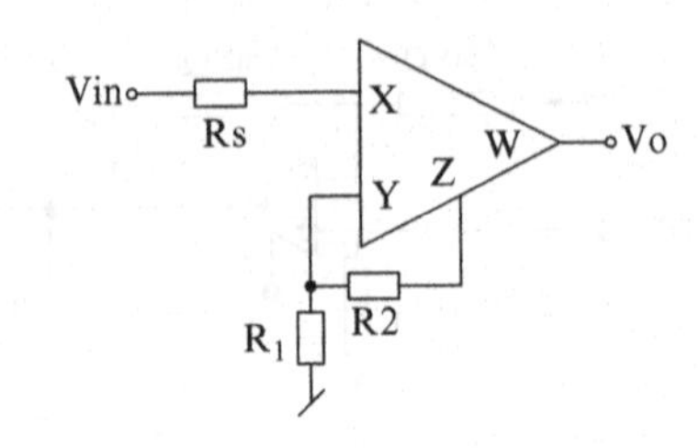

Fig. 6.4 Schmitt Trigger circuit using a CFOA (adapted from [2] © 1995 John Wiley & Sons Ltd.)

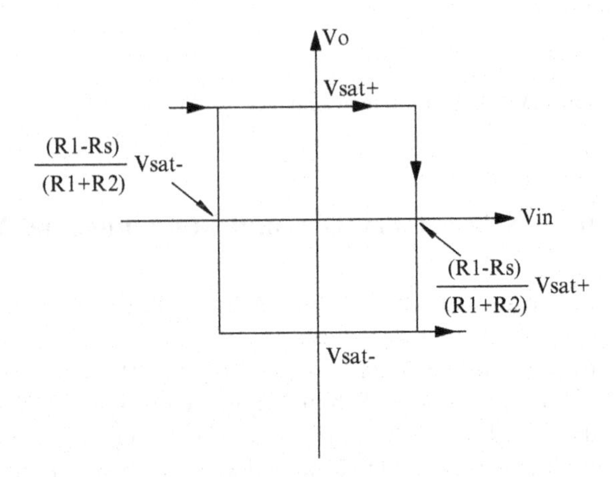

Fig. 6.5 Transfer characteristics of the Schmitt Trigger of Fig. 6.4 (adapted from [2] © 1995 John Wiley & Sons Ltd.)

Similarly, it can be found that the lower threshold level V_{TL} is given by

$$V_{TL} = \frac{R_1 - R_s}{R_1 + R_2} V_{sat-} \tag{6.5}$$

From the above analysis, the transfer characteristic of this Schmitt trigger can be drawn as shown in Fig. 6.5.

The circuit can be easily converted into a relaxation oscillator by connecting a capacitor from the input terminal-X to ground. With this addition, the circuit would generate a square wave output at V_o. Figure 6.6 shows the resulting relaxation oscillator incorporating the non-ideal model of the CFOA AD844 where the non-ideal parameter values are typically given by $r_x = 50\ \Omega$, $R_y = 2\ \text{M}\Omega$, $R_p = 3\ \text{M}\Omega$, $C_x = C_y = 2$ pF and $C_p = 4.5$ pF. In reference [3], it has been shown that the oscillation period of the waveform generated by this circuit is given by

$$T = 2C_T r_x l_n\left(2\frac{R_1}{r_x} - 1\right); \quad \text{where } C_T = C + C_x \tag{6.6}$$

Thus, the time period T is a function of the external capacitor C and the resistors r_x and R_1.

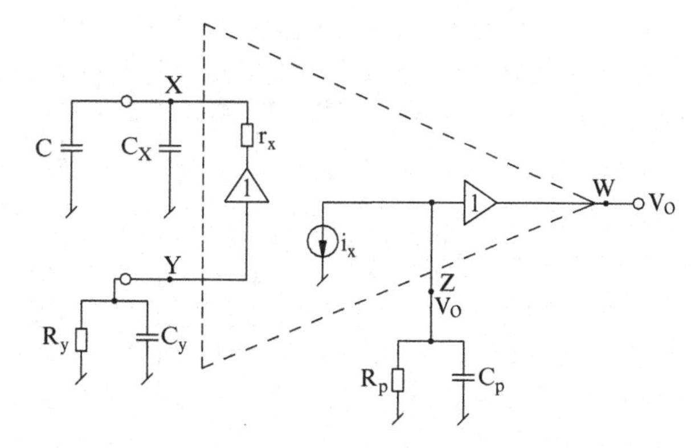

Fig. 6.6 A relaxation oscillator incorporating non-ideal model of the CFOA showing various parasitic impedances (adapted from [3] © 2005 Taylor & Francis)

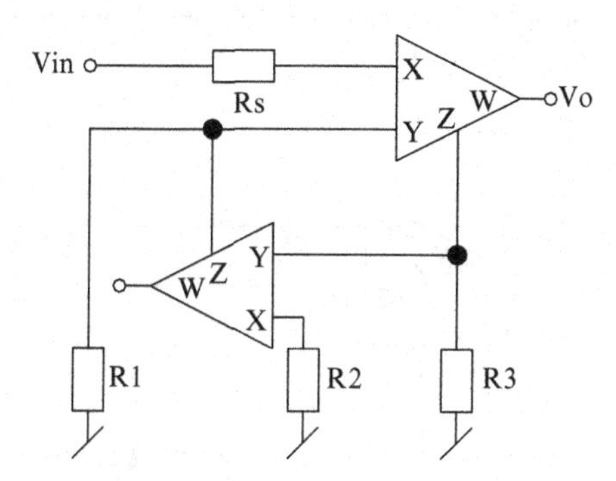

Fig. 6.7 An improved CFOA-version of the CCII-based circuit of Schmitt Trigger using two CFOAs (adapted from [4] © 2011 John Wiley & Sons Ltd.)

An improved CFOA-version of the CCII-based Schmitt trigger proposed by Srinivasulu [4] is shown here in Fig. 6.7.

In this circuit, the two threshold voltage levels are given by

$$V_{TH} = \left(1 - \frac{R_2 R_s}{R_1 R_3}\right) V_{sat+} \tag{6.7}$$

$$V_{TL} = -\left(1 - \frac{R_2 R_s}{R_1 R_3}\right) V_{sat-} \tag{6.8}$$

Based upon the above, the transfer characteristics of the circuit can be drawn as follows (Fig. 6.8).

A square wave/triangular wave generator using the Schmitt trigger of Fig. 6.7 is shown in Fig. 6.9.

In this circuit, the resistors R and R_4 together with the capacitor C constitute an integrator. A straight forward analysis of this circuit shows that the time period (T) of the waveforms generated (a square wave at V_{01} and triangular wave at V_{02}) is given by

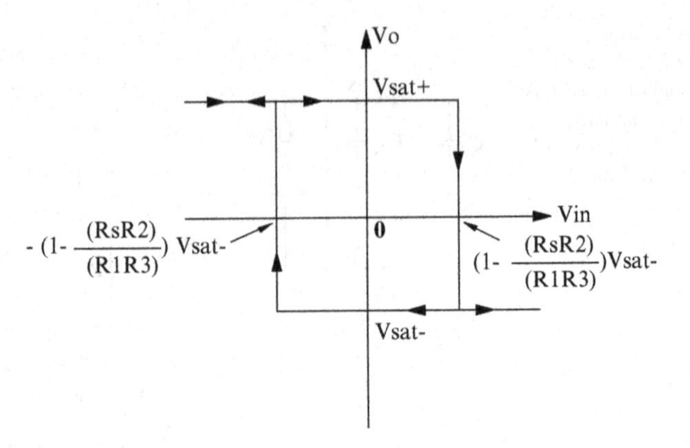

Fig. 6.8 Transfer characteristic of the Schmitt trigger of Fig. 6.7 (adapted from [4] © 2011 John Wiley & Sons Ltd.)

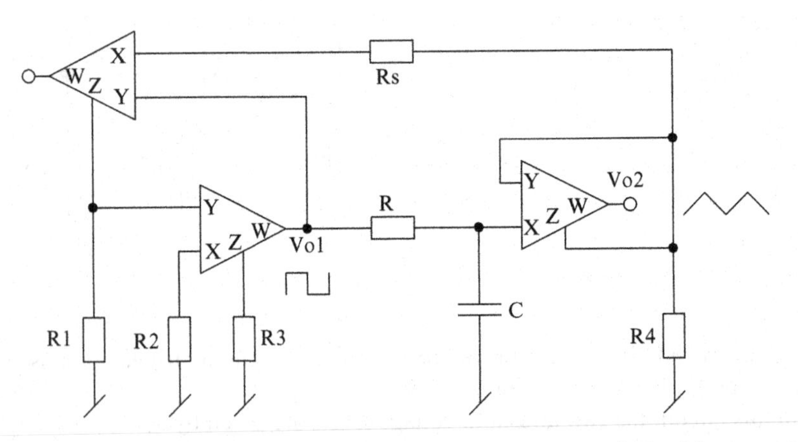

Fig. 6.9 A square/triangular wave generator using Schmitt Trigger of Fig. 6.7 proposed by Srinivasulu (adapted from [4] © 2011 John Wiley & Sons Ltd.)

$$T = 2\pi RC\left[1 - \frac{R_2 R_s}{R_1 R_3}\right] \tag{6.9}$$

Another two-CFOA-based triangular/square wave generator was advanced by Haque et al. in [5]. The circuit, however, requires two CFOAs, four resistors, one capacitor and 2n number of diodes to stabilize the Schmitt trigger output levels at $\pm nV_{D(on)}$.

The frequency of oscillation for triangular/square wave generator by the circuit of Fig. 6.10 is given by

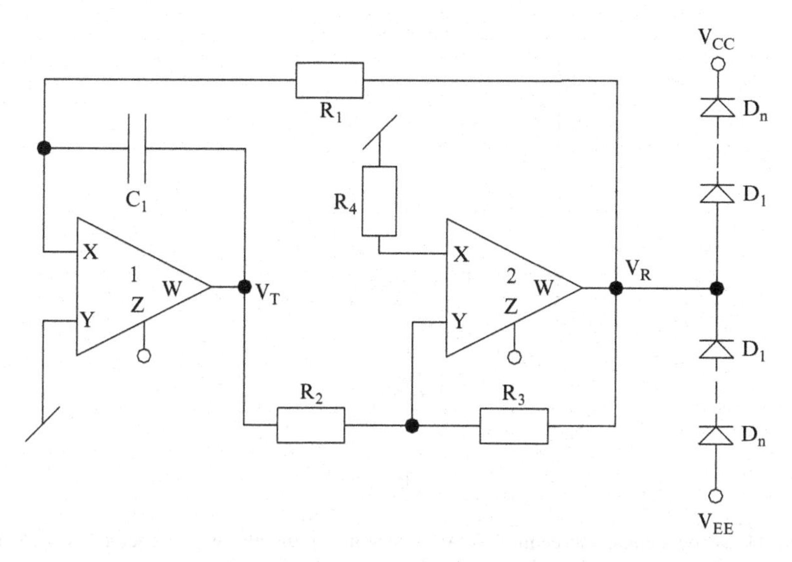

Fig. 6.10 Another two CFOA based triangular/square wave generator (adapted from [5] © 2008 IEEE)

$$f_O = \frac{\frac{V_R}{R_1} - \frac{V_N}{Z_T}\left(1 + \frac{R_2}{R_3}\right) + \frac{V_R R_2}{R_3 Z_T}}{4C_1\left[V_N\left(1 + \frac{R_2}{R_3}\right) - \frac{V_R R_2}{R_3}\right]} \tag{6.10}$$

where Z_T is the open loop transimpedance of the CFOA, V_N is the peak voltage at X-input terminal of $CFOA_2$ and V_R is the peak voltage of the square waveform.

A novel two CFOA and one grounded capacitor based square/triangular wave generator was proposed by Minaei and Yuce in [6]. This circuit is shown in Fig. 6.11.

The operation of this circuit can be explained as follows. Both the CFOAs in this circuit operate as voltage saturated elements. If we assume $Vs_{quare} = V_{sat+}$, the capacitor charges by a constant current $V_{sat}+/R_3$ so that a positive ramp appears at the output of $CFOA_2$ consequently, current flowing through R_1 decreases. When i_x becomes $\leq i_z$ then output voltage of $CFOA_1$ switches to other stable state V_{sat-}. Accordingly, we can write

$$\frac{V_{sat+}}{R_2} = \frac{V_{Sat+} - V_{tri(Peak+)}}{R_1} \tag{6.11}$$

From the above equation, the positive peak voltage of the triangular wave (higher threshold voltage) and the negative peak voltage (lower threshold voltage) are respectively given by

$$V_{tri(Peak+)} = \left(1 - \frac{R_1}{R_2}\right) V_{Sat+} \tag{6.12}$$

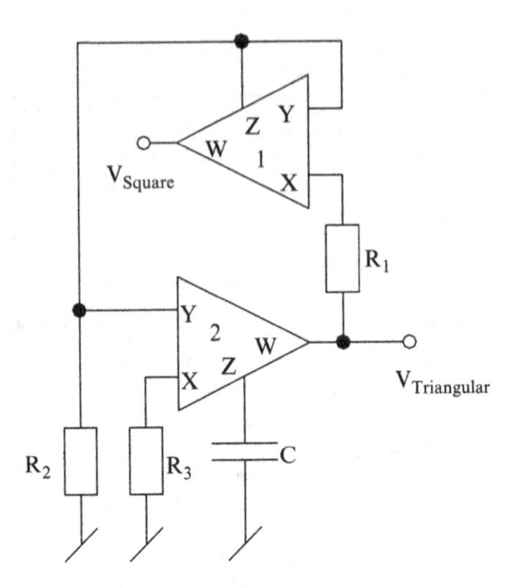

Fig. 6.11 A low-component-count CFOA-based square/triangular wave generator proposed by Minaei and Yuce (adapted from [6] © 2012 Springer)

$$V_{tri(Peak-)} = \left(1 - \frac{R_1}{R_2}\right) V_{Sat-} \tag{6.13}$$

Assuming the two saturation voltages to be equal in magnitude, the time period of the waveforms generated by this circuit is given by

$$T = 4CR_3\left(1 - \frac{R_1}{R_2}\right) \tag{6.14}$$

Finally, we show a triangular/square wave generator made from a single CFOA as shown in Fig. 6.12a. This circuit was proposed by Abuelma'atti and Al-Shahrani in [7]. In this circuit, the CFOA behaves as a Schmitt trigger with the input–output characteristic shown in Fig. 6.12b where the two threshold voltages are given by

$$V_{TH} = \frac{R_1 - r_x}{R_1 + R_2} V_{sat+} \text{ and } V_{TL} = \frac{R_1 - r_x}{R_1 + R_2} V_{sat-} \tag{6.15}$$

where V_{sat+} and V_{sat-} are two stable states decided by the DC biasing power supply voltages of the CFOA and r_x is the input resistance of the CFOA looking into terminal-X of the CFOA.

The circuit can be analyzed by starting from any one of the two stable states of the output voltage V_0 (for details, the reader is referred to [7]). The circuit generates a square wave signal at V_0 and a triangular wave signal at V_x. The frequency of the generated waveforms is given by

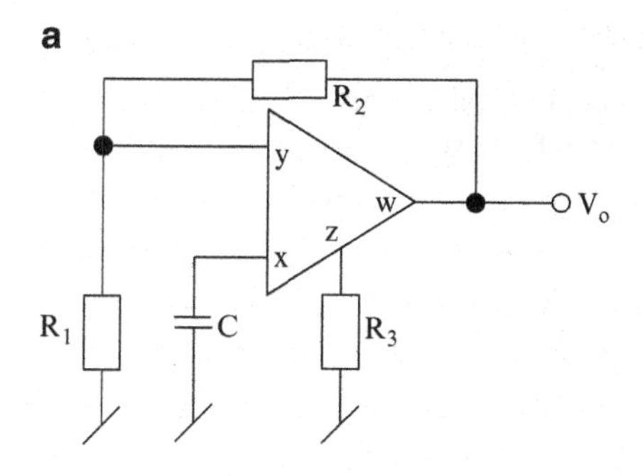

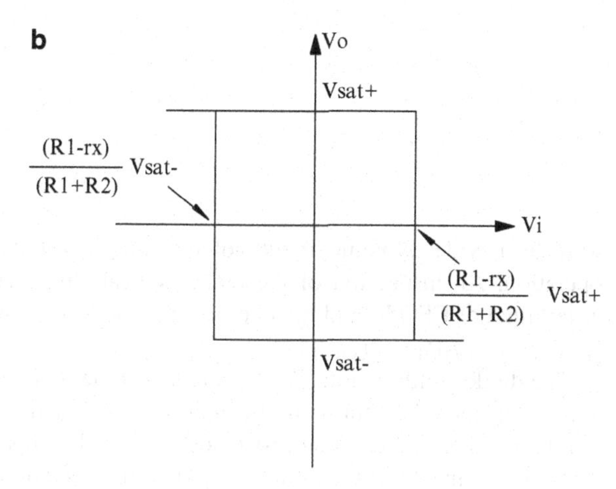

Fig. 6.12 Relaxation oscillator proposed by Abuelma'atti and Al-Shahrani (adapted from [7] © 1998 Taylor & Francis). (**a**) Triangular/square wave generator. (**b**) Transfer characteristic of the Schmitt Trigger composed of CFOA along with R_2 and R_1

$$f = 1 \Big/ \left(2CR_3 \left(\frac{R_1 - r_x}{R_1 + R_2} \right) \right) \cong \frac{1}{2CR_3} \left(1 + \frac{R_2}{R_1} \right); \quad \text{for } R_1 >> R_x \tag{6.16}$$

Out of the various circuits presented, the one in Fig. 6.12 is appealing due to its lowest-component-count whereas those of Figs. 6.10 and 6.11 have the advantage of providing low-output impedance outputs for both square and triangular wave outputs.

6.6 Precision Rectifiers

There have been several attempts of making precision current-mode full wave rectifiers using current conveyors quite often realized with AD844-type CFOAs. Here, we present a typical design of a simple full wave precision rectifier circuit proposed by Khan et al. in [8]. This circuit is shown in Fig. 6.13 and is claimed to

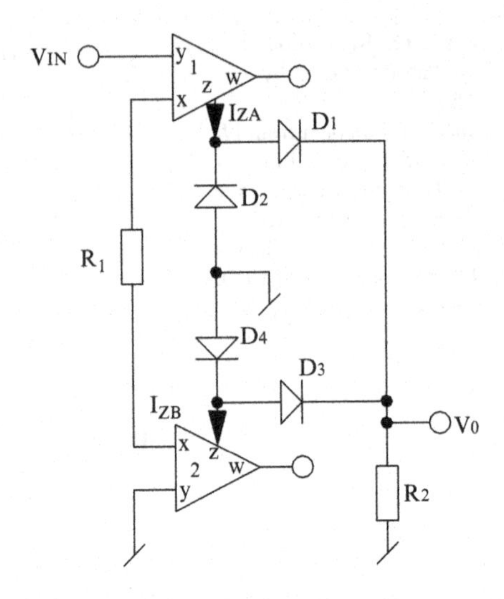

Fig. 6.13 Full wave rectifier proposed by Khan et al. (adapted from [8] © 1995 Taylor & Francis)

provide a wide dynamic input voltage range over a wide frequency range of operation. An inspection of the circuit reveals that the current flowing into the X-terminal of $CFOA_2$ and the one flowing out of Z-terminal $CFOA_1$ are given by: $i_z = i_x = V_{in}/(R_1 + 2r_x)$.

The diode combinations D_1–D_2 and D_3–D_4 are connected in such a manner that depending upon the polarity of the input voltage V_{in} the output current of the CFOA will flow either into the load resistance R_2 or will be bypassed to the ground. Thus, when V_{in} is positive, i_x, i_{ZA} and i_{ZB} have the directions such that diodes D_1–D_2 allow i_{ZA} to be flowing into the load resistance R_2 whereas at the same time, diodes D_3–D_4 allow i_{ZB} to go the ground. On the other hand, when V_{in} is negative, i_x, i_{ZA} and i_{ZB} reverse their directions and as a consequence, now i_{ZB} flows into the load while i_{ZA} flows to the ground. In view of this, it is clear that current through the load R_2 will be uni-directional thereby resulting in an output voltage given by

$$V_0 = i_{ZA}R_2 = i_{ZB}R_2 = (R2/\ (R_1 + 2r_x))V_{in} \qquad (6.17)$$

It is worth pointing out that by reversing the connections of all the four diodes, one can obtain a full wave rectified signal with negative sign (i.e. $-V_0$). Lastly, it must be mentioned that as compared to VOA-based precision rectifiers, which generally require four or more matched resistors, the circuit described here uses a bare minimum of (only two) resistors and no resistor-matching is needed.

6.7 Analog Squaring Circuit

Figure 6.14 shows a CFOA-version of the CCII-based squaring circuit presented in [9]. This circuit can be regarded as the simplest squaring circuit using only a single CFOA, a pair of identical (matched) MOSFETs and a single resistor.

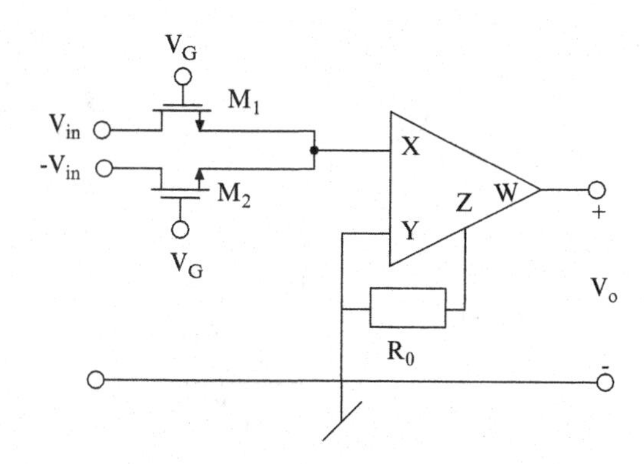

Fig. 6.14 A voltage squaring circuit proposed by Liu (adapted from [9] © 1995 IEE)

Assuming MOSFETs M_1 and M_2 to be operating in saturation mode an analysis of this circuit shows that the current entering into the terminal-X is given by

$$I_X = -KV_{in}^2 \tag{6.18}$$

where K is the transconductance parameter of the MOSFETs.

Since $I_y = 0$, $V_x = V_y$, $I_Z = I_x$ and $V_w = V_z$, for an ideal CFOA, the output voltage of the squaring circuit can be given as

$$V_O = KR_0V_{in}^2 \tag{6.19}$$

The transconductance parameter K of the MOSFETs is given by $K = \frac{\mu_s C_{OX}}{2}\left(\frac{W}{L}\right)$, where the symbols have their usual meanings.

Thus, the circuit of Fig. 6.14 gives an output voltage V_O which is proportional to the square of the input voltage V_{in}.

6.8 Analog Divider

Among various non-linear applications of CFOAs evolved so far, an interesting application is that of realizing an analog divider using CFOAs and MOSFETs. One such circuit having all MOSFETs operating in triode region is shown in Fig. 6.15 and was proposed by Liu and Chen [10].

Assuming the input signals V_x and V_y to be small and assuming all MOSFETs to be identical and (i.e. same K) operating in triode region a straight forward analysis of the circuit of Fig. 6.15 gives the output voltage as

$$v_O = (V_{GA} - V_{GB})\left(\frac{v_y}{v_x}\right) \tag{6.20}$$

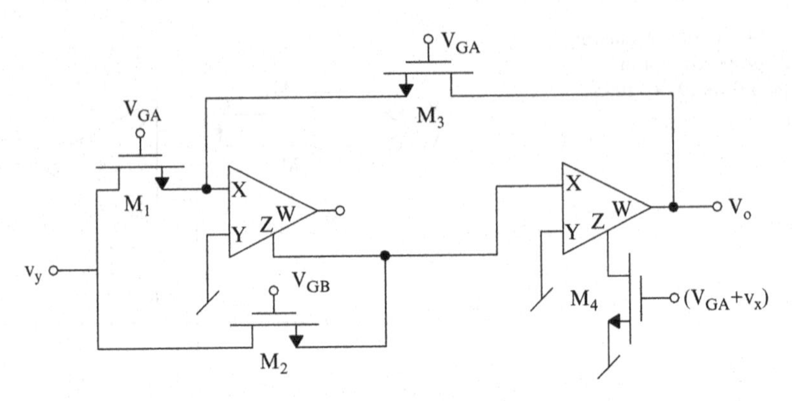

Fig. 6.15 An analog divider using CFOAs proposed by Liu and Chen (adapted from [10] © 1995 IET)

from where it is seen that the circuit functions as an analog divider with input signals as v_x and v_y where the scale factor $(V_{GA} - V_{GB})$ is controllable through external voltages V_{GA} and V_{GB}.

6.9 Pseudo-exponential Circuits

Pseudo-exponential functions play an important role in allowing wide gain control changes with a control parameter in a number of communication and signal processing systems [11]. Although, the exponential v–i characteristics of a BJT or a MOSFET operating in week inversion mode can be exploited to design exponential circuits, an alternative way is to employ a Pseudo-exponential function given by

$$e^{2\beta x} \approx \frac{1+\beta x}{1-\beta x} \tag{6.21}$$

Two circuits to realize the above function are shown in Fig. 6.16.

Consider first the circuit of Fig. 6.16a. In this circuit, the voltage at node-Y with identical resistors $R_a = R_b$ is given by

$$V_y = \frac{V_{in} + V_o}{2} \tag{6.22}$$

Because of the voltage buffer between terminals Y and X of the CFOA, the same voltage is transferred to the X-terminals of both the CFOAs. By a straight forward analysis, it can be proved that

$$V_o = \frac{1 + R_2\left(\frac{1}{R_1} - \frac{1}{R_3}\right)}{1 - R_2\left(\frac{1}{R_1} - \frac{1}{R_3}\right)} V_{in} \tag{6.23}$$

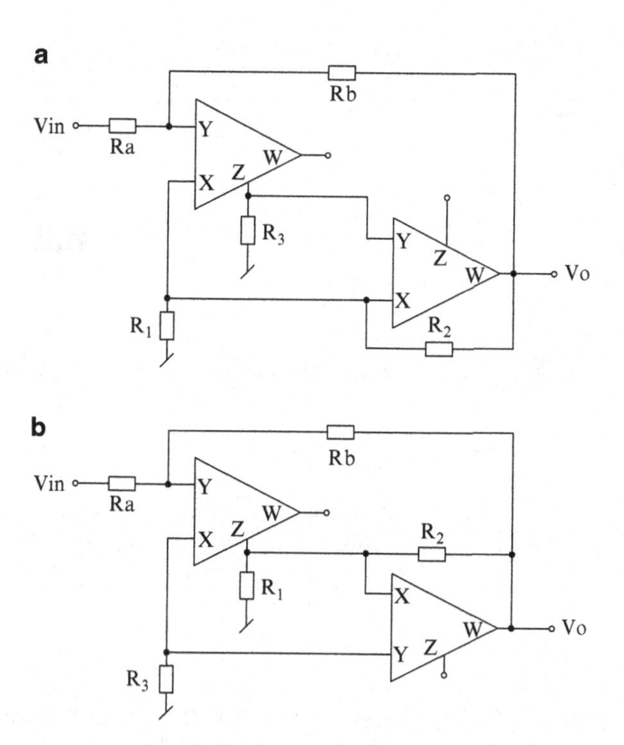

Fig. 6.16 CFOA-based realizations of Pseudo exponential function proposed by Maundy and Gift (adapted from [11] © 2005 IEEE)

which simplifies to

$$V_o = \frac{1+x}{1-x} V_{in} \text{ where; } x = R_2\left(\frac{1}{R_1} - \frac{1}{R_3}\right) \tag{6.24}$$

A similar analysis of the second circuit of Fig. 6.16b shows that this circuit also realizes the same function as in (6.24). The condition required for both the circuits to realize Pseudo exponential circuit is given by $x < 0$, if $R_1 > R_3$ or alternatively, $x > 0$ if $R_1 < R_3$.

6.10 Chaotic Oscillators Using CFOAs

Chua's circuit [12] has been a very active topic of research in the study of non-linear dynamical circuits and systems during the past two decades. There has been considerable interest in devising inductor-less realizations of Chua's oscillator. A variety of circuit configurations have been evolved employing traditional VOAs, current conveyors and CFOAs as building blocks. The advantage of using CFOAs is the ease with which current state variables can be readily made available

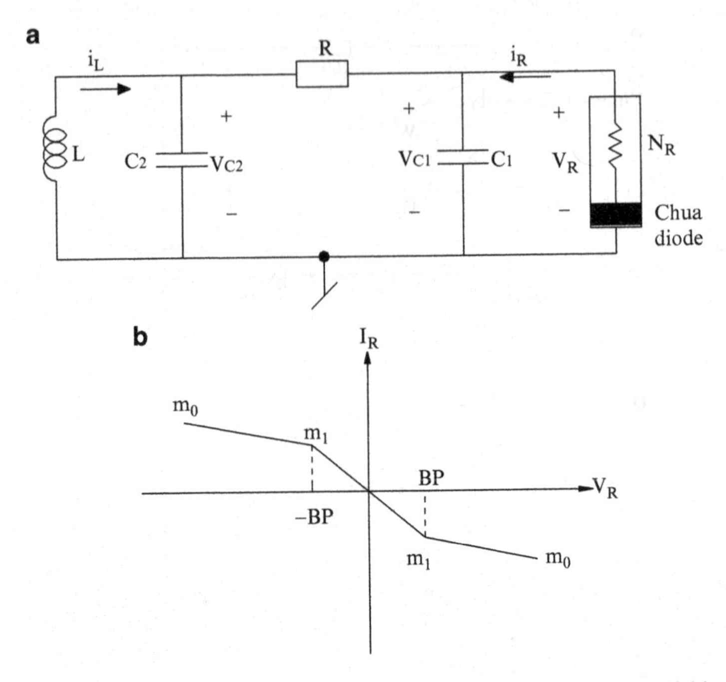

Fig. 6.17 (**a**) Chua's oscillator $L = 28.53$ mH, $C_1 = 5$nF, $C_2 = 50$ nF, $R =$ variable, (**b**) V–I characteristics of Chua diode, $m_o = -0.5$, $m_1 = -0.8$, $BP = -1$

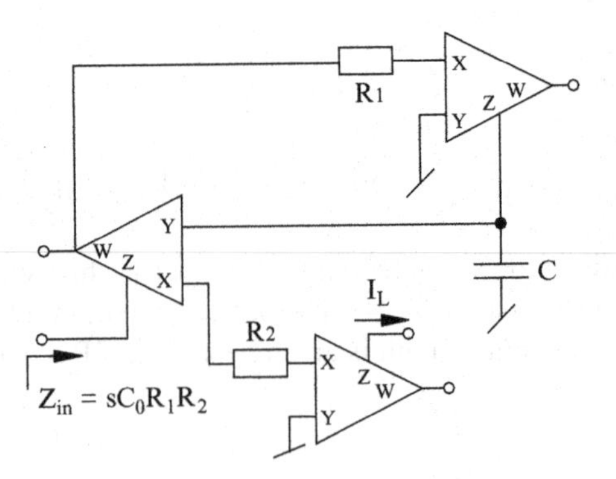

Fig. 6.18 The simulated inductor with inductor current as output (adapted from [13] © 1998 IEE)

as output. This was demonstrated by Senani and Gupta in [13] and by Elwakil and Kennedy in [14] in 1998 and 2000 respectively.

It is well known that Chua's oscillator requires a two segments piece-wise nonlinear negative resistor called Chua's diode (see Fig. 6.17). An interesting circuit for the simulation of the grounded inductor is shown in Fig. 6.18 which

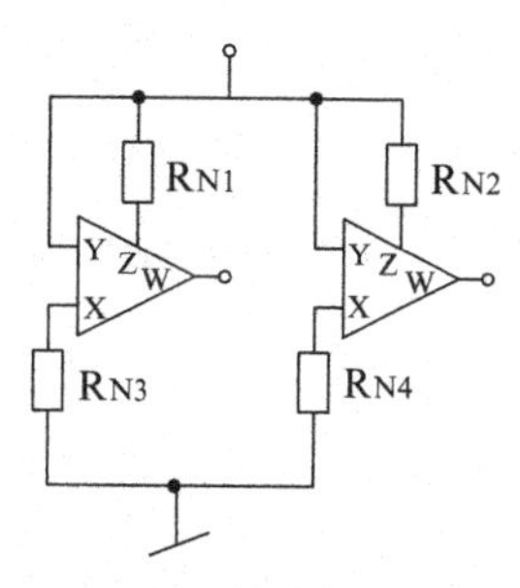

Fig. 6.19 The CFOA-based Chua diode (adapted from [13] © 1998 IEE)

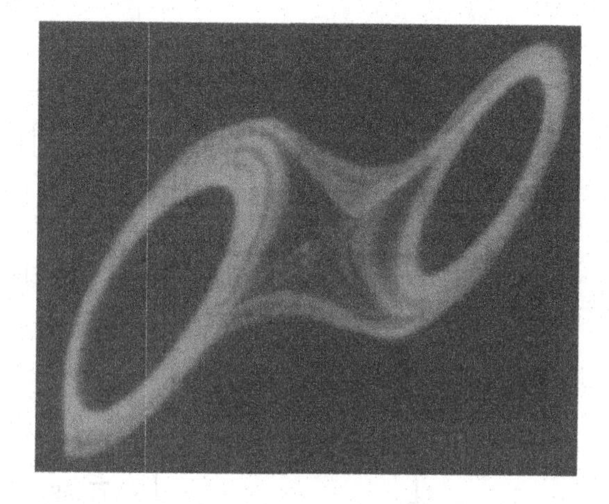

Fig. 6.20 Double scroll attractor obtained from the entirely CFOA-based hardware implementation of the circuit of Fig. 6.17

has a special feature that the current flowing into the simulating inductor is available explicitly from the Z-terminal of one of the CFOAs.

The CFOA-based Chua diode is shown in Fig. 6.19 whereas the double scroll chaotic attractor obtained from the complete implementation obtained by the placement of the sub-circuits of Figs. 6.19 and 6.18 into the main circuit of Fig. 6.17 is shown in Fig. 6.20.

The advantage of this implementation is that other than the availability of the two capacitor voltages as state variables, the third state variable namely, the inductor current i_L, is also available explicitly from the Z-output terminal of one of the CFOAs.

An alternative Chua's oscillator implementation proposed by Elwakil and Kennedy [14] using CFOAs is shown in Fig. 6.21 which also provides a current-mode output.

Other than autonomous chaotic oscillators, several researchers have also investigated the implementation and application of non-autonomous (derived) chaotic circuits and mixed-mode chaotic circuits (containing both autonomous and non-autonomous circuit realizations). A popular mixed-mode chaotic circuit is shown in Fig. 6.22 [15] in which through a digitally controlled switch, it becomes

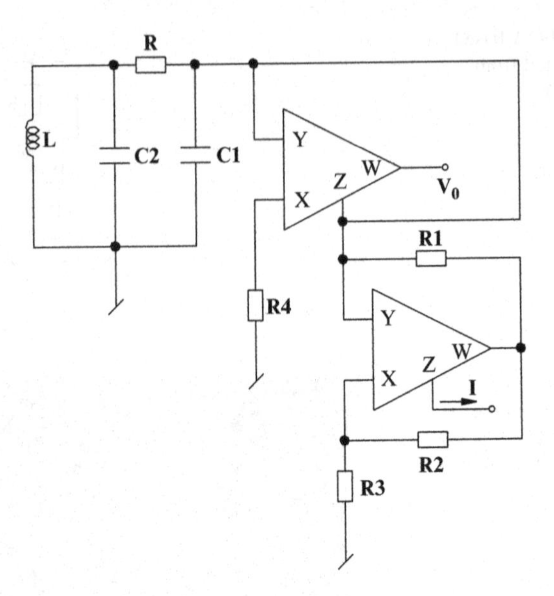

Fig. 6.21 CFOA-based Chua's Oscillator proposed by Elwakil and Kennedy (adapted from [14] © 2000 IEEE)

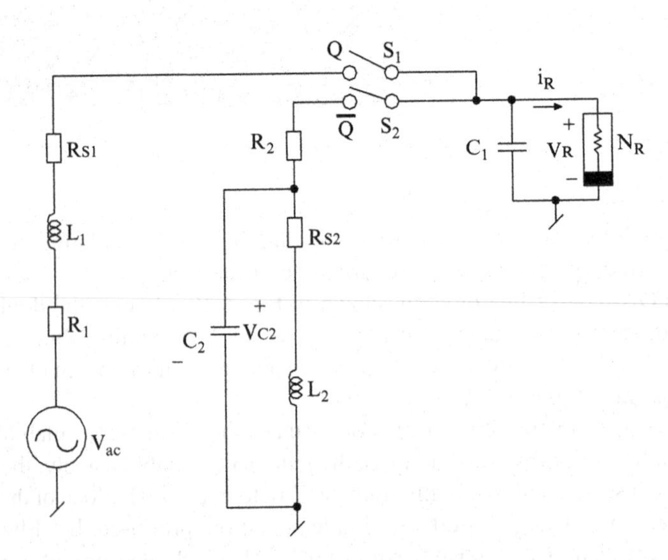

Fig. 6.22 Mixed-mode chaotic circuit proposed by Cam (adapted from [15] © 2004 Elsevier)

possible to realize an autonomous as well as non-autonomous chaotic circuit from the same general structure. When switch S_1 is closed, the circuit becomes a derived chaotic circuit; on the other hand, when S_2, closed, the circuit assumes the form of an autonomous Chua's oscillator.

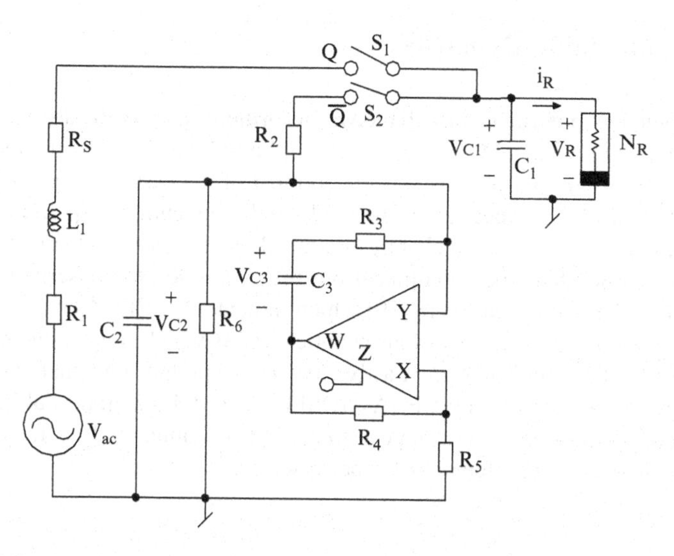

Fig. 6.23 Mixed-mode chaotic circuit with Wien bridge configuration proposed by Kilic (adapted from [16] © 2007 Elsevier)

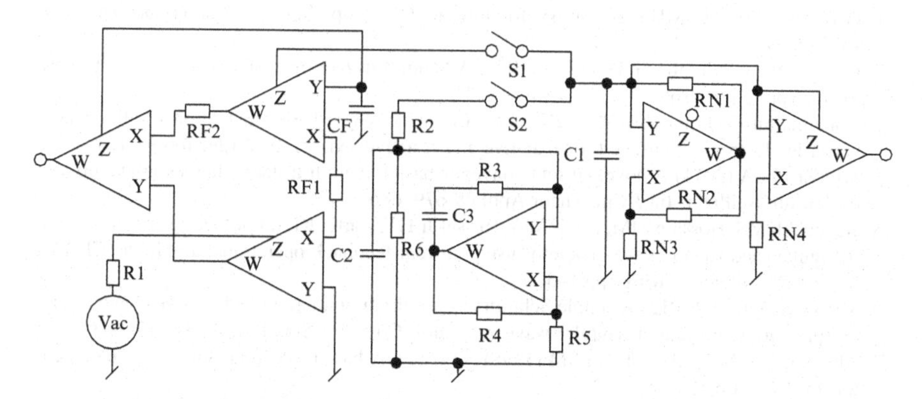

Fig. 6.24 Inductorless Wien bridge-based mixed-mode chaotic circuit configuration proposed by Kilic (adapted from [16] © 2007 Elsevier)

There have been several development of chaotic oscillators based upon classic Wien bridge oscillator. A mixed-mode chaotic circuit made from the Wien bridge configuration is shown here in Fig. 6.23 and its complete CFOA implementation using CFOA-based Wien Bridge oscillator, CFOA-based Chua's diode and CFOA-based lossless floating inductance, is shown in Fig. 6.24 [16].

It may be pointed out that a three-CFOA-based floating inductance simulator was proposed earlier by Senani in [17].

6.11 Concluding Remarks

This chapter has presented miscellaneous linear/nonlinear applications of CFOA which demonstrate that CFOAs are useful building blocks in realizing a variety of non-linear functions and circuits as well. In most of the cases, the type of circuits possible with CFOAs either cannot be made with conventional op-amps with the same advantages. In the area of chaotic circuits, there are a lot of other CFOA-based realizations which have been omitted from the discussion to conserve space; the interested reader is referred to [18–39] which indicate that CFOAs have been used quite prominently in realizing various chaotic circuits which confirm their utility in nonlinear chaotic circuit design. Lastly, it may be pointed out that, to the best knowledge of the authors, any CFOA-MOSFET based 4-quadrant multipliers and Square-rooting circuits do not appear to have been attempted so far and thus, constitute interesting problems for further research.

References

1. Data sheet of 60 MHz 2000V/μS Monolithic Op Amp AD844. Analog Devices, Inc. 1983–2011
2. Di Cataldo G, Palumbo G, Pennisi S (1995) A Schmitt trigger by means of a CCII+. Int J Circ Theor Appl 23:61–165
3. Abuelma'atti MT, Al-absi MA (2005) A current conveyor-based relaxation oscillator as a versatile electronic interface for capacitive and resistive sensors. Int J Electron 92:473–477
4. Srinivasulu A (2011) A novel current conveyor-based Schmitt trigger and its application as a relaxation oscillator. Int J Circ Theor Appl 39:679–686
5. Haque AKMS, Hossain MM, Davis WA, Russel Jr HT, Carter RL (2008) Design of sinusoidal, triangular, and square wave generator using current feedback operational amplifier (CFOA). Region 5 Conference IEEE. pp 1–5
6. Minaei S, Yuce E (2012) A simple Schmitt trigger circuit with grounded passive elements and its application to square/triangular wave generator. Circ Syst Sign Process 31:877–888
7. Abuelma'atti MT, Al-Shahrani SM (1998) New CFOA-based triangular/square wave generator. Int J Electron 84:583–588
8. Khan AA, El-Ela MA, Al-Turaigi (1995) Current-mode precision rectification. Int J Electron 79:853–859
9. Liu SI (1995) Square-rooting and vector summation circuits using current conveyors. IEE Proc Circ Devices Syst 142:223–226
10. Liu SI, Chen JJ (1995) Realization of analogue divider using current feedback amplifiers. IEE Proc Circ Devices Syst 142:45–48
11. Maundy B, Gift S (2005) Novel pseudo-exponential circuits. IEEE Trans Circuits Syst-II 52:675–679
12. Chua LO (1992) The genesis of Chua's circuit. Archiv Elektronik Uebertragungstechnik 46:250–257
13. Senani R, Gupta SS (1998) Implementation of Chua's chaotic circuit using current feedback op-amps. Electron Lett 34:829–830
14. Elwakil AS, Kennedy MP (2000) Improved implementation of Chua's chaotic oscillator using current feedback Op Amp. IEEE Trans Circ Syst-I 47:76–79

15. Cam U (2004) A new high performance realization of mixed-mode chaotic circuit using current-feedback operational amplifiers. Comput Electr Eng 30:281–290
16. Kilic R (2007) Mixed-mode chaotic circuit with Wien-bridge configuration: the results of experimental verification. Chaos Solitons Fractals 32:1188–1193
17. Senani R (1998) Realization of a class of analog signal processing/signal generation circuits: novel configurations using current feedback op-amps. Frequenz 52:196–206
18. Elwakil AS, Kennedy MP (1999) A family of Colpitts-like chaotic oscillators. J Franklin Inst 336:687–700
19. Elwakil AS, Kennedy MP (1999) Chaotic oscillators derived from Saito's double-screw hysteresis oscillator. IEICE Trans Fundament E82-A:1769–1775
20. Mahmoud SA, Elwakil AS, Soliman AM (1999) CMOS current feedback op amp-based chaos generators using novel active nonlinear voltage controlled resistors with odd symmetrical characteristics. Int J Electron 86:1441–1451
21. Elwakil AS, Kennedy MP (1999) Inductor less hyper chaos generator. Microelectron J 30:739–743
22. Elwakil AS, Kennedy MP (2000) Chua's circuit decomposition: a systematic design approach for chaotic oscillators. J Franklin Inst 337:251–265
23. Elwakil AS, Kennedy MP (2000) Novel chaotic oscillator configuration using a diode-inductor composite. Int J Electron 87:397–406
24. Elwakil AS, Kennedy MP (2001) Construction of classes of circuit-independent chaotic oscillator using passive-only nonlinear devices. IEEE Trans Circ Syst-I 48:289–307
25. Elwakil AS, Ozoguz S, Kennedy MP (2002) Creation of a complex butterfly attractor using a novel Lorenz-type system. IEEE Trans Circ Syst-I 49:527–530
26. Bernat P, Balaz I (2002) RC autonomous circuits with chaotic behavior. Radioengineering 11:1–5
27. Ozoguz S, Elwakil AS, Salama KN (2002) n-scroll chaos generator using nonlinear transconductor. Electron Lett 38:685–686
28. Ozoguz S, Elwakil AS, Kennedy MP (2002) Experimental verification of the butterfly attractor in a modified Lorenz system. Int J Bifurcation Chaos 12:1627–1632
29. Elwakil AS (2002) Non-autonomous pulse-driven chaotic oscillator based on Chua's circuit. Microelectron J 33:479–486
30. Kilic R (2003) On current feedback operational amplifier-based realizations of Chua's circuit. Circ Syst Sign Process 22:475–491
31. Yalcin ME, Suykens JAK, Vandewalle J (2004) True random bit generation from a double-scroll attractor. IEEE Trans Circ Syst-I 51:1395–1404
32. Elwakil AS (2004) Integrator-based circuit-independent chaotic oscillator structure. CHAOS 14:364–369
33. Ozoguz S, Elwakil AS (2004) On the realization of circuit-independent non-autonomous pulse-excited chaotic oscillator circuits. IEEE Trans Circuits Syst-II 51:552–556
34. Cam U, Kilic R (2005) Inductor less realization of non-autonomous MLC chaotic circuit using current-feedback operational amplifiers. J Circ Syst Comput 14:99–107
35. Cuautle ET, Hernandez AG, Delgado JG (2006) Implementation of a chaotic oscillator by designing Chua's diode with CMOS CFOAs. Analog Integr Circ Sign Process 48:159–162
36. Kilic R, Karauz B (2007) Implementation of a laboratory tool for studying mixed-mode chaotic circuit. Int J Bifurcation Chaos 17:3633–3638
37. Srisuchinwong B, Liou CH (2007) Improved implementation of Sprott's chaotic oscillators based on current-feedback op-amps. ECTI-CON:38–44
38. Petrzela J, Sotner R, Slezak J (2009) Electronically adjustable mixed-mode implementations of the jerk functions. Contemp Eng Sci 2:441–449
39. Elwakil AS, Ozoguz S (2006) On the generation of higher order chaotic oscillators via passive coupling of two identical or non-identical sinusoidal oscillators. IEEE Trans Circ Syst-I 53:1521–1532

Chapter 7
Realization of Other Building Blocks Using CFOAs

7.1 Introduction

Apart from the applications of the CFOAs as (pin-to-pin) replacements of voltage mode op-amps in which case they have been known to exhibit potential advantages as compared to the original VOA-based circuits, the CFOAs, as 4-terminal building blocks too, have yielded very interesting circuits, the type of which cannot be realized with conventional VOAs. This latter aspect has been amply demonstrated in the earlier chapters of this monograph.

In this chapter, we demonstrate how CFOAs have found potential applications in realizing a variety of other active building blocks too thereby quite often resulting in interesting topologies offering significant advantages.

7.2 Applications of the CFOAs in Realizing Other Building Blocks

Apart from being employed as four terminal building blocks in their own right, CFOAs have been employed to realize many other building blocks in the analog circuits literature, such as CCII+/−, unity gain voltage followers (VF) and unity gain current followers (CF), four terminal floating nullors (FTFN), Current differencing buffered amplifiers (CDBA), operational transresistance amplifiers (OTRA), Current differencing transconductance amplifiers (CDTA), third generation Current conveyors (CCIII), differential voltage second generation Current Conveyors (DVCC+), Current follower transconductance amplifiers (CFTA), current controlled current conveyor transconductance amplifier (CCCC-TA), differential-input buffered transconductance amplifier (DBTA), voltage differencing differential input buffered amplifier (VD-DIBA) etc.; see [1–146]. A few such equivalencies have been demonstrated earlier in [1, 2]. Also, a theorem facilitating replacement of CCIIs by CFOAs has been presented in [3].

R. Senani et al., *Current Feedback Operational Amplifiers and Their Applications*,
Analog Circuits and Signal Processing, DOI 10.1007/978-1-4614-5188-4_7,

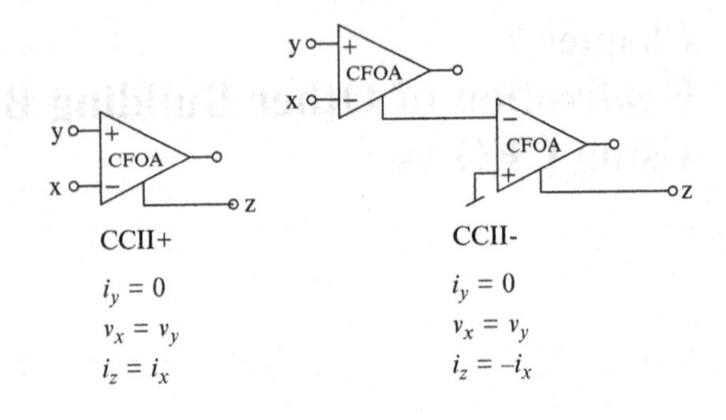

Fig. 7.1 Realization of CCII + and CCII− using CFOAs

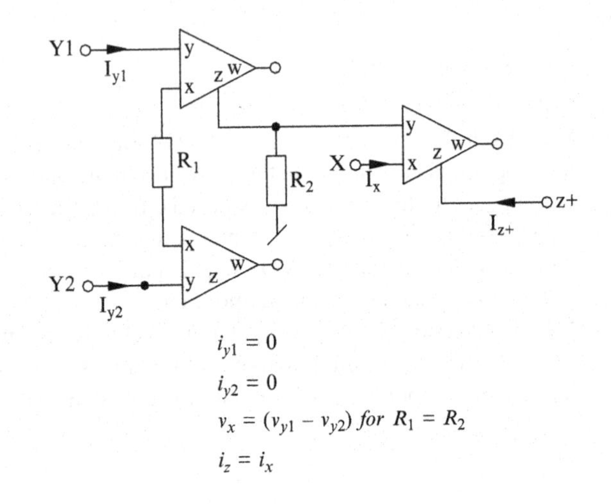

Fig. 7.2 Realization of a DVCC+ using CFOAs (based upon the idea given in [1])

7.2.1 CFOA Realizations of Various Kinds of Current Conveyors (CC)

A CCII+ is realizable with only a single CFOA while CCII− requires two of them as shown in Fig. 7.1.

$$\begin{array}{ll} i_y = 0 & i_y = 0 \\ v_x = v_y & v_x = v_y \\ i_z = i_x & i_z = -i_x \end{array} \tag{7.1}$$

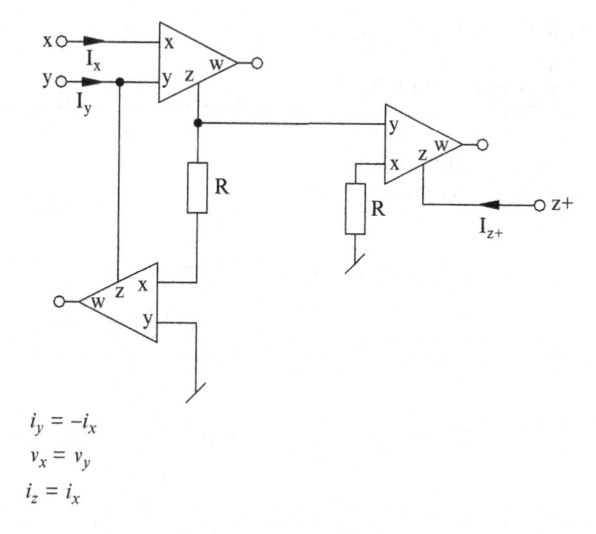

Fig. 7.3 Realization of a CCIII+ using CFOAs (based upon the idea given in [1])

An interesting variant of CCII, known as differential voltage CC (DVCC), can be realized with three CFOAs but needs two resistors as well (see Fig. 7.2).

$$\begin{aligned} i_{y1} &= 0 \\ i_{y2} &= 0 \\ v_x &= (v_{y1} - v_{y2}) \text{ for } R_1 = R_2 \\ i_z &= i_x \end{aligned} \tag{7.2}$$

Furthermore, yet another variant of CCs, known as a third generation Current Conveyor (CCIII) can also be realized with three CFOAs and two resistors as shown in Fig. 7.3.

$$\begin{aligned} i_y &= -i_x \\ v_x &= v_y \\ i_z &= i_x \end{aligned} \tag{7.3}$$

It is worth mentioning that CCII+, CCII−, DVCC or CCIII-based voltage-mode circuits would invariably require a voltage follower after the Z-terminal(s) of those CCs from which a voltage output is being taken since the Z-terminal being a current output terminal cannot be connected to the load impedance *directly* as this will modify and change the function realized by the circuit. Realizing the CC-based circuits by CFOAs will easily permit the Z-terminal voltage(s) to be available from the W-terminal(s) quite easily without requiring any additional external buffer because of the availability of an internal VF between the Z and W terminals in the CFOA(s) and thereby providing a remedy to this problem.

The CFOA-based CC implementations have been employed by many researchers for the verification of their CC-based circuit proposals, for example see [5–77] and the references cited therein.

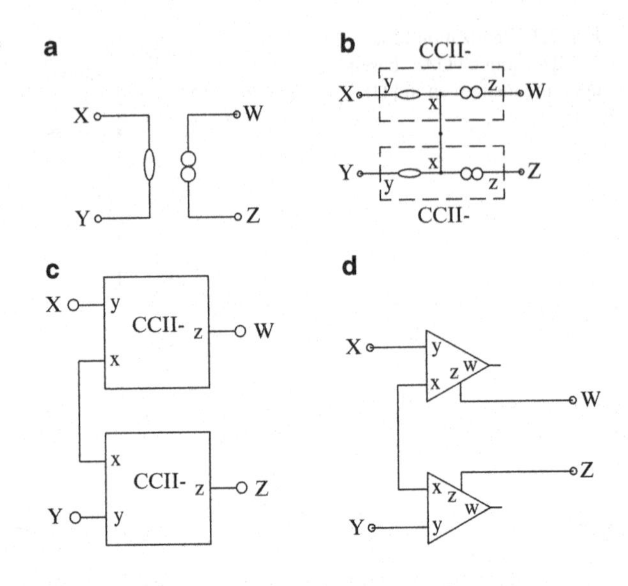

Fig. 7.4 Various steps in the implementation of FTFN using CFOAs (**a**) FTFN, (**b**) A FTFN using two three-terminal nullors, (**c**) FTFN using two CCII⊖, (**d**) FTFN using two CFOAs

7.2.2 *CFOA-Realization of the Four-Terminal-Floating-Nullors (FTFN)*

It was shown by a number of researchers (such as Nordholt [78], Stevenson [79], Huijsing [80] and Senani [81] (who coined the term 'four-terminal floating-nullor' (FTFN) to represent a fully floating nullor)) that fully floating versions of Op-amps (termed as operational floating amplifier (OFA)) and FTFNs are more versatile and flexible building blocks than the traditional op-amps in several applications.

It was suggested in [81] that a composite connection of two CCII− can be used to realize an FTFN (see Fig. 7.4). This follows from the fact that the representation of FTFN of Fig. 7.4a is equivalent to the combination of two 3-terminal nullors as shown in Fig. 7.4b where each 3-terminal nullor is equivalent to a CCII−, thus, finally, leading to the implementation of Fig. 7.4c. In fact, two CCII+ or two current feedback op-amps (CFOA) such as AD844, can be readily used to realize an FTFN using the same configuration (as in Fig. 7.4d).

A novel application of FTFNs has been in the area of floating impedance simulation; for instance, see [81, 91, 93].

In this context it may be noted that there are a number of realizations of floating impedances (FI) which call for two FTFNs and hence, four CFOAs, for instance, see [93]. In Fig. 7.5 we show a novel FI circuit[1] which is realizable with only a single FTFN and hence, only by two CFOAs. A routine circuit analysis shows that the floating inductance simulated by both the circuits is given by

[1] R. Senani (1987) Generation of new two-amplifier synthetic floating inductors. Electron Lett 23 (22):1202–1203

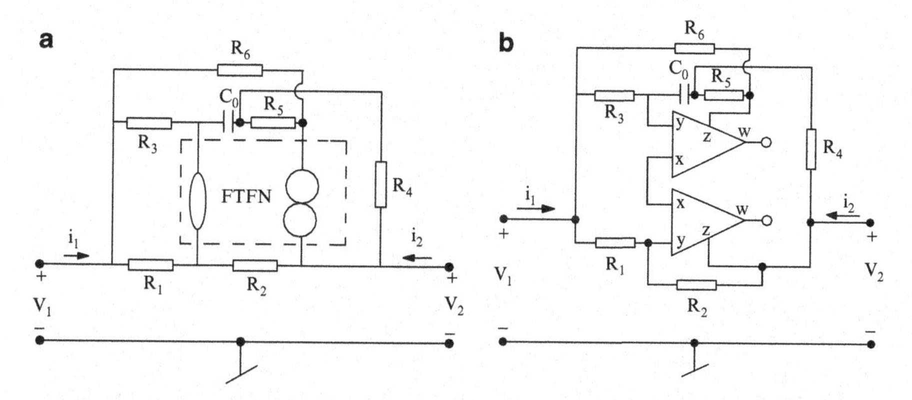

Fig. 7.5 (**a**) FTFN based lossless floating simulated inductance, (**b**) CFOA equivalent of a FTFN based floating simulated lossless inductance

$$L_{eq} = \frac{C_0 R_3 R_6 (1 + \frac{R_2}{R_1})}{(1 + \frac{R_5}{R_4})}; \text{ provided that } \frac{R_2}{R_4} = \left(\frac{R_1 + R_6}{R_5}\right) + \frac{R_1}{R_3}\left(1 + \frac{R_6}{R_5}\right) \quad (7.4)$$

From the discussions contained in Chap. 3, it may be seen that although a lossless FI was shown to be realizable with only two CFOAs as in the circuit of Fig. 3.15 but the circuit therein needs two matched capacitors. By contrast, the circuit of Fig. 7.5b has the novelty of employing only a single capacitor.

The CFOA-based FTFN implementation of Fig. 7.4d has been employed by many researchers for the verification of their FTFN-based propositions, for example, see [82–95] and the references cited therein.

7.2.3 *CFOA Realization of Operational Trans-resistance Amplifier (OTRA)*

Operational trans-resistance amplifier (OTRA) [119, 120] is characterized by the terminal equations

$$\begin{aligned} v_p &= 0 \\ v_n &= 0 \\ v_0 &= R_m(i_p - i_n); R_m \to \infty \end{aligned} \quad (7.5)$$

OTRAs have been employed as alternative building blocks to realize a number of functions such as all pass filters, inductance simulators, MOS-C biquads, sinusoidal oscillators and multivibrators. Although in several publications, CMOS OTRA architectures have been employed, in many others, the two-CFOA-based implementation of the OTRA as shown in Fig. 7.6 has been employed.

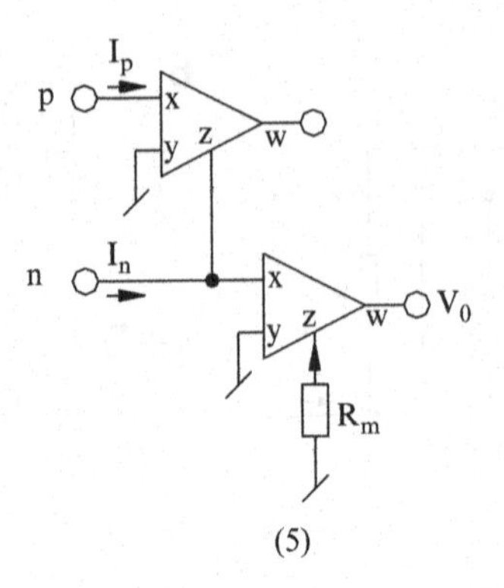

Fig. 7.6 Operational trans-resistance amplifier (OTRA) (adapted from [121] © 2004 Springer)

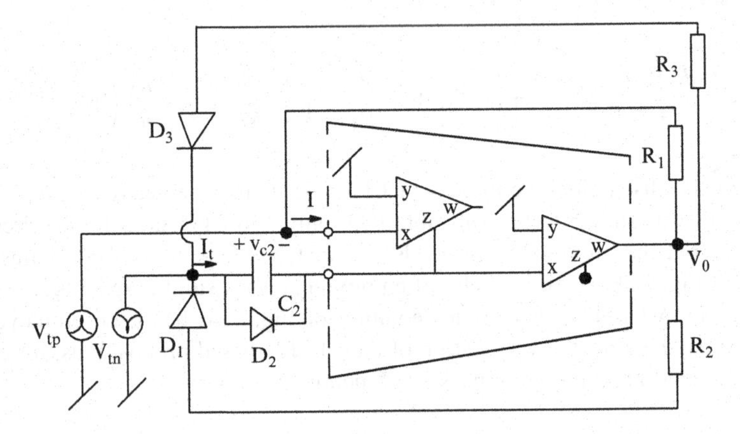

Fig. 7.7 The CFOA-version of the OTRA-based mono-stable multi-vibrator (adapted from [123] © 2006 IEEE)

Out of the various applications of OTRAs reported till date, a particularly interesting application was proposed by Lo and Chien [123] who realized a mono-stable multivibrator featuring both positive and negative triggering modes. A CFOA-version of this circuit is shown in Fig. 7.7, where the dotted box contains a two-CFOA realization of the OTRA.

The time period T during which the circuit remains in quasi-stable state and the recovery time T_r are given by

$$\begin{aligned} T &= R_2C_2 ln\left[\tfrac{R_1}{R_2}(1+K)\right] \\ T_r &= R_3C_2 ln\left(\tfrac{2-R_2/R_1}{1-K}\right), \quad \text{where } K = \tfrac{V_{D2}}{|V_{sat}|} \end{aligned} \tag{7.6}$$

The novelty of the structure of Fig. 7.7 is that no such circuit has been explicitly proposed in the existing literature using CFOAs.

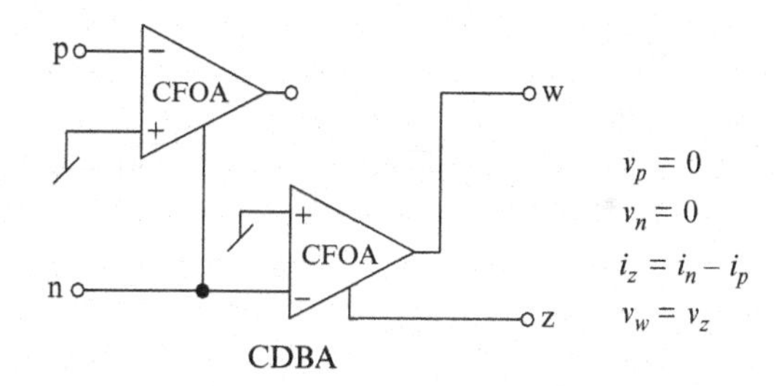

Fig. 7.8 Current differencing buffered amplifier (CDBA) (adapted from [96] © 1999 Elsevier)

The CFOA-based OTRA implementation has been employed by many researchers for the verification of their propositions for example, see [119–131] and the references citied therein.

7.2.4 CFOA Realization of Current Differencing Buffered Amplifier (CDBA) Based Circuits

Acar and Ozoguz in 1999 introduced current differencing buffered amplifier (CDBA) as a new versatile building blocks for analog signal processing [96]. Since then, CDBAs have been employed to realize a variety of linear and non-linear functions. Although fully integratable circuit implementations of CDBAs have been proposed by a number of researchers, CFOAs have been found to be quite handy in realizing them. Since CDBA is characterized by the terminal equations $V_p = V_n = 0$, $i_z = (i_p - i_n)$, and $V_w = V_z$ it was found that it could be readily implemented by two CFOAs as follows (Fig. 7.8).

$$\begin{aligned} v_p &= 0 \\ v_n &= 0 \\ i_z &= i_n - i_P \\ v_W &= v_z \end{aligned} \tag{7.7}$$

As an exemplary application of this equivalence we show in the following CDBA-based analog multiplier circuit and its CFOA-based implementation.

Assuming MOSFETs M_1 and M_2 to be matched and operating in triode region, the analysis of this circuit shows that the output voltage is given by

$$V_0 = \frac{K_{12}}{K_{34}(V_+ - V_T)} \cdot V_x V_y; \ |V_Z| = (V_+ - V_T) \tag{7.8}$$

where K is the transconductance parameter of MOSFETs M_1 and M_2 and is given by

$$K_{12} = \mu_s C_{ox} \left(\frac{W}{L}\right)_1 \tag{7.9}$$

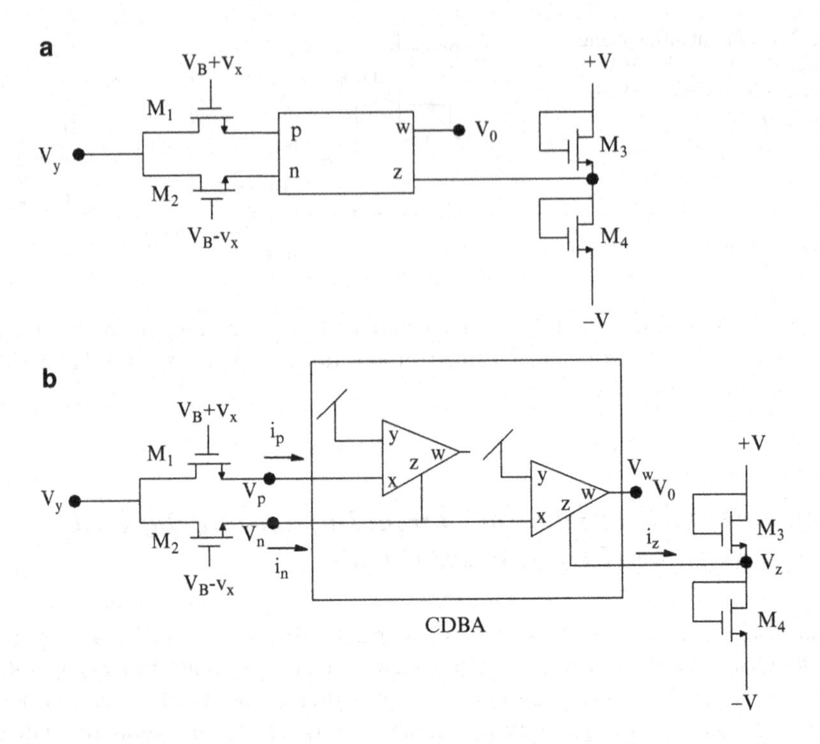

Fig. 7.9 (**a**) A CDBA-based analog multiplier (adapted from [101] © 2004 Springer). (**b**) CFOA-based implementation

and K_{34} is the transconductance parameter of the matched pair of MOSFETs M_3 and M_4 (both operating in saturation) and V_T is the threshold voltage of the MOSFETs.

It is worth pointing out that although two-CFOA based analog divider using four MOSFETs has been known in literature [4], any circuit for realizing a 4-quadrant analog multiplier using an exactly the same number of CFOAs has not been known. This application demonstrates how by realizing a CDBA through its two-CFOA-based implementation such a circuit (as in Fig. 7.9b) becomes possible. It must be emphasized that any such circuit using conventional VOAs and only four MOSFETs is not known to exist.

The CFOA-based CDBA implementation has been employed by many researchers for the verification of their propositions, for example, see [96–110] and the references cited therein.

7.2.5 *CFOA Realization of Circuits Containing Unity Gain Cells*

A large number of active circuit building blocks of varying complexity have been introduced by various researchers quite often having three or more external terminals. Interest in unity gain voltage follower (VF) and unity gain current

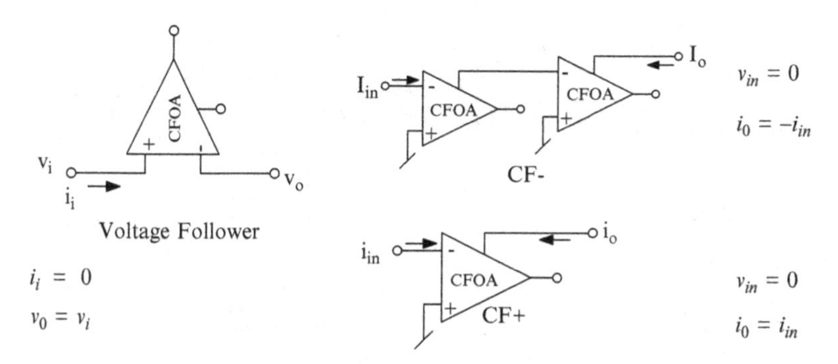

Fig. 7.10 Realization of voltage and current followers using CFOA(s)

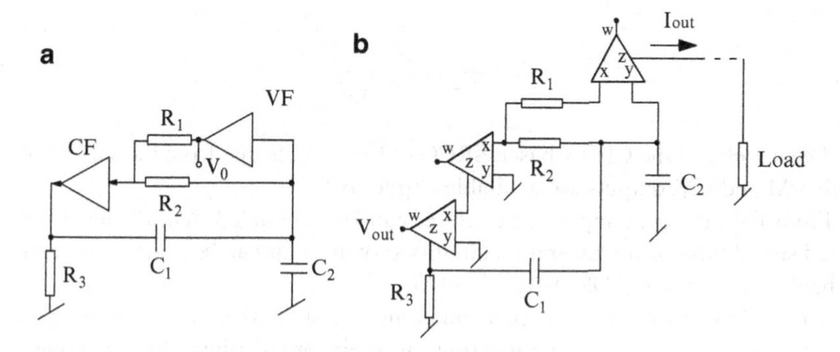

Fig. 7.11 (**a**) A VF-CF based SRCO, (**b**) CFOA implementation

follower (CF) is primarily attributed to the relatively larger bandwidth offered by them as well as the theoretical novelty that from several building blocks, VFs and CFs can be realized without requiring any external resistors. For instance, a unity gain VF is known to be realizable by a single VOA with its inverting terminal shorted to the output terminal. Likewise, a non-inverting VF and non-inverting/inverting CFs are also realizable with CFOAs without requiring any external resistances (see Fig. 7.10). In fact, these CFOA-based implementations of non-inverting VF and non-inverting/inverting CF have already been used by a number of researchers to prove the workability of their VFs/CFs based analog signal processing/signal generating circuits.

As an exemplary application, we demonstrate a VF-based sinusoidal oscillator and its CFOA implementation here in Fig. 7.11.

The condition of oscillation (CO) and the frequency of oscillation (FO) for both the circuits are given by

$$\mathrm{CO}: \ R_1 = \frac{C_1 R_3}{C_1 + C_2} \tag{7.10}$$

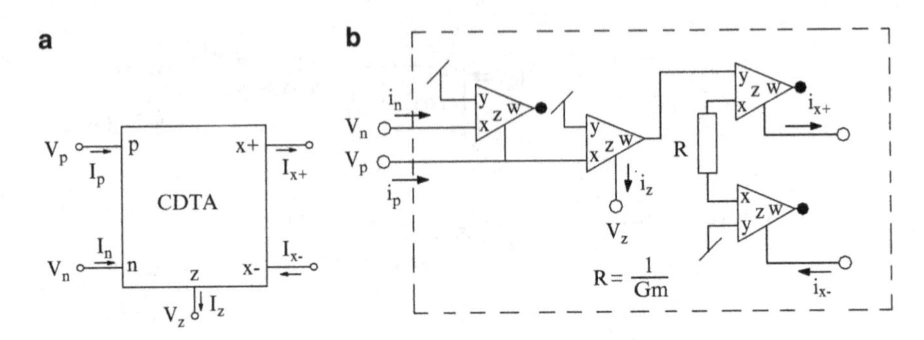

Fig. 7.12 (**a**) Symbolic notation, (**b**) CFOA implementation

and FO:

$$f_0 = \frac{1}{2\pi\sqrt{C_1 C_2 R_2 R_3}} \tag{7.11}$$

The novelty of the CFOA-based SRCO of Fig. 7.11 is that from the same circuit, both VM and CM outputs are available explicitly.

From the above example it, therefore, can be concluded that all the VFs/CFs based signal processing/generation circuits known so far can be practically realized in hardware by using AD844-type CFOAs.

The CFOA-based VF/CF implementations of Fig. 7.10 have been employed by many researchers for the verification of their propositions for example see [111–118] and the references cited therein

7.2.6 Current Differencing Transconductance Amplifier (CDTA)

The current differencing transconductance amplifier was introduced by Biolek in [135] as a new building block suitable for current-mode analog signal processing. The symbolic notation of the CDTA is shown in Fig. 7.12.

A CDTA is characterized by the following matrix equation.

$$\begin{pmatrix} I_z \\ I_{x+} \\ I_{x-} \\ V_p \\ V_n \end{pmatrix} = \begin{pmatrix} 0 & 0 & 0 & 1 & -1 \\ g_m & 0 & 0 & 0 & 0 \\ -g_m & 0 & 0 & 0 & 0 \\ 0 & 0 & 0 & 0 & 0 \\ 0 & 0 & 0 & 0 & 0 \end{pmatrix} \begin{pmatrix} V_z \\ V_{x+} \\ V_{x-} \\ I_p \\ I_n \end{pmatrix} \tag{7.12}$$

where g_m is the transconductance of the CDTA. An entirely CFOA-based implementation of the CDTA, based upon the idea given in [135], is shown in Fig. 7.12.

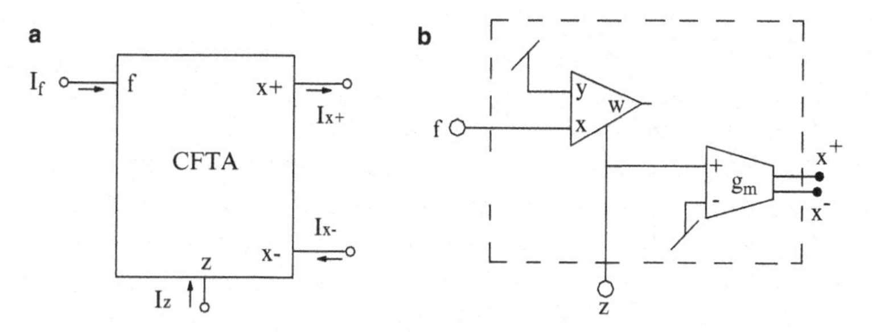

Fig. 7.13 (**a**) Symbolic notation, (**b**) CFOA implementation of the CFTA

CDTAs have received considerable attention in realizing various types of filters, oscillators, impedance simulators and other applications. For those cases where CDTAs have been implemented with CFOAs, the reader is referred to [136–140] and the references cited therein for further details.

7.2.7 Current Follower Transconductance Amplifiers (CFTA)

The current follower transconductance amplifier was introduced by Herencsar et al. in [141]. This current input, current output building block has been shown to be particularly useful in realizing analog signal processing functions requiring explicit current outputs. The symbolic notation of CFTA is shown in Fig. 7.13a and its characterizing matrix equation is given by (7.13).

$$\begin{pmatrix} I_z \\ I_{x+} \\ Ix- \\ V_f \end{pmatrix} = \begin{pmatrix} 0 & 0 & 0 & 1 \\ g_m & 0 & 0 & 0 \\ -g_m & 0 & 0 & 0 \\ 0 & 0 & 0 & 0 \end{pmatrix} \begin{pmatrix} V_z \\ V_{x+} \\ V_{x\_} \\ I_f \end{pmatrix} \tag{7.13}$$

A discrete version of CFTA can be implemented using one AD844 type CFOA and one balanced output transconductance amplifier such as MAX 435 and is shown in Fig. 7.13b.

This CFOA based implementation of the CFTA has been employed in the Realization of current-mode KHN-equivalent biquad using CFTAs presented in [142].

7.2.8 Current Controlled Current Conveyor Transconductance Amplifier (CCCC-TA)

The current controlled current conveyor transconductance amplifier (CCCC-TA) was introduced as a building block for analog signal processing by Siripruchyanun and Jaikla in [143]. A CFOA-based implementation of this was devised by Maheshwari et al. in [144] and is shown here in Fig. 7.14.

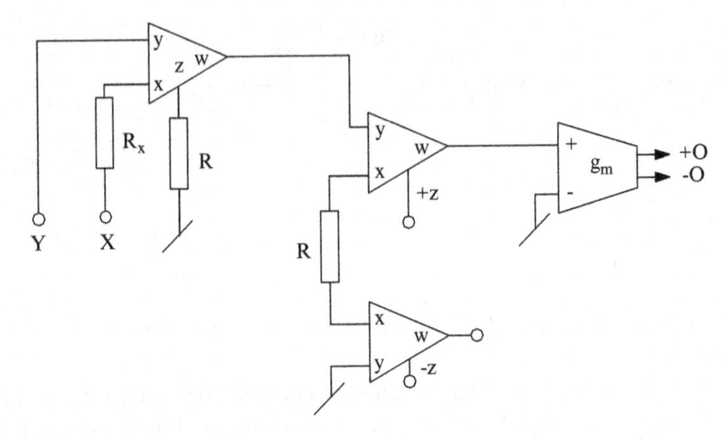

Fig. 7.14 A CFOA-based implementation of CCCC-TA (adapted from [144] © 2011 IET)

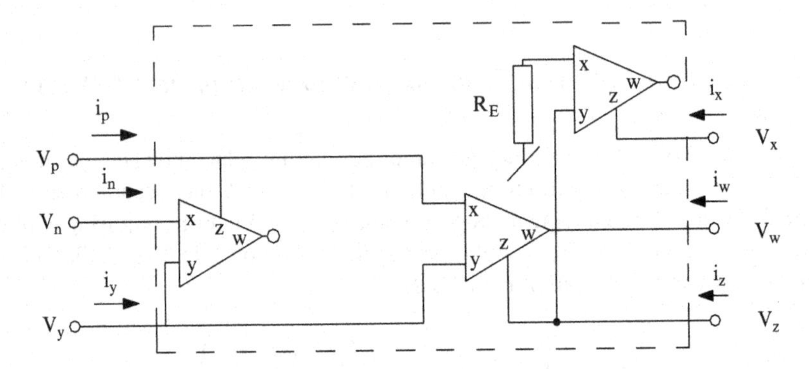

Fig. 7.15 A CFOA implementation of DBTA (adapted from [133] © 2009 IEICE)

7.2.9 *Differential Input Buffered Transconductance Amplifier (DBTA)*

The differential input buffered transconductance amplifier (DBTA) was introduced by Herencsar et al. in [132]. DBTA has been found to be an useful building block in realizing sinusoidal oscillators, quadrature oscillators and universal filters. Quite often, all these functions can be carried out effectively using only a single DBTA as in [133, 134]. A DBTA is a six port building block characterizing by the following equation:

$$v_p = v_y,\ v_n = v_y,\ i_y = 0,\ i_z = (i_p - i_n),\ v_w = v_z,\ i_x = g_m v_z \quad (7.14)$$

where $g_m = 1/R_E$

A CFOA implementation of this building block is shown in Fig. 7.15 and has been used in [133] to verify their proposed quadrature oscillator.

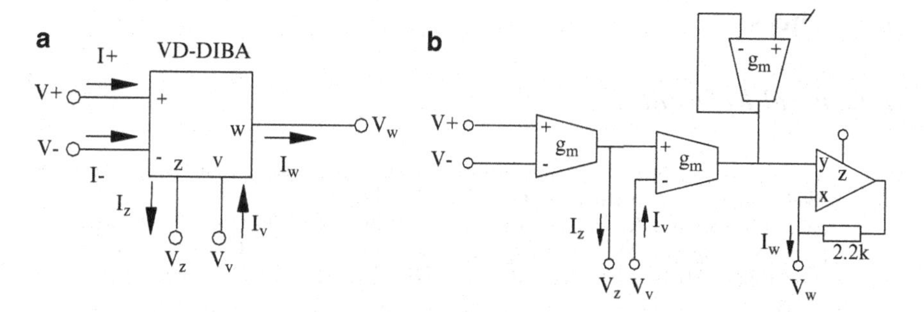

Fig. 7.16 (**a**) Schematic symbol, (**b**) CFOA implementation (adapted from [146] © 2011 Hindawi Publishing Corporation)

7.2.10 Voltage Differencing Differential Input Buffered Amplifier (VD-DIBA)

The voltage differencing differential input buffered amplifier (VD-DIBA) was introduced by Biolek et al. in [145]. Although some applications of VD-DIBAs have been reported in the open literature but for the very first time an application implementing VD-DIBA using OTAs and CFOA was presented in [146].

The schematic symbol of the VD-DIBA is shown in Fig. 7.16a.

A VD-DIBA is characterized by the following matrix equation.

$$\begin{pmatrix} I_+ \\ I_- \\ I_z \\ I_v \\ V_w \end{pmatrix} = \begin{pmatrix} 0 & 0 & 0 & 0 & 0 \\ 0 & 0 & 0 & 0 & 0 \\ g_m & -g_m & 0 & 0 & 0 \\ 0 & 0 & 0 & 0 & 0 \\ 0 & 0 & 1 & -1 & 0 \end{pmatrix} \begin{pmatrix} V_+ \\ V_- \\ V_z \\ V_v \\ I_w \end{pmatrix} \tag{7.15}$$

The CFOA-implementation of the VD-DIBA as proposed in [146] is shown in Fig. 7.16b

7.3 Concluding Remarks

In this chapter, we have demonstrated that CFOAs (sometimes only CFOAs and sometimes in conjunction with additional resistors and/or OTA(s)) have potential applications in realizing a variety of other active building blocks such as CCII+/− CCIII, unity gain VF and CF, FTFN, CDBA, OTRA, CDTA, DVCC+, CFTA, CCCC-TA, DBTA and VD-DIBA-all proposed and being employed in recent technical literature.

It was also demonstrated that when CFOAs are used to realize the circuits employing the considered building blocks, in several cases, quite versatile and efficient functional circuits result. These applications, therefore, further establish the flexibility and versatility of CFOAs in analog circuit design.

References

I. General References

1. Yuce E, Minaei S (2007) Realization of various active devices using commercially available AD844s and external resistors. Electron World 113:46–49
2. Senani R, Bhaskar DR, Gupta SS, Singh VK (2012) Current-feedback op-amp, their applications, Bipolar/CMOS implementation and their variants, Chapter 2. In: Esteban Tlelo-Cuautle (ed) Integrated circuits in analog signal processing. Springer, New York, pp 61–84
3. Aronhime P, Wang K, Qian T (2001) Generalization of a theorem for replacing CCIIs by CFOAs in current-mode circuits. Analog Integr Circ Sign Process 28:27–33
4. Liu SI, Chen JJ (1995) Realisation of analogue divider using current feedback amplifiers. IEE Proc Circ Devices Syst 142:45–48

II. Use of CFOAs in Realizing Various Types of Current Conveyors

5. Hou CL, Chen RD, Wu YP, Hu PC (1993) Realization of grounded and floating immittance function simulators using current conveyors. Int J Electron 74:917–923
6. Svoboda JA (1994) Comparison of RC op-amp and RC current conveyors filters. Int J Electron 76:615–626
7. Chang C, Lee MS (1994) Universal voltage-mode filter with three inputs and one output using three current conveyors and one voltage follower. Electron Lett 30:2112–2113
8. Vosper JV, Heima M (1996) Comparison of single- and dual-element frequency control in a CCII-based sinusoidal oscillator. Electron Lett 32:2293–2294
9. Horng JW, Tsai CC, Lee MH (1996) Novel universal voltage-mode biquad filter with three inputs and one outputs using only two current conveyors. Int J Electron 80:543–546
10. Wang HY, Lee CT (1997) Immittance function simulator using a single current conveyor. Electron Lett 33:574–576
11. Liu SI, Lee JL (1997) Voltage-mode universal filters using two current conveyors. Int J Electron 82:145–149
12. Horng JW, Lee MH, Cheng HC, Chang CW (1997) New CCII-based voltage-mode universal biquadratic filter. Int J Electron 82:151–155
13. Chang CM (1997) Multifunction biquadratic filters using current conveyors. IEEE Trans Circ Syst II 44:956–958
14. Lee JY, Tsao HW (1992) True RC integrators based on current conveyors with tunable time constants using active control and modified loop technique. IEEE Trans Instrum Meas 41 (5):709–714
15. Liu S-I, Kuo J-H, Tsay J-H (1992) New CCII-based current-mode biquadratic filters. Int J Electron 72:243–252
16. Wilson B (1992) Trends in current conveyor and current-mode amplifier design. Int J Electron 73:573–583
17. Brunn E, Olesen OH (1992) Conveyor implementations of generic current mode circuits. Int J Electron 73:129–140
18. Svoboda JA (1994) Transfer function synthesis using current conveyors. Int J Electron 76:611–614
19. Fabre A, Dayoub F, Duruisseau L, Kamoun M (1994) High input impedance insensitive second-order filters implemented from current conveyors. IEEE Trans Circ Syst I 41:918–921

20. Martinez PA, Celma S, Gutiérrez I (1995) Wien-type oscillators using $CCII^+$. Analog Integr Circ Sign Process 7:139–147
21. Hwang YS, Liu SI, Wu DS, Wu YP (1995) Linear transformation all-pole filters based on current conveyors. Int J Electron 79(4):439–445
22. Soliman AM (1996) New inverting-non-inverting bandpass and lowpass biquad circuit using current conveyors. Int J Electron 81:577–583
23. Cajka J, Lindovsky D (1997) Universal RC-Active network using CCII+. J Electr Eng 48:98–100
24. Soliman AM (1997) Generation of current conveyors-based all-pass filters from op-amp-based circuits. IEEE Trans Circ Syst II 44:324–330
25. Al-Walaie SA, Alturaigi MA (1997) Current mode simulation of lossless floating inductance. Int J Electron 83:825–829
26. Vrba K, Cajka J, Zeman V (1997) New RC-active networks using current conveyors. Radioengineering 6:18–21
27. Vrba K, Cajka J (1997) High-order one port elements for lowpass filter realization. J Electr Eng 48:31–34
28. Cajka J, Dostal T, Vrba K (1997) Realization of Nth-order voltage transfer function using current conveyors CCII. Radioengineering 6:22–25
29. Cicekoglu O (1998) New current conveyor based active-gyrator implementation. Microelectron J 29:525–528
30. Cicekoglu MO (1998) Active simulation of grounded inductors with CCII + s and grounded passive elements. Int J Electron 4:455–462
31. Ozoguz S, Acar C (1998) On the realization of floating immittance function simulators using current conveyors. Int J Electron 85:463–475
32. Soliman AM (1999) Synthesis of grounded capacitor and grounded resistor oscillators. J Franklin Inst 336:735–746
33. Cicekoglu O, Ozcan S, Kuntman H (1999) Insensitive multifunction filter implemented with current conveyors and only grounded passive elements. Frequenz 53:158–160
34. Chang CM, Tu SH (1999) Universal voltage-mode filter with four inputs and one output using two CCII + s. Int J Electron 86:305–309
35. Abuelma'atti MT, Tasadduq NA (1999) New negative immittance function simulators using current conveyors. Microelectron J 30:911–915
36. Soliman AM, Elwakil AS (1999) Wien oscillators using current conveyors. Comput Electr Eng 25:45–55
37. Abuelma'atti MT (2000) Comment on: Active simulation of grounded inductors with CCII + s and grounded passive elements. Int J Electron 87:177–181
38. Cicekoglu O (2000) Reply to comment on: Active simulation of grounded inductors with CCII + s and grounded passive elements. Int J Electron 87:183–184
39. Abuelma'atti MT, Tasadduq NA (2000) Current-mode lowpass/bandpass and highpass filter using CCII + s. Frequenz 54:162–164
40. Abuelma'atti MT (2000) New sinusoidal oscillators with fully uncoupled control of oscillation frequency and condition using three CCII+s. Analog Integr Circ Sign Process 24:253–261
41. Ozoguz S, Acar C, Toker A, Gunes EO (2001) Derivation of low-sensitivity current-mode CCII-based filters. IEE Proc Circ Devices Syst 148:115–120
42. Horng JW (2001) A sinusoidal oscillator using current-controlled current conveyors. Int J Electron 88:659–664
43. Cicekoglu O, Toker A, Kuntman H (2001) Universal immittance function simulators using current conveyors. Comput Electr Eng 27:227–238
44. Biolek D, Cajka J, Vrba K, Zeman V (2002) Nth-order allpass filters using current conveyors. J Electr Eng 53:50–53

45. Hwang YS, Hung PT, Chen W, Liu SI (2002) Systematic generation of current-mode linear transformation filters based on multiple outputs CCIIs. Analog Integr Circ Sign Process 32:123–134
46. Aksoy M, Ozcan S, Cicekoglu O, Kuntman H (2002) High output impedance current-mode third-order Butterworth filter topologies employing unity gain voltage buffers and equal-valued passive components. Int J Electron 90:589–598
47. Shah NA, Malik MA (2005) High impedance voltage- and current-mode multifunction filters. Int J Electron Commun (AEU) 59:262–266
48. Kumar P, Pal K (2005) Variable Q all-pass, notch and band-pass filters using single CCII. Frequenz 59:235–239
49. Horng JW (2004) High input impedance voltage-mode universal biquadratic filters with three inputs using plus-type CCIIs. Int J Electron 91:465–475
50. Gift SJG (2004) New simulated inductor using operational conveyors. Int J Electron 91:477–483
51. Abuelma'atti MT, Bentrcia A, Al-Shahrani SM (2004) A novel mixed-mode current conveyor-based filter. Int J Electron 91:191–197
52. Horng JW (2004) Voltage-mode universal biquadratic filters using CCIIs. IEICE Trans Fundam E-87-A:406–409
53. Horng JW, Hou CL, Chang CM, Chung WY, Tang HW, Wen YH (2005) Quadrature oscillators using CCIIs. Int J Electron 92:21–31
54. Fongsamut C, Fujii N, Surakampontorn W (2005) Two new RC oscillators using CCIIs. Proc ISCIT 2:1138–1141
55. Khan AA, Bimal S, Dey KK, Roy SS (2005) Novel RC sinusoidal oscillator using second-generation current conveyors. IEEE Trans Instrum Meas 54:2402–2406
56. Horng JW (2005) Current conveyors based allpass filters and quadrature oscillators employing grounded capacitors and resistors. Comput Electr Eng 31:81–92
57. Abuelma'atti MT, Shahrani SMA, Al-Absi MK (2005) Simulation of a mutually coupled circuit using plus-type CCIIs. Int J Electron 92:49–54
58. Keskin AU (2005) Single CFA-based NICs with impedance scaling properties. J Circ Syst Comput 14:195–203
59. Horng JW, Hou CL, Chang CM, Chung WY, Wei HY (2005) Voltage-mode universal biquadratic filters with one input and five outputs using MOCCIIs. Comput Electr Eng 31:190–202
60. Horng JW, Hou CL, Chang CM, Chung WY (2006) Voltage-mode universal biquadratic filters with one input and five outputs. Analog Integr Circ Sign Process 47:73–83
61. Yuce E, Cicekoglu O (2006) The effects of non-idealities and current limitations on the simulated inductances employing current conveyors. Analog Integr Circ Sign Process 46:103–110
62. Pandey N, Paul SK, Bhattacharyya A, Jain SB (2006) A new mixed mode biquad using reduced number of active and passive elements. IEICE Electron Express 3:115–121
63. Metin B, Cicekoglu O (2006) A novel floating lossy inductance realization topology with NICs using current conveyors. IEEE Trans Circ Syst II 53:483–486
64. Maundy B, Gift S, Aronhime P (2007) Realization of a GIC using hybrid current conveyors/operational amplifier circuits. 50th Midwest Symp Circ Syst (MWSCAS 2007), pp 163–166, DOI:10.1109/MWSCAS.2007.4488562
65. Maundy B, Gift S, Aronhime P (2007) A novel hybrid active inductor. IEEE Trans Circ Syst II 54:663–667
66. Kumar P, Pal K, Rana S (2008) High input impedance universal biquadratic filters using current conveyors. J Active Passive Electron Devices 3:17–27
67. Kumar P, Pal K (2008) Universal biquadratic filter using single current conveyor. J Active Passive Electron Devices 3:7–16
68. Pandey N, Paul SK, Jain SB (2008) Voltage mode universal filter using two plus type CCIIs. J Active Passive Electron Devices 3:165–173

69. Yuce E (2008) Negative impedance converter with reduced nonideal gain and parasitic impedance effect. IEEE Trans Circ Syst I 55:276–283
70. Yuce E (2008) Grounded inductor simulators with improved low-frequency performances. IEEE Trans Instrum Meas 57:1079–1084
71. Maundy B, Gift S, Aronhime P (2008) Practical voltage/current-controlled grounded resistor with dynamic range extension. IET Circ Devices Syst 2:201–206
72. Ferri G, Guerrini N, Silverii E, Tatone A (2008) Vibration damping using CCII-based inductance simulators. IEEE Trans Instrum Meas 57(5):907–914
73. Pal K, Nigam MJ (2008) Novel active impedances using current conveyors. J Active Passive Electron Devices 3:29–34
74. Yuce E, Minaei S (2008) Electronically tunable simulated transformer and its application to Stagger-tuned filter. IEEE Trans Instrum Meas 57:2083–2088
75. Senani R, Bhaskar DR (2008) Comment: Practical voltage/current-controlled grounded resistor with dynamic range extension. IET Circ Devices Syst 2:465–466
76. Skotis GD, Psychalinos C (2010) Multiphase sinusoidal oscillator using second generation current conveyors. Int J Electron Commun (AEU) 64:1178–1181
77. Maheshwari S (2010) Current-mode third-order quadrature oscillator. IET Circ Devices Syst 4:188–195

III. Realization of Four Terminal Floating Nullors (FTFN) Using CFOAs

78. Nordholt EH (1982) Extending op-amp capabilities by using a current-source power supply. IEEE Trans Circ Syst 29:411–412
79. Stevenson JK (1984) Two-way circuits with inverse transmission properties. Electron Lett 20:965–967
80. Huijsing JH (1990) Operational amplifier. IEE Proc Circ Devices Syst 137:131–136
81. Senani R (1987) A novel application of four-terminal floating nullor. IEEE Proc 75:1544–1546
82. Hou CL, Yean R, Chang CK (1996) Single-element controlled oscillators using single FTFN. Electron Lett 32:2032–2033
83. Liu SI (1997) Single-resistance-controlled sinusoidal oscillators using two FTFNs. Electron Lett 33:14
84. Abuelma'atti MT, Al-Zaher HA (1998) Current-mode sinusoidal oscillator using two FTFNs. Proc Natl Sci Counc Repub China A 22:758–764
85. Wang HY, Lee CT (1998) Realization of R-L and C-D immittances using single FTFN. Electron Lett 34:502–503
86. Bhaskar DR (1999) Single resistance controlled sinusoidal oscillator using single FTFN. Electron Lett 35:190
87. Abuelma'atti MT, Al-Zaher HA (1999) Current-mode quadrature sinusoidal oscillators using two FTFNs. Frequenz 53:27–30
88. Abuelma'atti MT, Al-Zaher HA (1999) Current-mode sinusoidal oscillator using single FTFN. IEEE Trans Circ Syst II 46:69–74
89. Gunes EO, Anday F (1999) Realization of voltage/current-mode filters using four-terminal floating nullors. Microelectron J 30:211–216
90. Cam U, Toker A, Cicekoglu O, Kuntman H (2000) Current-mode high output impedance sinusoidal oscillator configuration employing single FTFN. Analog Integr Circ Sign Process 24:231–238
91. Cam U, Cicekoglu O, Kuntman H (2000) Universal series and parallel immittance simulators using four terminal floating nullors. Analog Integr Circ Sign Process 25:59–66

92. Lee CT, Wang HY (2001) Minimum realization for FTFN based SRCO. Electron Lett 37:1207–1208
93. Cam U, Cicekoglu O, Kuntman H (2001) Novel lossless floating immittance simulator employing only two FTFNs. Analog Integr Circ Sign Process 29:233–235
94. Wang HY, Chung H, Huang WC (2002) Realization of an nth-order parallel immittance function employing only $(n - 1)$ FTFNs. Int J Electron 89:645–650
95. Bhaskar DR (2002) Grounded-capacitor SRCO using only one PFTFN. Electron Lett 38(20):1156–1157

IV. CFOA-Based of Current Differencing Buffered Amplifier (CDBA)

96. Acar C, Ozoguz S (1999) A new versatile building block: current differencing buffered amplifier for analog signal processing filters. Microelectron J 30:157–160
97. Acar C, Ozoguz S (2000) nth-order current transfer function synthesis using current differencing buffered amplifier: signal-flow graph approach. Microelectron J 31:49–53
98. Ozcan S, Toker A, Acar C, Kuntman H, Cicekoglu O (2000) Single resistance-controlled sinusoidal oscillators employing current differencing buffered amplifier. Microelectron J 31:169–174
99. Ozcan S, Kuntman H, Cicekoglu O (2002) Cascadable current mode multipurpose filters employing current differencing buffered amplifier (CDBA). Int J Electron Commun (AEU) 56:67–72
100. Horng JW (2002) Current differencing buffered amplifiers based single resistance controlled quadrature oscillator employing grounded capacitors. IEICE Trans Fundam E85-A:1416–1419
101. Keskin AU (2004) A four quadrant analog multiplier employing single CDBA. Analog Integr Circ Sign Process 40:99–101
102. Keskin AU (2005) Voltage-mode notch filters using single CDBA. Frequenz 59:1–4
103. Tangsrirat W, Surakampontorn W (2005) Realization of multiple-output biquadratic filters using current differencing buffered amplifiers. Int J Electron 92:313–325
104. Keskin AU (2006) Multi-function biquad using single CDBA. Electr Eng 88:353–356
105. Keskin AU, Aydin C, Hancioglu E, Acar C (2006) Quadrature oscillator using current differencing buffered amplifiers (CDBA). Frequenz 60:21–23
106. Koksal M, Sagbas M (2007) A versatile signal flow graph realization of a general transfer function by using CDBA. Int J Electron Commun (AEU) 61:35–42
107. Tangsrirat W, Pisitchalermpong S (2007) CDBA-based quadrature sinusoidal oscillator. Frequenz 61:102–104
108. Tangsrirat W, Pukkalanun T, Surakampontorn W (2008) CDBA-based universal biquad filter and quadrature oscillator. Active Passive Electron Comp: Article ID 247171
109. Tangsrirat W, Prasertsom D, Piyatat T, Surakampontorn W (2008) Single-resistance-controlled quadrature oscillator using current differencing buffered amplifiers. Int J Electron 95:1119–1126
110. Pathak JK, Singh AK, Senani R (2010) Systematic realization of quadrature oscillators using current differencing buffered amplifiers. IET Circ Devices Syst 5:203–211

V. Unity Gain VF and CF Based Circuits Realized with CFOAs

111. Celma S, Sabadell J, Martinez P (1995) Universal filter using unity-gain cells. Electron Lett 31:1817–1818

112. Senani R, Gupta SS (1997) Universal voltage-mode/current-mode biquad filter realized with current feedback op-amps. Frequenz 51:203–208
113. Abuelma'atti MT, Daghreer HA (1997) New single-resistor controlled sinusoidal oscillator circuit using unity-gain current followers. Active Passive Electron Comp 20:105–109
114. Weng RM, Lai JR, Lee MH (2000) New universal biquad filters using only two unity gain cells. Int J Electron 87(1):57–61
115. Kuntman H, Cicekoglu O, Ozcan S (2002) Realization of current-mode third order Butterworth filters employing equal valued passive elements and unity gain buffers. Analog Integr Circ Sign Process 30:253–256
116. Gupta SS, Senani R (2004) New single resistance controlled oscillators employing a reduced number of unity-gain cells. IEICE Electron Express 1:507–512
117. Keskin AU, Toker A (2004) A NIC with impedance scaling properties using unity gain cells. Analog Integr Circ Sign Process 41:85–87
118. Nandi R, Kar M (2009) Third order lowpass Butterworth filters using unity gain current amplifiers. IEICE Electron Express 6:1450–1455

VI. Use of CFOAs in Realizing Operational Trans-resistance Amplifiers (OTRA)

119. Chen JJ, Tsao HW, Chen CC (1992) Operational transresistance amplifier using CMOS technology. Electron Lett 28:2087–2088
120. Salama KN, Elwan HO, Soliman AM (2001) Parasitic-capacitance-insensitive voltage-mode MOSFET-C filters using differential current voltage conveyor. Circ Syst Sign Process 20:11–26
121. Cam U, Kacar F, Cicekoglu O, Kuntman H, Kuntman A (2004) Novel two OTRA-based grounded immittance simulator topologies. Analog Integr Circ Sign Process 39:169–175
122. Hou CL, Chien HC, Lo YK (2005) Square wave generators employing OTRAs. IEE Proc Circ Devices Syst 152:718–722
123. Lo YK, Chien HC (2006) Current-mode monostable multivibrators using OTRAs. IEEE Trans Circ Syst II 53:1274–1278
124. Kilinc S, Salama KN, Cam U (2006) Realization of fully controllable negative inductance with single operational transresistance amplifier. Circ Syst Sign Process 5(1):47–57
125. Chen JJ, Tsao HW, Liu SI, Chiu W (1995) Parasitic-capacitance-insensitive current-mode filters using operational transresistance amplifiers. IEE Proc Circ Devices Syst 142:186–192
126. Lo YK, Chien HC (2007) Switch-controllable OTRA-based square/triangular waveform generator. IEEE Trans Circ Syst II 54:1110–1114
127. Lo YK, Chien HC (2007) Single OTRA-based current-mode monostable multivibrator with two triggering modes and a reduced recovery time. IET Circ Devices Syst 1:257–261
128. Lo YK, Chien HC, Chiu HJ (2008) Switch-controllable OTRA-based bistable multivibrator. IET Circ Devices Syst 2:373–382
129. Lo YK, Chien HC, Chiu HJ (2010) Current-input OTRA Schmitt trigger with dual hysteresis modes. Int J Circ Theory Appl 38:739–746
130. Sanchez-Lopez C, Martinez-Romero E, Tlelo-Cuautle E (2011) Symbolic analysis of OTRAs-based circuits. J Appl Res Technol 9:69–80
131. Gupta A, Senani R, Bhaskar DR, Singh AK (2011) OTRA-based grounded-FDNR and grounded-inductance simulators and their applications. Circ Syst Sign Process 31:489–499

VII. Use of CFOA in Realizing Differential Input and Buffered Trans-Conductance Amplifier (DBTA)

132. Herencsar N, Vrba K, Koton J, Lattenberg I (2009) The conception of differential-input buffered and transconductance amplifier (DBTA) and its application. IEICE Electron Express 6(6):329–334
133. Herencsar N, Koton J, Vrba K, Lahiri A (2009) New voltage-mode quadrature oscillator employing single DBTA and only grounded passive elements. IEICE Electron Express 6:1708–1714
134. Herencsar N, Koton J, Vrba K, Lattenberg I (2010) New voltage-mode universal filter and sinusoidal oscillator using only single DBTA. Int J Electron 97:365–379

VIII. Current Differencing Transconductance Amplifier (CDTA) Using CFOAs

135. Biolek D (2003) CDTA-building block for current-mode analog signal processing. Proc ECCTD'03, Krakow, Poland III, pp 397–400
136. Bekri AT, Anday F (2005) nth-order low-pass filter employing current differencing transconductance amplifiers. Proc 2005 European Conf Circ Theor Appl 2:II/193-II/196
137. Tangsrirat W (2007) Current differencing transconductance amplifier-based current-mode four-phase quadrature oscillator. Indian J Eng Mater Sci 14:289–294
138. Prasad D, Bhaskar DR, Singh AK (2008) Realisation of single-resistance-controlled sinusoidal oscillator: a new application of the CDTA. WSEAS Trans Electron 5:257–259
139. Silapan P, Siripruchyanum M (2011) Fully and electronically controllable current-mode Schmitt triggers employing only single MO-CCCDTA and their applications. Analog Integr Circ Sign Process 68:111–128
140. Lahiri A (2010) Resistor-less mixed-mode quadrature sinusoidal oscillator. Int J Comput Electr Eng 2:63–66

IX. Current Follower Transconductance Amplifier (CFTA) Realized with CFOAs

141. Herencsar N, Koton J, Vrba K, Misurec J (2009) A novel current-mode SIMO type universal filter using CFTAs. Contemp Eng Sci 2:59–66
142. Herencsar N, Koton J, Vrba K (2010) Realization of current-mode KHN-equivalent biquad using current follower transconductance amplifiers (CFTAs). IEICE Trans Fundam E93:1816–1819

X. CFOA Realizations of Current-Controlled Current Conveyor Transconductance Amplifier (CCCC-TA)

143. Siripruchyanun M, Jaikla W (2007) Current controlled current conveyor transconductance amplifier (CCCCTA): a building block for analog signal processing. Electr Eng 19:443–453
144. Maheshwari S, Singh SV, Chauhan DS (2011) Electronically tunable low-voltage mixed-mode universal biquad filter. IET Circ Devices Syst 5(3):149–158

XI. Voltage-Differencing Differential-Input Buffered Amplifier (VD-DIBA) Realized with CFOAs

145. Biolek D, Senani R, Biolkova V, Kolka Z (2008) Active elements for analog signal processing: classification, review, and new proposals. Radioengineering 17:15–32
146. Prasad D, Bhaskar DR, Pushkar KL (2011) Realization of new electronically controllable grounded and floating simulated inductance circuits using voltage differencing differential input buffered amplifiers. Active passive Electron Compon: Article ID 101432

Chapter 8
Advances in the Design of Bipolar/CMOS CFOAs and Future Directions of Research on CFOAs

8.1 Introduction

Motivated by the widespread and potential applications of the CFOAs, as evidenced by the publication of several hundred research papers, most of which have been cited in this monograph, various researchers have worked towards evolving new bipolar or CMOS architectures for CFOAs possessing one or more of the several desirable features such as reduced input impedance at X-input terminal, better accuracy of unity current gain between Z and X terminals and unity voltage gain between Y and X terminals, higher slew rates, increased CMRR, enhanced gain-bandwidth-products, lower DC offset voltage, better current drive capability and reduced operating voltages etc. In this chapter, we will outline major developments which have taken place on the improvement in the design of Bipolar/CMOS/BiCMOS CFOAs and will also make some comments on future directions of research on CFOAs and their applications.

8.2 Progress in the Design of Bipolar CFOAs

Although there have been hundreds of publications on improving the design of Current Conveyors, surprisingly, in spite of the wide spread applications of the CFOAs, as exemplified in this monograph and in the references cited in the various chapters, there has been comparatively a much lesser effort [1–24] on improving the design of bipolar and CMOS CFOAs.

8.2.1 Bipolar CFOA with Improved CMRR

Conventional CFOAs generally exhibit a poor CMRR. In [1], Tammam et al. have carried out an analysis of CMRR of a conventional CFOA and have identified the mechanism primarily responsible for the CMRR. They have then presented a

R. Senani et al., *Current Feedback Operational Amplifiers and Their Applications*, Analog Circuits and Signal Processing, DOI 10.1007/978-1-4614-5188-4_8,

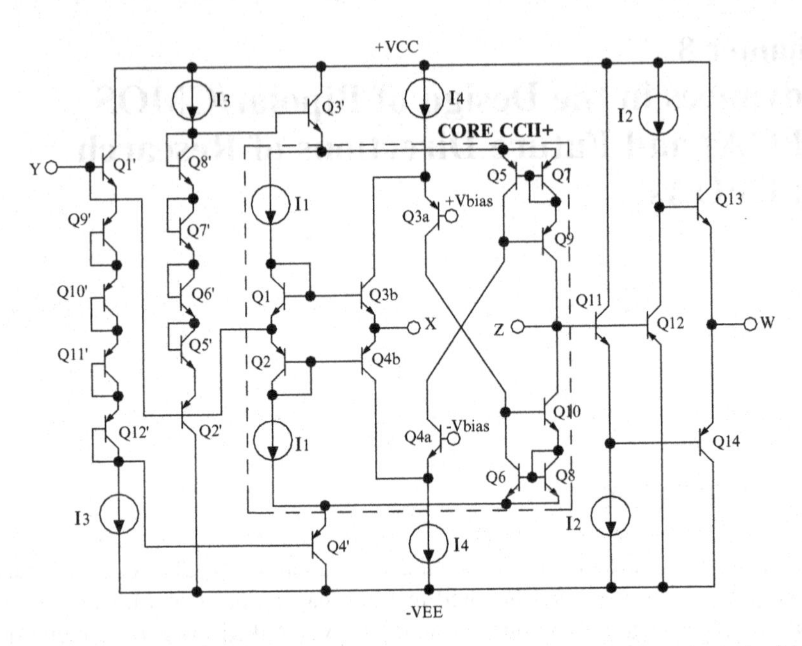

Fig. 8.1 CFOA architecture proposed by Tammam et al. employing *bootstrapping* and a *folded-cascode* in the input stage (adapted from [1] © 2003 Taylor & Francis)

modified CFOA input stage circuit design by introducing a combination of a *bootstrapping* technique and *folded-cascode* transistors resulting in a new CFOA architecture which is shown to result in significant improvement in the CMRR and gain accuracy. The circuit proposed in [1] is shown in Fig. 8.1.

This architecture has been shown to achieve a CMRR increased by some 62 dB as compared to the conventional CFOA along with an improvement in the AC gain error and input dynamic range although at the expense of reduction of slew rate and slight increase in the supply currents by about 15%.

8.2.2 *Bipolar CFOA with Higher Gain Accuracy, Lower DC Offset Voltage and Higher CMRR*

Hayatleh et al. in [2] presented a bipolar CFOA architecture based upon a new type of input stage incorporating both forward and reverse bootstrapping technique resulting in the architecture shown in Fig. 8.2. This novel topology was shown to provide higher gain accuracy; lower DC offset voltage and higher CMRR. It was shown that in comparison with the conventional CFOA, the CMRR increases by

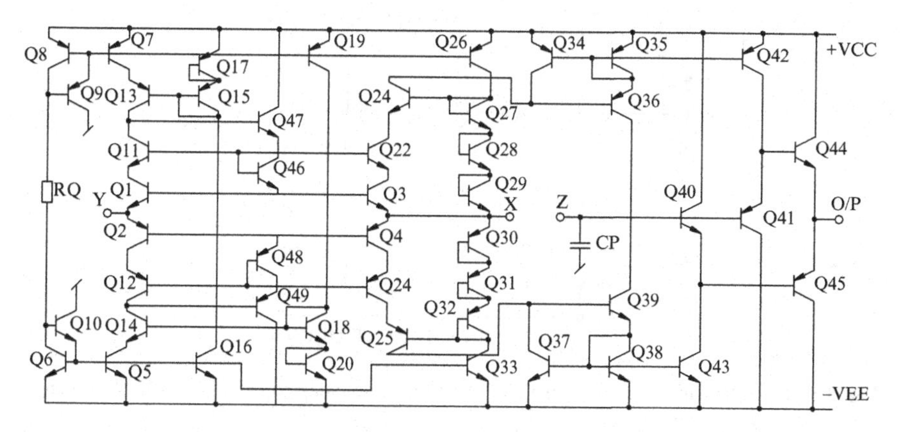

Fig. 8.2 CFOA using forward and reverse bootstrapping proposed by Hayatleh-Tammam et al. (adapted from [2] © 2007 Taylor & Francis)

about 46 dB and the input offset voltage reduces by a factor of two. While the majority of other characteristics are also better, the price to be paid is a reduced output voltage swing because of vertical transistors stacking.

8.2.3 *Bipolar CFOA Architectures with New Types of Input Stages*

Hayatleh et al. in [3] considered six new input stages with the intention of improving the performance of Bipolar CFOAs over a CFOA made from the conventional input stage, taking three major characteristics to improve upon, namely, CMRR, offset voltage and slew rate.

Figure 8.3 shows the schematic of a traditional CFOA architecture in which the half circuit of the input stage has been shown in dotted box as half circuit-A. Apart from this basic stage, six new half circuit stages have been presented therein which are shown here in Fig. 8.4. Accordingly, six different bipolar CFOA formulations were made and their performance in terms of CMRR, AC gain accuracy, and frequency response for unity closed loop gain, transient response and input impedance were studied and compared.

Out of the six new CFOA architectures, the one based upon input stage **F** has been found to be superior as compared to the rest in respect of slew rate (950 V/μs), input offset voltage (±0.75 mV), input resistance (62.9 Ω), bandwidth (61.6 MHz) and AC gain error (1.1 mV). This circuit has been shown in Fig. 8.5.

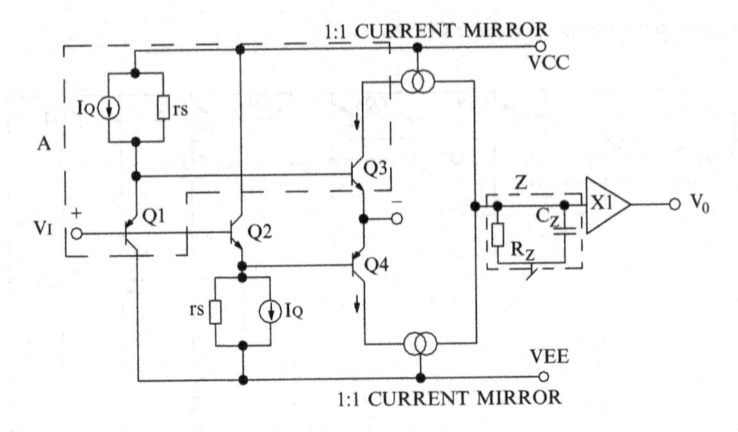

Fig. 8.3 The schematic of a traditional CFOA architecture (adapted from [3] © 2007 Springer)

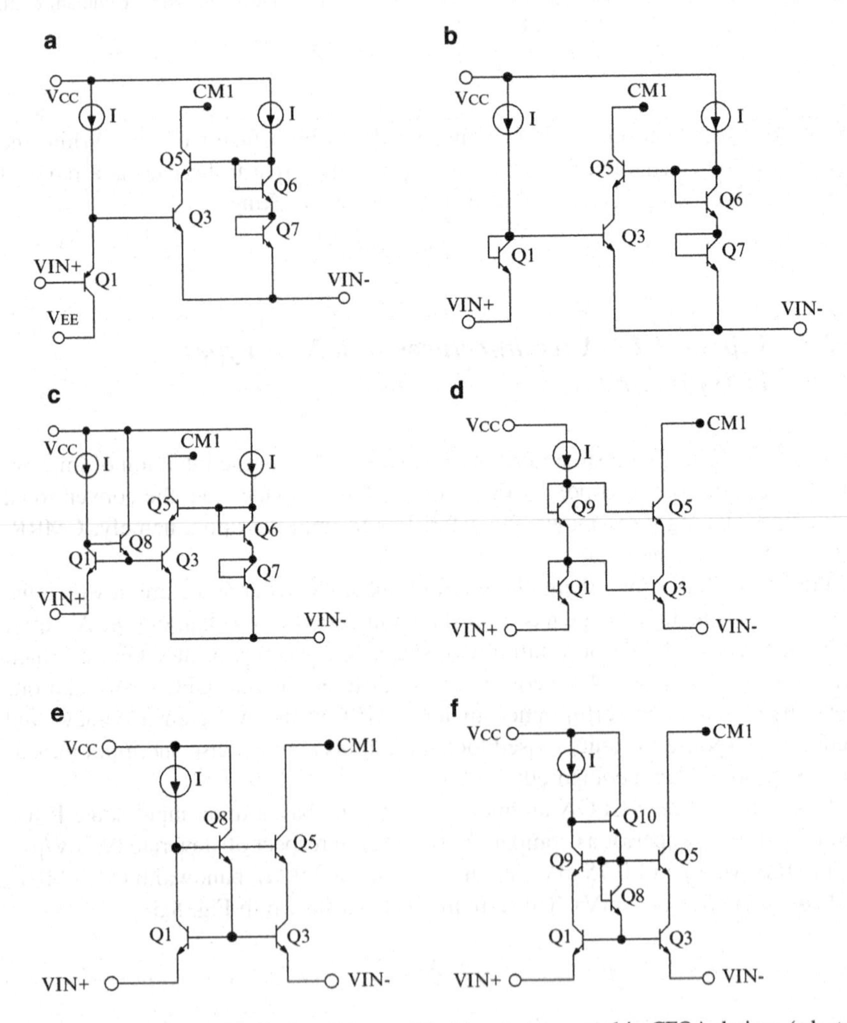

Fig. 8.4 The various half circuits of the modified input stages used in CFOA designs (adapted from [3] © 2007 Springer)

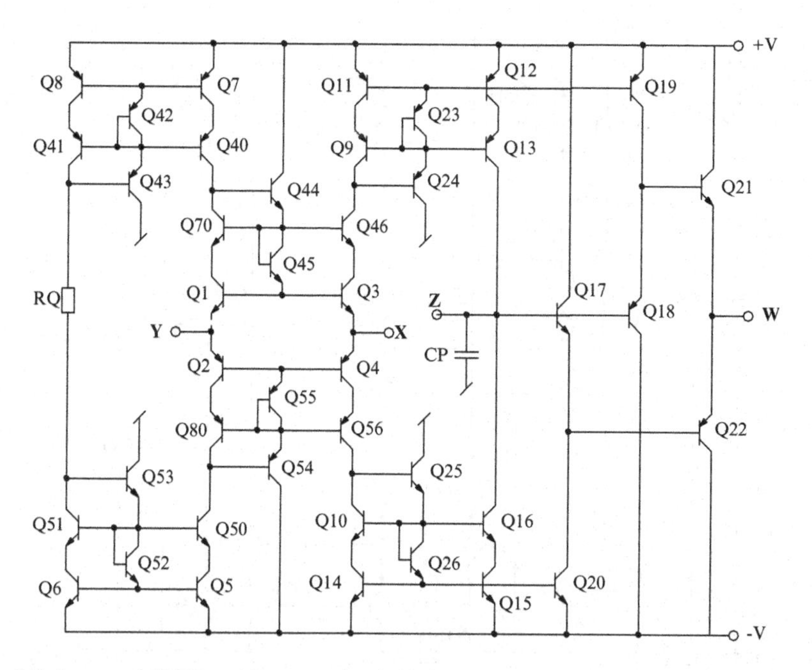

Fig. 8.5 Improved CFOA architecture using half circuit F (adapted from [3] © 2007 Springer)

8.2.4 Novel CFOA Architecture Using a New Current Mirror Formulation

Recently, a novel CFOA utilizing a new current cell for biasing the CFOA has been presented in [4] which is shown in Fig. 8.6. Although a drawback of this circuit is its moderately higher power supply operation ($\pm V = \pm 3$ V), this CFOA exhibits performance characteristics, superior to those obtained with an established input architecture, in terms of higher CMRR (91 dB), smaller offset voltage (<26 mV) with an acceptable high slew rate and gain accuracy.

8.3 The Evolution of CMOS CFOAs

In this section, we will highlight some significant contributions made in the design of CMOS CFOAs.

One of the early attempts in this direction was made by Bruun [5] who presented a CMOS CFOA based upon a bipolar counterpart and demonstrated that this CMOS CFOA has performance characteristics comparable to that of bipolar CFOA. This structure is shown in Fig. 8.7.

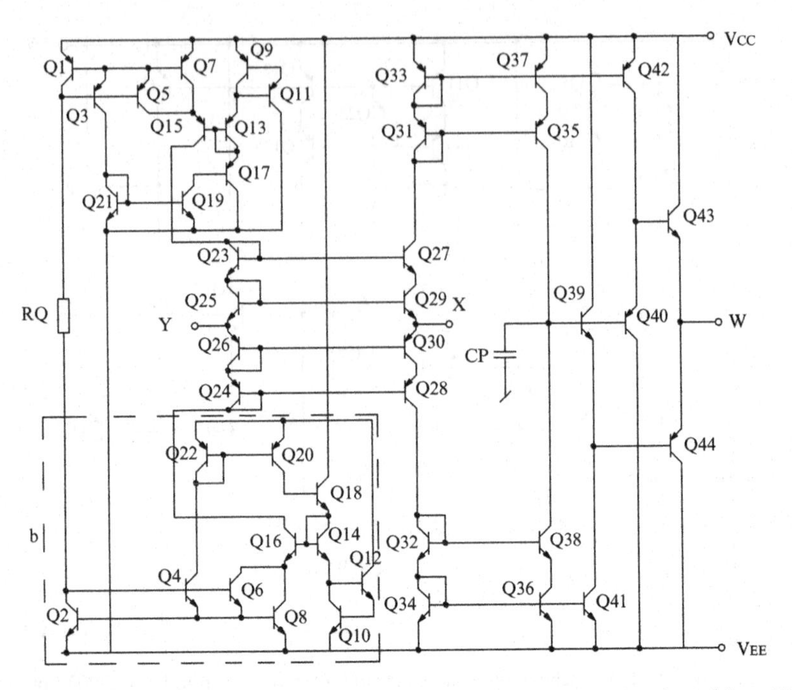

Fig. 8.6 New CFOA utilizing new current cell proposed by Tammam et al. (adapted from [4] © 2012 World Scientific Publishing Company)

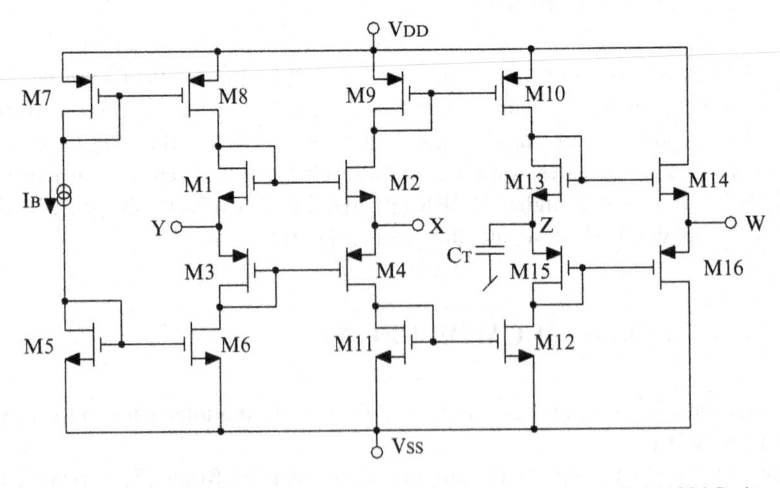

Fig. 8.7 CMOS Current feedback op-amp due to Bruun (adapted from [5] © 1994 Springer)

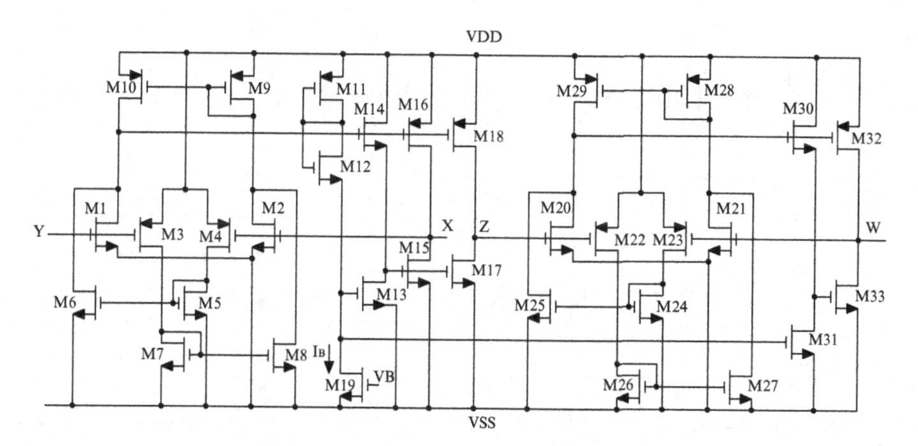

Fig. 8.8 The rail-to-rail CMOS CFOA proposed by Mahmoud et al. (adapted from [6] © 2000 Springer)

8.3.1 CMOS CFOA with Rail-to-Rail Swing Capability

Mahmoud et al. in [6], presented a CMOS structure with rail-to-rail swing capability. This circuit is shown in Fig. 8.8. The circuit operates as class AB, is capable of operating with ±1.5 V DC power supplies and has a standby current of 200 μA. This circuit appears to be suitable for low-voltage, low-power applications.

8.3.2 CMOS CFOA for Low-Voltage Applications

Two different variants of CMOS CFOA architectures were advanced by Maundy et al. in [8]. It was shown that the two variants exhibited overall gain bandwidth products in excess of 59 and 102 MHz respectively for a gain of −10 and with a 3.3 V DC power supply. One of these circuits is shown in Fig. 8.9.

8.3.3 Fully-Differential CMOS CFOAs

Mahmoud and Awad in [9] introduced a CMOS fully-differential CFOA (FDCFOA) using 0.35 μm technology which is shown in Fig. 8.10. This implementation was evolved with a view to be used for the synthesis of fully differential integrators and filters. This FDCFOA architecture is based upon a novel class AB fully differential buffer circuit. The circuit is capable of being operated with a supply voltage of ±1.5 V and has a total standby current of 400 μA.

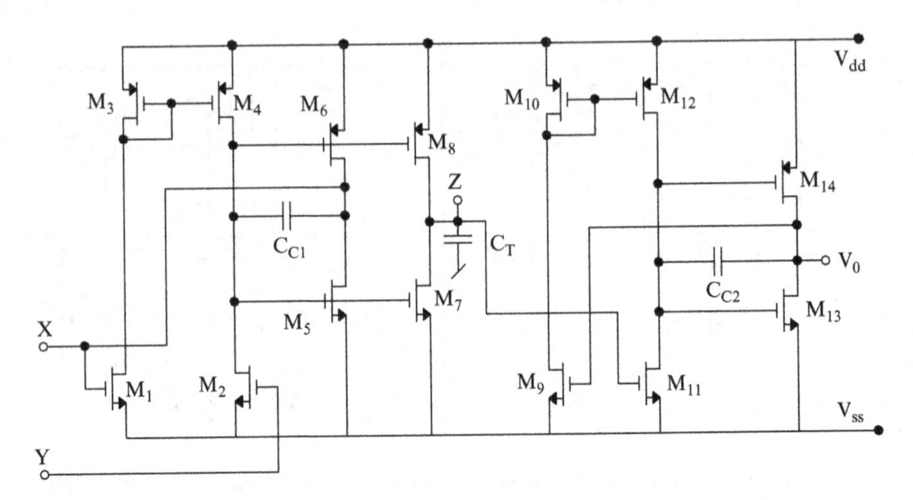

Fig. 8.9 CMOS CFOA proposed by Maundy et al. (adapted from [8] © 2002 Springer)

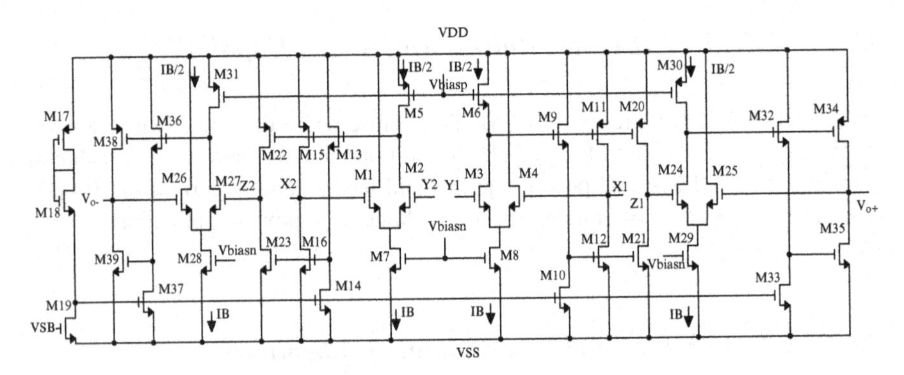

Fig. 8.10 Fully differential current feedback operational amplifier proposed by Mahmoud and Awad (adapted from [9] © 2005 Springer)

Another CMOS fully-differential CFOA was presented in [15] by Madian et al. providing a wide range controllable bandwidth of 57–500 MHz without changing the feedback resistance. The circuit had a standby current of 320 μA using a dual power supply voltages of ±0.75 V.

8.3.4 *CMOS CFOAs with Increased Slew Rate and Better Drive Capability*

Two internally-compensated CMOS CFOAs were introduced by Cataldo et al. in [11]. The circuits were made from class AB stages thereby obtaining increased slew rate and better drive capability. Experimental results on a prototype implemented in

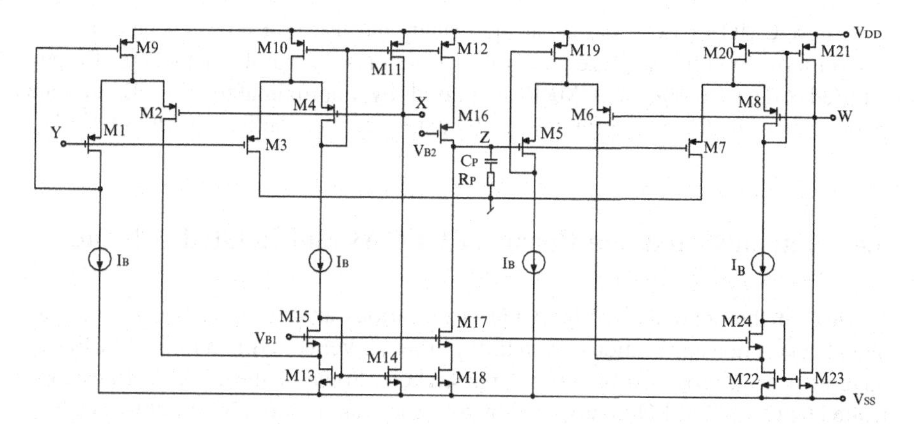

Fig. 8.11 CMOS CFOA due to Cataldo et al. (adapted from [11] © IEEE 2007)

0.35 μm process demonstrated a slew rate of 35 V/μs and a constant bandwidth of the order of 2 MHz in an inverting amplifier configuration with a 10 kΩ feedback resistor. One of these circuits is shown in Fig. 8.11.

8.3.5 *Other CMOS CFOA Architectures*

A CMOS CFOA was proposed by Ismail and Soliman in [7] which was implemented in 0.5 μm CMOS technology. This circuit exhibited frequency range of operation up to around 180 MHz using a DC bias power supply ±2.5 V and offered input impedance of the order of 2 Ω at the x-input terminal.

The CMOS CFOA presented in [10] is based upon connecting two high performance CCIIs and is aimed at achieving good input/output swing and drive capability. The circuit, however, has the drawback of having poor slew rate performance which, however, was shown to be substantially improved by adopting a class AB differential stage in the CCIIs.

The CMOS CFOA of [12] exhibited a bandwidth of 34.6 MHz, CMRR of 41 dB, input impedance of X-terminal as 1.65 kΩ, that of Y-terminal as 227.16 MΩ , the output impedance of 125.51 MΩ at Z-terminal and of W-terminal 1.65 kΩ.

Mahmoud et al. in [14] presented a low-voltage CMOS CFOA which allows rail-to-rail input/output operation with high drive capability using a supply voltage of ±0.75 V and a total standby current of 304 μA. The circuit had a bandwidth of 120 MHz and a current drive capability of ±1 mA.

In [23], Maundy et al. proposed a new topology for designing a CMOS CFOA by employing a CMOS CCII+ and a CMOS op-amp in an unconventional manner. The workability of the configuration was verified making an IC manufactured in 0.18 μm digital CMOS process.

A new CMOS CFOA based on the design and use in a repeated pattern of a current transfer cell was presented in [24] which resulted in the reduction of input referred offset voltage, a CMRR increased by approximately 53 dB and an improvement in AC gain accuracy as compared to a conventional CMOS CFOA.

8.4 Various Modified Forms of CFOAs and Related Advances

In view of the popularity of the CFOAs in various analog signal processing and signal generation applications, from time to time, several modified forms of CFOAs have also been proposed by various researchers. Some of these are: current controlled CFOA (CC-CFOA) proposed by Siripruchyanun et al. [16], modified CFOA (MCFOA) proposed by Yuce and Minaei [17], differential voltage CFOA (DVCFOA) proposed by Gunes and Toker [18], differential difference complimentary current feedback amplifier (DDCCFA) proposed by Gupta and Senani [19].

8.4.1 The Modified CFOA

The so-called MCFOA [17], when closely examined, turns out to be exactly same as the 'Composite Current Conveyor' proposed by Smith and Sedra in one of their very early publications on current conveyors [25]. In fact, its CMOS realization also can be identified to be a composite connection of a CCII+ and CCII−. Furthermore, its implementation in terms of normal kind CFOAs has three CFOAs out of which the first one is used as a CCII+ while the remaining two together are configured as CCII−. Thus, any single MCFOA-based circuit is actually a circuit involving a CCII + and a CCII− and hence, in the opinion of the present authors, the MCFOA as a building block is not distinctly different than the composite current conveyor of [25].

8.4.2 Current-Controlled CFOA

BiCMOS CC-CFOA architecture was proposed by Siripruchyanun et al. in [16] which is shown in Fig. 8.12.

The BiCMOS CC-CFOA of [16] consists of two blocks out of which the first is a CC-CCII+ and the other is a modified voltage follower. The CC-CCII stage consists of mixed translinear loop consisting of transistors Q_6–Q_9 which is biased by a current I_B through the current mirrors consisting of M_1–M_3 and M_{10}–M_{11}. The input resistance looking into the terminal-X is given by $r_x = V_T/2I_B$. A replica of the current i_x is generated and conveyed to the Z terminal using the CMOS transistors M_4–M_5 and M_{12}, M_{13}. On the other hand, the voltage buffer is realized using bipolar transistors Q_{10}–Q_{22} and M_{14}–M_{21}. Since the structure uses only NPN

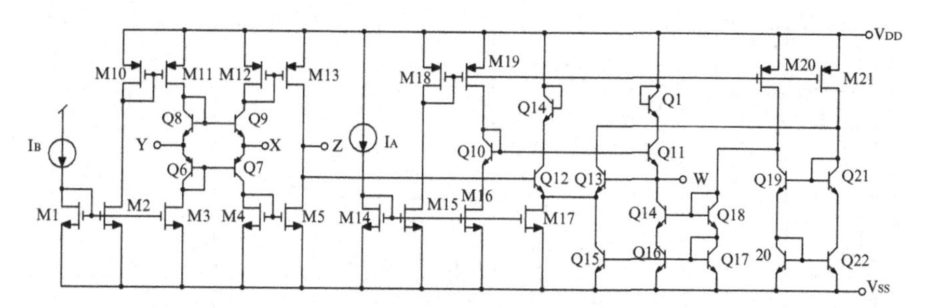

Fig. 8.12 A BiCMOS CFOA architecture proposed by Siripruchyanun et al. (adapted from [16] © 2008 WSEAS)

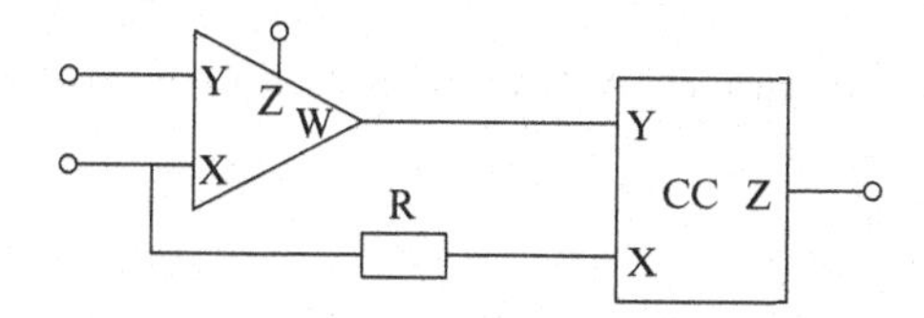

Fig. 8.13 Current feedback conveyor proposed by Gift and Maundy (adapted from [20] © 2008 John Wiley & Sons)

transistors, good high frequency behavior is expected. SPICE Simulations, using PMOS and NMOS transistors with parameters of 0.35 μm TSMC CMOS technology, have shown that r_x is controllable from 53 Ω to 12.62 kΩ by varying I_B from 1 μA to 300 μA. The circuit when operated from ±1.5 V biasing shows a power consumption of 4.16 mW with 3 dB bandwidth of 76 MHz for i_z/i_x, 342 MHz for v_x/v_y and 260 MHz for v_w/v_z.

8.4.3 Current Feedback Conveyor

A novel circuit element utilizing a CCII+ (realizable from a CFOA) with its input circuit in the feedback loop of a CFOA was advanced by Gift and Maundy [20] who preferred to call it a Current feedback conveyor (CFC). A CFC is shown in Fig. 8.13.

It was shown to be particularly attractive for realizing Current amplifiers and can be implemented from variety of CFOAs such as AD844, OPA 603 and OPA 623 and a variety of CCs such as AD844 and OPA660.

8.4.4 The Differential Voltage Current Feedback Amplifier

The differential voltage current feedback amplifier (DVCFA) introduced by Gunes and Toker in [18] provides a differential Y input (consisting of two terminals Y_1

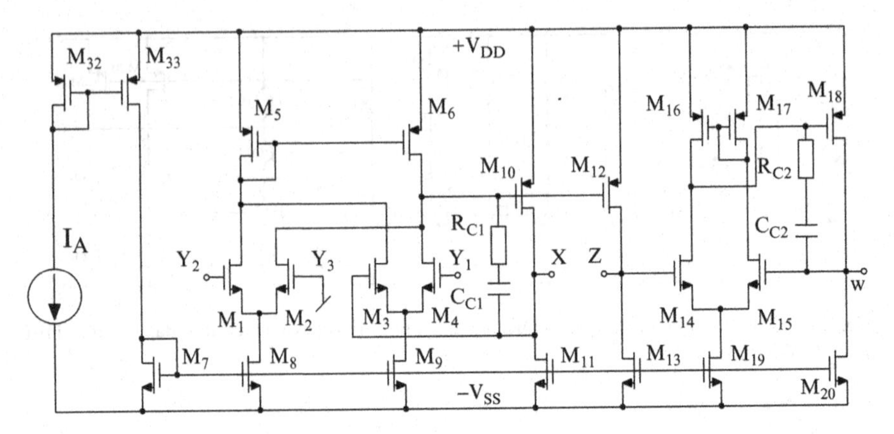

Fig. 8.14 CMOS implementation of DVCFA by Gunes and Toker (adapted from [18] © 2002 Elsevier)

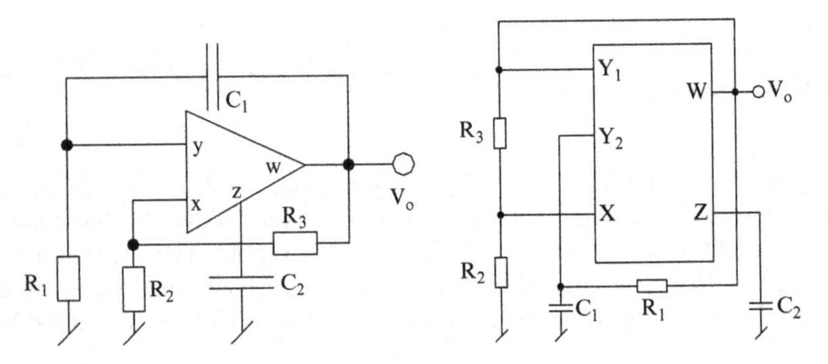

Fig. 8.15 A CFOA-based SRCO and its DVCFA-based counterpart employing both GCs as proposed by Gunnes and Toker (adapted from [18] © 2002 Elsevier)

and Y_2), has the characterization $I_{yi} = 0$, $i = 1, 2$; $V_x = (V_{y1} - V_{y2})$, $I_z = I_x$ and $V_w = V_z$ and is shown in Fig. 8.14.

DVCFAs have been shown to be particularly useful building blocks for synthesizing SRCOs employing grounded capacitors (GC). In this context, it may be noticed that while it has been amply demonstrated by a number of researchers that single resistance controlled oscillators (SRCO) can be realized using only a single CFOA however, none of the circuits known so far is able to employ both GCs as desirable for integrated circuit implementation. A DVCFA is particularly useful in removing this difficulty and it makes GC-SRCOs realizable from a single DVCFA. A family of eight such GC-oscillators has been derived by Gunes and Toker in [18]. An exemplary realization therefrom is shown here in Fig. 8.15.

Both the circuits are characterized by exactly the same CO and FO which are given by

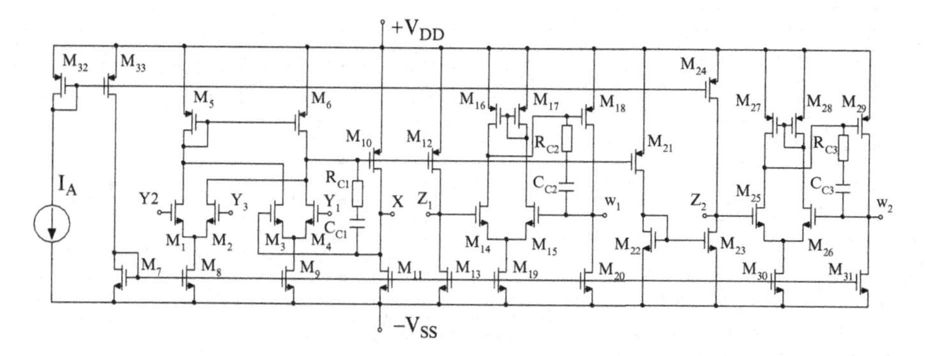

Fig. 8.16 An exemplary CMOS implementation of the DDCCFA [26]

$$C_1R_1 = C_2R_2 \text{ (adjustable by } R_2) \tag{8.1}$$

$$\omega_0 = \sqrt{\frac{1}{C_1C_2R_1R_3}} \text{ (adjustable by } R_3) \tag{8.2}$$

8.4.5 *Differential Difference Complementary Current Feedback Amplifier*

Another extension of the CFOA named Differential Difference Complementary Current Feedback Amplifier (DDCCFA) obtained by appropriate modification of the DVCFA of Gunes and Toker [18], was proposed by Gupta and Senani in [19, 26]. It is an active eight-port building block defined by the following characterizing equations: $i_{yk} = 0$; $k = 1$–3, $v_x = v_{y1} - v_{y2} + v_{y3}$, $i_{z1} = i_x$, $i_{z2} = -i_x$, $v_{w1} = v_{z1}$ and $v_{w2} = v_{z2}$. An exemplary implementation of DDCCFA is shown in Fig. 8.16.

It was shown in [19] and [26] that a single DDCCFA is sufficient to generate SRCO circuits possessing the following properties simultaneously: (a) use of a single active building block, (b) employment of two GCs along with a minimum number (only three) of resistors, (c) non-interacting controls of CO and FO, (d) a simple condition of oscillation (i.e. not more than one condition) and (e) availability of current-mode and voltage-mode outputs *both explicitly*. An exemplary SRCO using a DDCCFA is shown in Fig. 8.17.

This circuit is characterized by the following CO and FO

$$\text{CO}: \quad \frac{C_1}{C_2} = \frac{R_3}{R_1} \tag{8.3}$$

$$\text{FO}: \quad f_0 = \sqrt{\frac{1}{C_1C_2R_2R_3}} \tag{8.4}$$

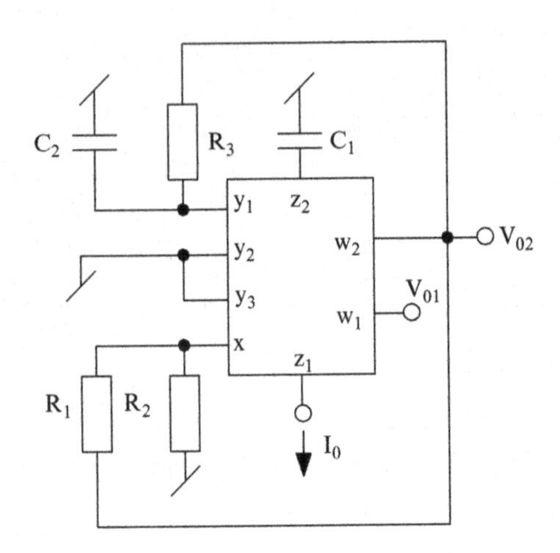

Fig. 8.17 An exemplary SRCO using a DDCCFA [26]

from where it is seen that CO is adjustable by R_1 while FO is independently variable through R_2. Furthermore, both VM and CM outputs are available and circuit employs both GCs as preferred for IC implementation.

It has also been demonstrated in [19] and [26] that employing a single DDCCFA, a large number of previously known building blocks can be derived as special cases as shown in the following:

(a) *First generation current conveyor minus type* (CCI^-) can be realized by grounding terminals y_2, y_3, interconnecting terminals y_1, z_1 and leaving terminals w_1, w_2 unused.
(b) *Second Generation Current Conveyor* ($CCII^+$) can be realized by grounding y_2, y_3, z_2 terminals while leaving terminals w_1, w_2 unused.
(c) *Second Generation Current Conveyor* ($CCII^-$) can be realized by grounding terminals y_2, y_3, z_1 and leaving terminals w_1, w_2 unused.
(d) *Dual Output Current Conveyor* (*DOCC*) can be realized by grounding y_2, y_3 terminals and leaving w_1, w_2 terminals unused.
(e) *Differential Voltage Current Conveyor* ($DVCC^+$) [27] can be realized by grounding terminals y_3, z_2 while w_1, w_2 terminals are unused.
(f) *Differential Voltage Current Conveyor* ($DVCC^-$) [27] can be realized by grounding terminals y_3, z_2 and leaving terminals w_1, w_2 unused.
(g) *Differential Voltage Complementary Current Conveyor* (*DVCCC*) [27, 28] can be realized by grounding terminal y_3 and terminals w_1, w_2 are left unused.
(h) *Current Feedback Operational Amplifier* (*CFOA*) can be realized by grounding y_2, y_3, z_2 terminals while leaving terminal w_2 unused.
(i) *Differential Difference Complementary Current Conveyor* (*DDCCC*) [27, 29] can be realized by leaving terminals w_1 and w_2 unused.
(j) *Differential Difference Complementary Current Conveyor* (*DDCCC*) [27, 29] can be realized by grounding y_3, z_2 terminals and leaving terminal w_2 unused.
(k) *Inverting Second Generation Current Conveyor* ($IVCCII^+$) [29] can be realized by connecting terminals y_1, y_3, z_2 to ground and terminals w_1, w_2 are left unused.

(l) *Inverting Second Generation Current Conveyor* (*IVCCII*$^-$) [29] can be realized by connecting terminals y_1, y_3, z_1 to ground and terminals w_1, w_2 are left unused.

(m) *Third Generation Current Conveyor* (*CCIII*$^+$) [30] can be realized by grounding terminals y_2, y_3 , interconnecting terminals y_1, z_2 and leaving terminals w_1, w_2 unused.

With more than one DDCCFA, other building blocks such as first generation CCI+, third generation CCIII$^-$, current differencing buffered amplifier (CDBA) [31], four terminal floating nullor (FTFN) [32, 33], operational trans-resistance amplifier (OTRA) [34] and fully differential second generation current conveyors (FDCCII) [35] can also be realized. Thus, the DDCCFA can be considered to be a generalized building block.

8.5 Future Directions of Research on CFOAs and Their Applications

While a considerable number of applications of CFOAs have been elaborated in this monograph, a number of applications such as those of realizing precision rectifiers, relaxation oscillators and analog multipliers/dividers etc. are, in the opinion of the authors, far from being completely investigated and a number of new configurations might be still waiting to be discovered. Thus, there is ample scope for evolving new configurations for the above mentioned as well as many newer applications of the CFOAs, not attempted so far.

Another area in which research work is still continuing is the development of improved bipolar/CMOS/BiCMOS circuit architectures of the CFOA itself. In this respect, one can also add a number of recent innovations like several varieties of so-called modified CFOAs (MCFOA), DVCFOA, and DDCCFA which were shown to be useful and versatile building blocks capable of providing many advantages in analog circuit design over the traditional type of CFOAs.

Among the continued efforts on improving the design of Bipolar and CMOS CFOAs, the systematic synthesis of CFOAs advanced by Torres-Papaqui and Tlelo-Cuautle [22] through manipulation of voltage followers and current followers looks very promising.

8.6 Epilogue

In spite of covering various aspects of the design and applications of CFOAs in this monograph, it can be concluded that the area still holds a lot of promise for the discovery of further new applications and the design of improved CFOAs. We hope that this exposition should be able to provide to the general readers, circuit designers and researchers, the needed stimulus to carry out further work on newer possibilities in improving the design of the CFOA and in searching newer applications of the CFOAs. For further studies, an additional list of references has been provided at the end.

It is interesting to note that the existing literature on CFOAs and their applications spread over dozens of international journals resulting in the publication of several hundred research papers on CFOAs have curiously focused their attention quite dominantly only on one specific type of CFOA namely, the AD844, which is uniquely different from the variety of other CFOAs from various manufacturers in that, this happens to be the only CFOA which has its Z-pin accessible outside the chip. As a consequence of this, researchers and academicians have found this CFOA to be particularly flexible and versatile because it (1) can be used as an op-amp, (2) can be used to realize current conveyor based circuits, (3) can be used to realize circuits based upon many other building blocks of more recent origin and at the same time, (4) can also be used as a 4-terminal building block in its own right.

In view of the wide spread use and applications of the CFOA with an external accessible Z-pin, it is, therefore, extremely surprising that none of the manufacturers have turned their attention to produce any more CFOAs of this kind and have limited themselves without exception, to only three terminal CFOAs which can be used only as a replacement of the traditional VOAs with the only advantage of offering superior performance in the same topologies (as compared to their VOA-based counterparts).

In view of the forgoing, the authors strongly feel that to tap the full potential of CFOAs with external accessible Z-terminal, the leading ICs manufacturers should produce improved versions of bipolar/CMOS/BiCMOS CFOAs and should necessarily provide one or more (if two, then complimentary) current-mode outputs and as many buffered voltage-mode outputs which will definitely make such a building block much more capable, flexible and versatile for various analog signal processing and signal generation applications in both linear and non-linear modes.

References

1. Tammam AA, Hayatleh K, Lidgey FJ (2003) High CMRR current-feedback operational amplifier. Int J Electron 90:87–97
2. Hayatleh K, Tammam AA, Hart BL, Lidgey FJ (2007) A novel current-feedback op-amp exploiting bootstrapping techniques. Int J Electron 94:1157–1170
3. Hayatleh K, Tammam AA, Hart BL (2007) Novel input stages for current feedback operational amplifiers. Analog Integr Circ Sign Process 50:163–183
4. Tammam AA, Ben-Esmael M, Abazab MR (2012) Current feedback op-amp utilizes new current cell to enhance the CMRR. J Circ Syst Comput 21:1250038-1–13
5. Bruun E (1994) A dual current feedback Op amp in CMOS technology. Analog Integr Circ Sign Process 5:213–217

6. Mahmoud SA, Elwan HO, Soliman AM (2000) Low voltage rail to rail CMOS current feedback operational amplifier and its applications for analog VLSI. Analog Integr Circ Sign Process 25:47–57
7. Ismail AM, Soliman AM (2000) Novel CMOS current feedback op-amp realization suitable for high frequency applications. IEEE Trans Circ Syst-I 47:918–921
8. Maundy BJ, Finvers IG, Aronhime P (2002) Alternative realizations of CMOS current feedback amplifiers for low voltage applications. Analog Integr Circ Sign Process 32:157–168
9. Mahmoud SA, Awad IA (2005) Fully differential CMOS current feedback operational amplifier. Analog Integr Circ Sign Process 43:61–69
10. Mita R, Palumbo G, Pennisi S (2005) Low-voltage high-drive CMOS current feedback op-amp. IEEE Trans Circ Syst-II 52:317–321
11. Cataldo GD, Grasso AF, Pennisi S (2007) Two CMOS current feedback operational amplifiers. IEEE Trans Circ Syst-II 54:944–948
12. Perez AP, Cuautle ET, Mendez AD, Lopez CS (2007) Design of a CMOS compatible CFOA and its application in analog filtering. IEEE Latin Am Trans 5:72–76
13. Maundy B, Gift S, Magierowski S (2007) Constant bandwidth current feedback amplifier from two operational amplifiers. Int J Electron 94:605–615
14. Mahmoud SA, Madian AH, Soliman AM (2007) Low-voltage CMOS current feedback operational amplifier and its application. ETRI J 29:212–218
15. Madian AH, Mahmoud SA, Soliman AM (2007) Low voltage CMOS fully differential current feedback amplifier with controllable 3-dB bandwidth. Analog Integr Circ Sign Process 52:139–146
16. Siripruchyanun M, Chanapromma C, Silapan P, Jaikla W (2008) BiCMOS current-controlled current feedback amplifier (CC-CFA) and its applications. WSEAS Trans Electron 5:203–219
17. Yuce E, Minaei S (2008) A modified CFOA and its applications to simulated inductors, capacitance multipliers, and analog filters. IEEE Trans Circ Syst-I 55(2):266–275
18. Gunes EO, Toker A (2002) On the realization of oscillators using state equations. Int J Electron Commun (AEU) 56:317–326
19. Gupta SS, Senani R (2005) Grounded-capacitor SRCOs using a single differential difference complimentary current feedback amplifier. IEE Proc Circ Devices Syst 152:38–48
20. Gift SJG, Maundy B (2008) A novel circuit element and its application in signal amplification. Int J Circ Theor Appl 36:219–231
21. Madian AH, Mahmoud SA, Soliman AM (2008) Configurable analog block based on CFOA and its application. WSEAS Trans Electron 5:220–225
22. Papaqui LT, Cuautle ET (2004) Synthesis of CCs and CFOAs by manipulation of VFs and CFs. Proc 2004 I.E. international behavioral modeling simulation conference. pp 92–96
23. Maundy BJ, Sarkar AR, Gift SJ (2006) A new design topology for low-voltage CMOS current feedback amplifiers. IEEE Trans Circ Syst-II 53:34–38
24. Tammam AA, Hayatleh K, Hart B, Lidgey FJ (2003) Current feedback operational amplifier with high CMRR. Electron Lett 39:2
25. Smith KC, Sedra A (1970) Realization of the Chua family of new nonlinear network elements using the current conveyor. IEEE Trans Circ Theor 17:137–139
26. Gupta SS (2005) Realization of some classes of linear /nonlinear analog electronic circuits using current-mode building blocks. PhD thesis (Supervisor Raj Senani), Faculty of Technology, University of Delhi, Ch 4. pp 128–188
27. Elwan HO, Soliman AM (1997) Novel CMOS differential voltage current conveyor and its applications. IEE Proc Circ Devices Syst 144:195–200
28. Gupta SS, Senani R (2000) Grounded-capacitor current-mode SRCO: novel application of DVCCC. Electron Lett 36:195–196
29. Awad IA, Soliman AM (1999) Inverting second generation current conveyors: the missing building blocks, CMOS realizations and applications. Int J Electron 86:413–432
30. Fabre A (1995) Third-generation current conveyor: a new helpful active element. Electron Lett 31:338–339

31. Acar C, Ozoguz S (1999) A new versatile building block: current differencing buffered amplifier suitable for analog signal-processing filters. Microelectron J 30:157–160
32. Senani R (1987) Generation of new two-amplifier synthetic floating inductors. Electron Lett 23:1202–1203
33. Senani R (1987) A novel application of four terminal floating nullors. Proc IEEE 75:1544–1546
34. Chen JJ, Tsao HW, Chen CC (1992) Operational transresistance amplifier using CMOS technology. Electron Lett 28:2087–2088
35. El-Adawy AA, Soliman AM, Elwan HO (2000) A novel fully differential current conveyor and applications for analog VLSI. IEEE Trans Circ Syst-II 47:306–313

References for Additional Reading

1. Smith SO (1993) The good, the bad and the ugly: current feedback-technical contributions and limitations. IEEE Int Sympos ISCAS 93:1058–1061
2. Bruun E (1993) Feedback analysis of trans-impedance operational amplifier circuits. IEEE Trans Circ Syst-I 40(4):275–278
3. Hart BL (1993) Current and voltage feedback op.-amps: a unified approach. Int J Electron 75(4):715–718
4. Higashimura M (1995) Filters and immittances using current-feedback amplifiers. Proceedings of the 20th International Conference on Microelectronics. 2: NIS SERBIA: 737742
5. Mahattanakul J, Toumazou C (1996) A theoretical study of the stability of high frequency current feedback Op-amp integrators. IEEE Trans Circ Syst-I 43(1):2–12
6. Fabre A, Amrani H, Saaid O (1996) Current-mode band-pass filters with Q-magnification. IEEE Trans Circ Syst-II 43(12): 839–842
7. Serrano L, Carlosena A (1997) Active RC impedances revisited. Int J Circ Theor Appl 25:289–305
8. Dostal T, Prokop R, Sarman R (1997) Functional blocks and biquadratic ARC filters using trans-impedance amplifiers. Radioengineering 6(1):9–15
9. Martinez PA, Aldea C, Sabadell J, Celma S (1998) Approach to the realization of state variable based oscillators. IEEE Int Conf Electron Circuits Syst 3:139–142
10. Palumbo G (1999) Bipolar current feedback amplifier: compensation guidelines. Analog Integr Circuits Signal Process 19:107–114
11. Yang Z (1999) Circuit transformation method from OTA-C circuits into CFA-based RC circuits. IEE Proc Circuits Devices Syst 146(2):99–100
12. Schmidt H (2000) Approximating the universal active element. IEEE Trans Circ Syst-II 47(11):1160–1169
13. Weng RM, Lai JR, Lee MH (2000) Realization of nth-order series impedance function using only (n-1) current-feedback amplifiers. Int J Electron 87(1):63–69
14. Gift SJG (2001) Hybrid current conveyor-operational amplifier circuit. Int J Electron 88(12):1225–1235
15. Palumbo G, Pennisi S (2001) Current-feedback amplifiers versus voltage operational amplifiers. IEEE Trans Circ Syst-I 48(5):617–623
16. Bayard J (2001) CFOA based inverting amplifier bandwidth enhancement. IEEE Trans Circ Syst-II 48(12):1148–1150
17. Takagi S (2001) Analog circuit designs in the last decade and their trends toward the 21st century. IEICE Trans Fundam E84-A(1):68–79
18. Gilbert B (2001) Analog at milepost 2000: a personal perspective. Proc IEEE 89(3):289–304

R. Senani et al., *Current Feedback Operational Amplifiers and Their Applications*, Analog Circuits and Signal Processing, DOI 10.1007/978-1-4614-5188-4,

19. Bayard J, Ayachi M (2002) OTA-or CFOA-based LC sinusoidal oscillators-analysis of the magnitude stabilization phenomenon. IEEE Trans Circ Syst-I 49(8):1231–1236
20. Venkateswaran P, Nagaria RK, Sanyal SK, Nandi R (2003) Dual-input single-tunable integrators and differentiators using current-feedback amplifier. Int J Electron 90(2): 109–115
21. Wu R, Lidgey FJ, Hayatleh K (2004) Using the 'T' feedback network with the current feedback operational amplifier. Int J Electron 91(11):685–695
22. Mita R, Palumbo G, Pennisi S (2004) Effect of CFOA non-idealities in Miller integrator cells. IEEE Trans Circ Syst-II 51(5):249–253
23. Natarajan S (2004) A variable frequency oscillator using modern current feedback amplifiers (CFAs) for high frequency applications. Proceedings of the 36th southeastern symposium on system theory. pp 417–421
24. Gilbert B (2004) Current mode, voltage mode, or free mode? a few sage suggestions. Analog Integr Circuits and Signal Process 38:83–101
25. Natarajan S (2005) Inductance simulation using modern current feedback amplifiers (CFAs). Proceedings of the 37th southeastern symposium on system theory. pp 60–64
26. Natarajan S (2005) A narrow-band VHF-tuned amplifier using modern current feedback amplifiers (CFAs). Proceedings of the 37th southeastern symposium on system theory. pp 55–59
27. Gift JGS, Maundy Brent (2005) Improving the bandwidth gain-independence and accuracy of the current feedback amplifier. IEEE Trans Circ Syst-II 52(3):136–139
28. Lim H, Park J (2006) Frequency-domain analysis of effects of the location of a feedback resistor in a current feedback amplifier. IEEE Trans Circ Syst-II 53(8):687–691
29. Gupta SS, Sharma RK, Bhaskar DR, Senani R (2006) Synthesis of sinusoidal oscillators with explicit-current-output using current feedback Op-amps. WSEAS Trans Electron 3(7): 385–388
30. Alzaher HA (2006) A novel band pass filter based on current feedback amplifier. Analog Integr Circuits Signal Process 46:145–148
31. Pal K (2006) All pass/notch filters using operational amplifier and current conveyors. JAPED 1:289–294
32. Vochyan J (2006) Synthesis of new biquad filters using two CFOAs. Radioengineering 15(4):76–79
33. Nagaria RK (2008) On the new design of CFA based voltage controlled integrator/ differentiator suitable for analog signal processing. WSEAS Trans Electron 6(5):232–237
34. Bayard J (2008) Universal filter using a 'pseudo' gyrator. Int J Electron 95(2):77–83
35. Raikos G, Psycalinos C (2009) Low-voltage current feedback operational amplifiers. Circuits Syst Signal Process 28:377–388
36. Yuce E (2010) Fully-integrable mixed-mode universal biquad with specific application of the CFOA. Int J Electron Commun (AEU) 64:304–309
37. Pennisi S, Scotti G, Trifiletti A (2011) Avoiding the gain-bandwidth trade off in feedback amplifiers. IEEE Trans Circ Syst-I 58(9):2108–2113
38. Ananda Mohan PV (2011) Comments on avoiding the gain-bandwidth trade off in feedback amplifiers. IEEE Trans Circ Syst-I 58(9):2114–2116
39. Pennisi S, Scotti G, Trifiletti A (2011) Reply to 'Comments on Avoiding the gain-bandwidth trade off in feedback amplifiers. IEEE Trans Circ Syst-I 58(9):2117
40. Rybin YK (2012) The nonlinear distortions in the oscillatory system of generator on CFOA. Hindawi Publishing Corporation: active and passive electronic components. Article ID 908716: doi:10.1155/2012/908716

About the Authors

Raj Senani received B.Sc. from Lucknow University, B.Sc. Engg. from Harcourt Butler Technological Institute, Kanpur, M.E. (Honors) from Motilal Nehru National Institute of Technology (MNNIT), Allahabad and Ph.D. in Electrical Engg. from the University of Allahabad.

Dr. Senani held the positions of Lecturer (1975–1986) and Reader (1986–1988) at the EE Department of MNNIT, Allahabad. He joined the ECE. Department of the Delhi Institute of Technology (now named as Netaji Subhas Institute of Technology) in 1988 and became a full Professor in 1990. Since then, he has served as Head, ECE Department, Head Applied Sciences, Head, Manufacturing Processes and Automation Engineering, Dean Research, Dean Academic, Dean Administration, Dean Post Graduate Studies and Director of the Institute, a number of times.

Professor Senani's teaching and research interests are in the areas of Bipolar and CMOS Analog Integrated Circuits, Electronic Instrumentation and Chaotic Nonlinear Circuits. He has authored/co-authored 135 research papers in various international journals and 4 book chapters for monographs published by Springer. He is currently serving as Editor-in-Chief for IETE Journal of Education and as an

R. Senani et al., *Current Feedback Operational Amplifiers and Their Applications*, Analog Circuits and Signal Processing, DOI 10.1007/978-1-4614-5188-4,

Associate Editor for the Journal on Circuits, Systems and Signal Processing, Birkhauser Boston (USA) since 2003, besides being on the editorial boards of several other journals and acting as an editorial reviewer for 30 international journals.

Professor Senani is a Senior Member of IEEE and was elected a Fellow of the National Academy of Sciences, India, in 2008. He is the recipient of Second Laureate of the 25th Khwarizmi International Award for the year 2012. Professor Senani's biography has been included in several editions of Marquis' Who's Who series (published from NJ, USA) and a number of other international biographical directories.

D.R. Bhaskar received B.Sc. degree from Agra University, B.Tech. degree from Indian Institute of Technology (IIT), Kanpur, M.Tech. from IIT, Delhi and Ph.D. from University of Delhi. Dr. Bhaskar held the positions of Assistant Engineer in DESU (1981–1984), Lecturer (1984–1990) and Senior Lecturer (1990–1995) at the EE Department of Delhi College of Engineering and Reader in ECE Department of Jamia Millia Islamia (1995–2002). He became a full Professor in January 2002 and has served as the Head of the Department of ECE during 2002–2005.

Professor Bhaskar's teaching and research interests are in the areas of Analog Integrated Circuits and Signal Processing, Communication Systems and Electronic Instrumentation. He has authored/co-authored 64 research papers in various International journals and 3 book chapters for monographs published by Springer. He has acted/has been acting as a Reviewer for several international journals. Professor Bhaskar is a Senior Member of IEEE. His biography is included in a number of international biographical directories.

A.K. Singh received M.Tech. in Electronics and Communication Engineering from IASED and Ph.D., in the area of Analog Integrated Circuits and Signal processing, from Netaji Subhas Institute of Technology (NSIT), University of Delhi. Dr. Singh held the positions of Lecturer and Senior Lecturer (June 2000–August 2001) at the ECE Department, AKG Engineering College, Ghaziabad. He joined ECE Department of Inderprastha Engineering College, Ghaziabad, India as a Senior Lecturer in August 2001 where he became Assistant Professor in April, 2002 and Associate Professor in 2006.

At present, he is a full Professor at the ECE Department of HRCT Group of Institutions, Faculty of Engineering and Technology, Ghaziabad, India. His teaching and research interests are in the areas of Bipolar and MOS Analog Integrated Circuits and Signal Processing. Dr. Singh has authored/co-authored 36 research papers in various International journals and 2 book chapters for monographs published by Springer.

V.K. Singh obtained B.E. and M.E. degrees in Electrical Engineering from MNR Engineering College, Allahabad and Ph.D. in Electronics and Communication Engineering from Uttar Pradesh Technical University, India. He worked as a

Research Assistant (1979–1980) at EE Department of MNR Engineering College Allahabad, as Teaching Assistant (1980–1981) and Assistant Professor at EE Department of GB Pant University; as a Lecturer (1986–1992) and Assistant Professor at Institute of Engineering and Technology (IET) Lucknow (1992–2004) where he became a full Professor in 2004. He has served as Head of the ECE Department at IET Lucknow between 1986–1988 and 2007–2010 where currently he is functioning as Dean of Research and Development since 2007. His teaching and research interests are in the areas of Analog Integrated Circuits and Signal Processing and he has published 16 research papers in various international Journals.

Index

A

Active R biquad, 107–110, 126
Active-R oscillators, 139, 148, 151–152
Active-R single resistance controlled oscillators (SRCO), 148–152
Amplifiers, 2, 7, 9–11, 17, 20, 25, 26, 28, 30–31, 33, 34, 45, 53, 72, 113, 211, 233
Analog circuit, 1–3
  design, 213
Analog dividers, 191–192, 208
Analog filters, 110
Analog multipliers (AM), 164, 165, 168–173, 175, 181, 207, 208, 237
Analog signal processing, 1, 9, 12, 19

B

Biquads, 82–107, 109, 115–116, 118, 126, 205
Bootstrapping, 224, 225
Building blocks, 236–238

C

Cable deriver, 182
Canonic single resistance controlled oscillators (SRCO), 135–137, 157, 175
Capacitive feedback, 32, 36
CCCS. *See* Current-controlled-current-sources (CCCS)
CCVS. *See* Controlled source implementations (CCVS)
Chaotic oscillators, 193–197
Chua diode, 194, 195
Chua's oscillator, 193–197
Circuit building blocks, 9–20
CMRR, 227, 231, 232
Compensating capacitor, 28, 30, 42
Compensation, 25–28, 36, 38, 39
Controlled source implementations (CCVS), 35
Controlled sources, 2, 7, 30, 35
Current-controlled current conveyor (CCCC), 201, 211–213
Current-controlled current feedback operational amplifiers (CFOA), 232–233
Current-controlled-current-sources (CCCS), 35
Current conveyors, 11–14, 26, 62, 72, 81, 131, 132, 153, 189, 193, 201, 223, 232
Current difference buffered amplifiers (CDBA), 18, 201, 207–208
Current differencing transconductance amplifier (CDTA), 13, 19–20, 201, 210, 211, 213
Current feedback, 26, 30
  amplifiers, 153, 214, 233–234
  conveyor, 233
Current feedback operational amplifiers (CFOA), 1–3, 12–15, 20
  architecture, 224–229
  pole, 148–152
Current followers, 32, 36, 37
Current follower transconductance amplifier (CFTA), 201, 211, 213
Current mirror, 232
Current mode (CM)
  biquad, 83–107
  circuits, 11, 13, 152
  oscillators, 156

R. Senani et al., *Current Feedback Operational Amplifiers and Their Applications*, Analog Circuits and Signal Processing, DOI 10.1007/978-1-4614-5188-4,

D
DC precision, 31
Deboo's integrator, 38
Decoupling of gain and bandwidth, 30–31
Differential difference complementary current feedback amplifier (DDCCFA), 235–237
Differential input buffered transconductance amplifier (DBTA), 201, 212, 213
Differential voltage Current feedback amplifier (DVCFA), 233, 234
Differentiators, 2
Double scroll attractor, 195

E
Electronically-gain variable amplifier, 181, 182
Explicit current output, 152–157, 175

F
Floating generalized impedance convertor (GIC), 64
Floating generalized impedance simulator, 60–65
Floating impedances, 53, 54, 63–65, 74, 76, 77, 204
Four terminal floating nullor (FTFN), 13, 17, 201, 204–205, 213
Frequency dependent negative conductance, 49, 67, 70
Frequency dependent negative resistance, 49, 56–60, 65, 70, 73, 74, 76, 77
Frequency stability, 132, 139, 143, 144, 160, 161, 167
Full power bandwidth (FPBW), 9
Fully differential biquad, 115–116
Fully-differential current feedback operational amplifiers (CFOA), 229, 230
Fully differential integrator, 113–115
Fully uncoupled oscillator, 157, 158, 160

G
Gain bandwidth conflict, 7–8
Generalized positive impedance converters, 55
Generalized positive impedance inverters, 55, 62, 63
Grounded capacitors, 62, 64, 67–70, 132, 137–139, 144, 148, 156, 157, 171, 174–176
Gyrator, 52, 54–56

H
Higher order filter, 55, 62, 63, 68–71

I
IC implementation, 83, 94
Impedance converters, 55, 63
Impedance inverters, 54, 55, 63
Inductance simulation, 50, 52, 53, 61
Instrumentation amplifier, 2, 6–8
Integrated circuits, 15
Integrators, 2
Inverse active filters, 110–112
Inverse transformation, 5

K
Kerwin Huelsman Newcomb (KHN) biquad, 84, 85, 87, 97

L
LC Ladder prototype, 118–122, 125
Linear voltage-controlled oscillators (VCOs), 161, 163–167, 173, 175
Lossless floating inductance simulator, 52, 62, 65–68
Lossy floating inductance simulator, 65–68
Lossy grounded inductor, 56–60
Low component count, 188

M
Mixed mode, 195–197
biquad, 83, 107, 126
Mixed translinear cell (MTC), 14
Modified current feedback operational amplifiers (CFOA), 232, 237
Modular filter structure, 119, 121–125
Monostable, 206
MOSFET-C filters, 16, 112–118, 126
MOSFET-C sinusoidal oscillator, 173–175
Multifunction biquad, 86, 100, 101
Multiphase oscillator, 151, 152
Multivibrators, 2, 205, 206

N
Negative capacitance, 56, 60, 61, 77
Negative impedance inverter, 55
Negative inductance, 56, 60, 61, 77
Network transformation, 70
Nonlinearity cancellation, 73, 173
Norton amplifier, 16

O
Offset voltage, 26, 31, 32, 47, 224–225
Op-amp circuits, 2–9
Operational amplifiers, 2, 15, 20, 230
Operational floating amplifier (OFA), 204
Operational transconductance
amplifier, 9–11
Operational transresistance amplifiers
(OTRA), 13, 15–16, 201,
205–207, 213

P
Passive compensation, 89
Passive LCR filter, 103
Positive impedance inverter, 55
Precision rectifiers, 2, 156, 189–190, 237
Pseudo exponential, 192–193
PSRR, 32

Q
Quadrature oscillators, 147, 148, 174, 175
Quasi stable state, 206

R
Rail-to-Rail, 229
Rectifier, 189–190
Relaxation oscillator, 184, 185, 189

S
Saturation, 188, 191
Schmitt trigger, 2, 183–189
Second order filter, 68–69
Series CD, 57, 58, 60, 70
Series RL, 51, 53, 54, 57, 58,
67–70, 77
Signal flow graph (SFG), 118, 119, 124, 126
Simulated impedances, 68–71
Simulated inductance, 56, 204, 205
Single element controlled oscillator,
131–132, 143
Single resistance controlled oscillators
(SRCO), 131, 134–138
Sinusoidal oscillators, 10, 131–176
Slew rate, 2, 8–9, 14, 15
Squarer, 9
Square rooter, 9, 198
Square wave generators, 16, 186–189
Squaring circuit, 190–191
State variable methodology, 140, 142, 153, 175
State variable synthesis, 140–143, 161,
163–167
Stray capacitances, 32

T
Third generation current conveyors, 201, 203
Threshold voltage, 183, 185, 187–189
Transconductance, 201, 207, 208, 210–212
Transfer characteristics, 30, 184–186, 189
Transimpedance, 187
Translinear circuits, 11
Triangular wave generators, 185–188
Triggering recovery time, 206
Tunability, 87, 90, 93, 97, 100, 102, 103,
109, 118
Tuning laws, 143–145, 157

U
Unity gain cells, 208–210
Universal biquad, 81, 83, 92, 95, 97–107, 126

V
VCCS. *See* Voltage-controlled-current-sources
(VCCS)
VCVS, 31–34, 47
Very low frequency oscillator, 132, 143
Voltage buffer, 14, 15, 20
Voltage-controlled-current-sources
(VCCS), 35
Voltage controlled impedances, 71–77
Voltage controlled oscillators (VCO), 71, 132,
161–164, 166–173, 175
Voltage controlled resistance (VCR), 137, 139,
141, 159–161, 163–167
Voltage differencing differential input
buffered amplifier (VD-DIBA),
201, 213
Voltage follower, 8, 16
Voltage mode
biquad, 102
circuits, 203

W
Wave active filters, 119–122, 124
Wave equivalents, 122, 124, 125
Waveform generators, 181, 183–189
Wien bridge oscillator (WBO), 131, 133–134
Wilson current Mirror, 27